TA403 .S773
Stokes, Vernon L.   030101   000
Manufacturing materials / Vern

0 2078 0003463 0
PSTCC LIBRARY KNOXVILLE, TN

AF352405

MANUFACTURING

Charles E. Merrill Publishing Company
A Bell & Howell Company
Columbus Toronto London Sydney

MATERIALS

Vernon L. Stokes
Tarrant County Junior College

Published by
Charles E. Merrill Publishing Company
A Bell & Howell Company
Columbus, Ohio 43216

This book was set in Times Roman.
The production editors were Frances Margolin and Jo Ellen Diehl.
The cover was prepared by Will Chenoweth.

Copyright ©, 1977, by Bell & Howell Company. All rights reserved. No part of this book may be reproduced in any form, electronic or mechanical, including photocopy, recording, or any information storage and retrieval system, without permission in writing from the publisher.

International Standard Book Number: 0-675-08493-8

Library of Congress Catalog Card Number: 76-526-23

1 2 3 4 5 6 7 8 9 10 — 86 85 84 83 82 81 80 79 78 77

Printed in the United States of America

To the students of engineering who are tomorrow's technicians
and engineers.

Preface

As knowledge increases in less and less time, and as new methods of manufacturing become more complex, materials take an increasingly important role. Demands for safer materials have forced engineers and technicians to become more knowledgeable in the field of materials. Consequently, these persons *must* understand the chemical, physical, and mechanical properties of materials. Engineers and technicians must be able to plan, predict, and produce the required strengths in their products as these products function in everyday life and in extreme environments. They must be aware of microstructures of materials and the relationships to mechanical properties.

For all these reasons, and because of the pressing need for qualified personnel in the field of engineering materials, *Manufacturing Materials* has been written. No advanced mathematics or physics is necessary to understand the detailed discussion of materials and their uses. The story of materials is told, beginning with the smallest particles of matter and ending as raw materials ready for processing into many kinds of needed parts.

The discussion identifies many materials used in manufacturing and points out their physical and mechanical capabilities so that engineers and technicians can choose the proper material for a given design. All the essential mechanical testing procedures are discussed, both destructive and nondestructive, along with detailed metallurgical descriptions of the numerous metals. Microstructures and strengths of materials are correlated throughout the text to add meaning to the discussion. And then, the harsh environments of cryogenics and elevated temperatures are pointed out, as these temperatures relate to the materials which must effectively resist them. Finally, because many materials are used in fatigue and corrosion environments, the student is given a practical view of these intense operating conditions so that proper choice of material will allow the part to function safely.

Ideally, the student of engineering technology should be well equipped with a working knowledge of materials capabilities before moving into the numerous manufacturing processes. Equipped with the knowledge of these materials, the designer, engineer, and technician can plan the manufacturing process with a high degree of validity. Consequently, it is recommended that all design, architectural, engineering, and manufacturing students equip themselves with a reasonable level of knowledge of the chemical, physical, and mechanical properties of materials.

This book is the result of several years of experimenting with what should be taught to beginning students of engineering technology. I want to take this opportunity to thank my many friends, including my students, for their helpful information in formulating the manuscript. Also, recognition must be given to the many manufacturers noted throughout this text and to the several professional organizations for their assistance in furnishing data. A special recognition goes to Tarrant County Junior College where several photographs were made in the engineering technology department. For their helpful guidance in the text on atomic structure, I want to thank Thomas Taaffe and Jerry Brammer. Finally, in recognition for their review of the entire manuscript, along with their helpful comments, I want to give special thanks to Professor Joe Waldinsperger of Rochester Institute of Technology and Donald Rawlins of General Dynamics, Convoir, Fort Worth, and for her excellent editing of the manuscript and preparation for production I want to extend my great appreciation and thanks to Marilyn J. Neyman. For helping me procure the many photographs and illustrations and for typing the manuscript and assisting me in producing this text in final form, I am indeed grateful to my wife Dorothy.

Vernon L. Stokes

Contents

Manufacturing Materials

The Need for Manufactured Parts

At no time in the history of the world has the need for new designs been greater than now. Accelerating technology has moved into all segments of society. New inventions cause initial increases in unemployment, but they ultimately decrease this factor by creating new jobs and demands for unheard-of materials. As new materials are found the need for new ways of processing them arises. But while technological expansion helps satisfy society's basic needs, it sometimes causes severe cultural lags such as regional unemployment, water and air pollution, waste of valuable resources, and local and international material problems.

A GOAL OF MANUFACTURING

The primary goal of manufacturing is the making of things for profit, and increased technology bolsters profit (Fig. 1–1). In this regard, the installation of automated machines causes some people to lose their jobs, but many others find new jobs in design, electronics, machinery manufacture, parts production, and sales. As production of parts increases, more money becomes available through buying and selling, and the standards of living become better for more people. But the availability of tangible goods and the reinforcement of the good life are occasionally accompanied by the ills of pollution and material waste. While manufacturing increases in one part of a society, it often causes curtailment of some aspects of production in another part, thereby instigating job losses and a lowered standard of living, but in time the differences are reduced. While part of an advanced society efficiently consumes its newly manufactured goods, another part wastes these materials because of negligence and overabundance of that which is wasted. Billions of dollars worth of recoverable materials are annually buried in public garbage dumps. Negligence in some areas of manufacturing and inability to control certain types of pollution have brought serious problems to advanced cultures. Local and national

FIGURE. 1–1   These automatic machines produce thousands of parts daily and require little supervision. (Courtesy of Martin Sprocket and Gear)

raw material needs have expanded to international magnitudes; serious problems exist between the have and have-not countries. As the great abundance of newly manufactured goods is distributed throughout the world, however, solutions to cultural problems become more probable because the intensity of need finds the inventions to help solve these harmful secondary effects of mass production. An ultimate result of manufacturing is continued advancement in production techniques; however, this advancement is slowed somewhat because of the attendant ill effects. An ultimate objective of manufacturing, then, is the production of finished parts from raw materials whereby a reasonable profit is made while the harmful effects of increased technology are kept to a minimum.

MATERIALS SERVE THE SOCIETY

Even though technology is accompanied by disagreeable side effects, manufacturing continues to provide the essential supplies to sustain the good life. Explorations in space have provided numerous consumer items that would normally not be available at this time. Modifications in radio and television reception, control devices, and production of electronic items such as the sophisticated pocket calculator have resulted from new findings in aerospace technology. Factors relating to materials, such as density, cryogenic and elevated temperatures, penetrating radiation, and mechanical properties, have led to the invention of hundreds of new consumer items. And wars, even though dreaded, produce numerous items for consumer use, such as improvement in optics, electronic communications, earth-moving devices, and explosion-formed metals. As if moving in a circle, manufacturing technology helps cause the social problems, helps solve the problems, provides the economic goods for social satisfaction while it breeds foul air and contaminated water, provides employment and available money, accelerates the knowledge growth in the sciences, raises the standards of living for the affected people, and provides an abundance of materials for use. Manufacturing is the backbone of progress (Fig. 1–2). Manufacturing materials that have engineering uses provide the reason for

FIGURE 1–2 Molten iron is charged into a basic oxygen furnace for refinement into steel. Man's continuous advancement depends on availability of steel. (Courtesy of Bethlehem Steel Corporation)

writing this text. Materials which serve as the foundation for technology's existence require intimate discussion for both understanding and application (Fig. 1–3).

ORIGIN OF MANUFACTURING MATERIALS

The origin of materials for subsequent manufacturing is varied. Most raw materials come from the earth in the form of ores, oil, and gas, while some basic elements

FIGURE 1–3 The main engineering materials include rubber (left), plastic, wood, metal, ceramic, concrete, glass, and brick (rear).

are derived from sea water and the atmosphere. These primary materials, ores or elements, must be converted into secondary raw materials, gases, liquids, and solids, through special processes for use in manufacturing. These secondary materials (Fig. 1–4) must be finished into consumer items at acceptable prices for the manu-

FIGURE 1–4   These stacks of steel bars will be machined into thousands of sprockets and gears. (Courtesy of Martin Sprocket and Gear)

facturing process to be successful (Fig. 1–5). The processes of manufacturing the numerous primary and secondary raw materials into finished parts are discussed in *Manufacturing Processes*. Subsequent discussions in this text will be concerned with the chemical, physical, and mechanical properties of engineering materials used in the metals, plastics, ceramics, glass, concrete, rubber, and wood industries. These studies focus on materials capabilities.

THE NEED FOR SPECIAL ENVIRONMENTAL MATERIALS

As manufacturing technology advances, other areas of society advance also. For example, several transportation systems continue to provide their vehicles with the

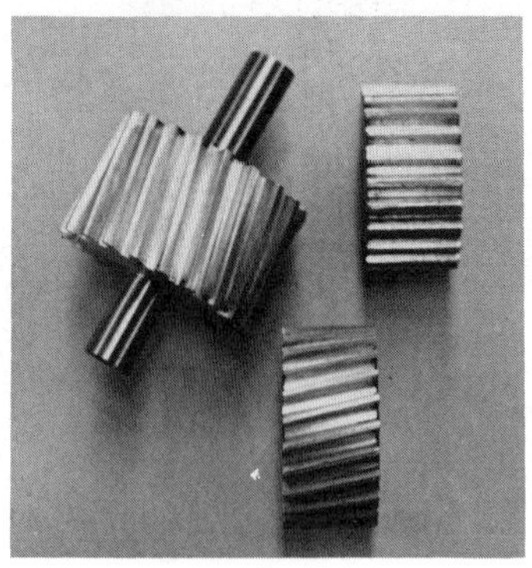

FIGURE 1–5   Gears which have been machined from bar stock, as shown in Figure 1–4.

capability of moving faster. Air transportation leads other systems in travel speeds mainly because of its uniqueness. But these speeds would not exist if it were not for the new materials only recently brought into use. Titanium and graphite are increasingly used where heat is a factor, and heat is very much a factor with objects moving at very high speeds in the atmosphere. Military aircraft (Fig. 1–6) require

FIGURE 1–6 The F-16 is a high performance combat aircraft powered with a single Pratt Whitney engine which gives it excellent maneuverability such as fast climbing. It is manufactured for the United States Air Force and several NATO countries. (Courtesy of General Dynamics, Fort Worth Division)

the best material available in strength, weight, and environmental applications. Spacecraft (Fig. 1–7) go one step further, especially in the use of cryogenic and elevated temperature resistant materials. Because space travel involves material exposure to a wide range of temperatures, a very limited number of materials are currently available for use in these environments where temperatures range from

FIGURE 1–7 Space radiator system for the space shuttle *Orbiter* is shown mounted on the vehicle's open payload bay doors. Some of these materials are discussed in following chapters. (Courtesy of Vought Corporation)

−350 °F (−598 °C) to earth entry temperatures as high as 4000 °F (7232 °C). But a unique material, silicon carbide, is now available to protect the leading edges of spacecraft as they collide with and move swiftly through the atmosphere (Fig. 1–8). Railroad trains are also moving faster, so locomotives must be soundly con-

FIGURE 1–8 Samples of heat shielding material which protects space vehicles entering the earth's atmosphere. The thin line of lightly shaded material is silicon carbide and this is diffused onto carbon.

structed. Rails and wheels therefore demand higher strength and safer metals in order to withstand the intermittent stresses (Fig. 1–9).

Adverse Environments Challenge Materials

In the field of nonmetals, such as carbides, oxides, and precious stones, a new need exists for special machining, for example, drilling very small vent holes and forming oddly shaped parts. Also, many very hard metals are being produced whereby

FIGURE 1–9 These 3600 hp (2.68 x 10^6 W) EMD locomotives are assembled at the Corwith Yard, Chicago. The metals constituting their makeup will be discussed. (Courtesy of the Santa Fe Railway)

machining must follow heat-treating processes. Ordinary tools will not cut hardened steels; therefore, new materials and new machining processes have been found. Special materials are therefore needed to make the machines and cutting tools. Ultrasonic cutting and electrical discharge machining, along with the powerful energies of the electron beam, plasma arc, and laser systems, provide the capacity to machine unusual shapes and selected materials, depending on the capability of the process.

Space vehicles moving into the severe temperature environment of space and adverse planetary surface conditions demand materials having universal usage, but these materials are few indeed. For example, spacecraft (Fig. 1–10) and missiles

FIGURE 1–10 This head-on view of the Apollo command and service modules was taken from the lunar module during the *Apollo IX* flight. Antenna at lower left was used to transmit deep space communications and television signals back to Earth. Unique materials are used in spacecraft's construction. (Courtesy of Rockwell International)

(Fig. 1–11) demand the best available materials, such as austenitic steels and titanium alloys. Rocket exhausts require alloys of tungsten and molybdenum, along with graphite. Ablative materials of carbon and resin have proven effective in nose cones of space vehicles. When all else fails, the compounds of aluminum oxide and silicon carbide, along with the element carbon, handle the ultimate in adverse temperature conditions.

THE NEED FOR SAFER MATERIALS

Because materials are being used in more adverse conditions and safety factors for selected designs are being reduced, materials must be safer with regard to freedom from defects. Cracked metals cannot be used when the pending fracture is projected

FIGURE 1–11   This Scout 181 missile is constructed of high strength but light-weight materials in order for it to endure the forces of flight and extreme temperature changes. (Courtesy of Vought Corporation)

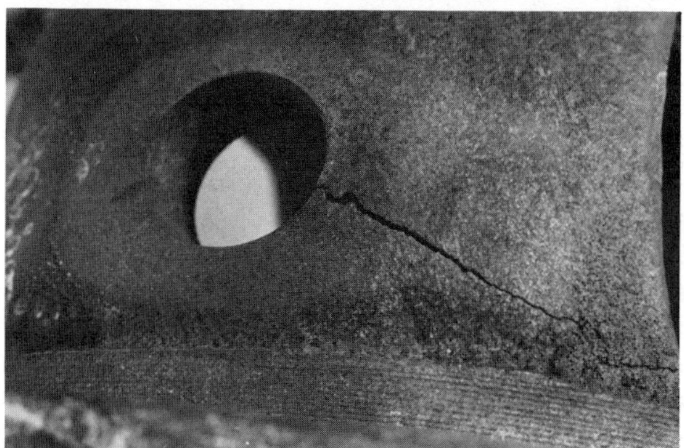

FIGURE 1–12   A cracked metal part headed for disaster. Most often, these cracks are not apparent to the naked eye.

for disaster (Fig. 1–12). Train wheels are inspected periodically to assure freedom from dangerous cracks (Fig. 1–13). Critical parts such as automobile camshafts are also tested before installation (Fig. 1–14). Materials in the ship in Figure 1–15 have been approved prior to construction. Manufacturing and fabrication processes use various testing and inspection techniques in their quality control processes. Inspections demand sophisticated searches for discontinuities in metals, while the replaced parts are made from cleaner metals. As the amount of material is decreased for a given load-holding design to reduce weight, material soundness is confirmed with a high degree of validity. Recent demands on materials require extreme tough-

FIGURE 1–13 A train wheel is inspected by ultrasonics for cracks.

FIGURE 1–14 An automotive camshaft is being inspected for surface cracks.

ness in the face of impact loading, some designs requiring exceptional toughness in a range from $-300\,°F$ $(-184\,°C)$ to $1200\,°F$ $(649\,°C)$. In the past certain exceptions have been allowed with regard to use of structural materials. However, manufacturers must now comply with procedures which lead to safe operating conditions. In this respect, knowledge of OSHA (Occupational Safety and Health Act) laws is essential for engineers and technicians involved in manufacturing.

MANUFACTURING MATERIALS AND PROCESSES

Manufacturing materials and processes go hand in hand, and at no time can they be separated during the processing cycle (Fig. 1–16). An automatic machine, for example, accepts a material and forms it according to predetermined specifications. Because the material is a tangible thing, it offers resistance to being formed, and this resistance is reflected in the capability of the machine. Consequently, it is essential to know the relationships between material and machine. Machines of all types effectively process diverse manufacturing materials to produce products of various shapes and strengths. Some of these machines are hand operated and some

FIGURE 1–15 Various types of structural steels are formed and welded into a large ship, the *Alaskan Mail* (bow view). (Courtesy of Newport News Shipbuilding)

FIGURE 1–16 An automatic cutting machine processes a large sprocket blank by cutting teeth along the periphery. (Courtesy of Martin Sprocket and Gear)

are automatic; the type of process and economic factors influence the choice of the particular machine to use. Numerous kinds of support and materials-handling equipment besides machines are essential in the finishing of the part. Because manufacturing is technical in nature, certain qualified persons guide the production activities from raw material to finished part (Fig. 1–17).

FIGURE 1–17 These sprocket teeth are fabricated with the use of an automatic cutting torch, the finish needing no further processing. (Courtesy of Martin Sprocket and Gear)

THE MANUFACTURING TEAM

A team consisting of engineers, technicians, operators, supervisors, and inspectors is primarily responsible for production of finished parts, while the whole plant personnel component has certain responsibilities. The process engineer and technicians, along with key persons such as designers and metallurgists, are charged with materials procurement and processing. Production results must pay the accumulated expenses and allow a reasonable profit. But part of the profit must be returned to the operating complex in order to sustain a long-term operation. And there must be an integration of people, machines, and materials, along with high morale. Certain knowledge and skills are essential in handling the several kinds and shapes of materials. In this respect, metals constitute by far the majority of manufacturing materials; however, plastics are used for a vast quantity of products. Large and diverse manufacturing plants also employ chemists, medical doctors, and psychologists, along with the normal array of support and sales personnel. In order to help assure smooth operating conditions, a training section is quickly established to provide the skills needed to process the materials. But ultimately, it is the manufacturing team that guides the material through the manufacturing processes.

CHOICE OF MATERIAL MUST BE CHALLENGED

Because a profit must be made, there is a constant watch of manufacturing activities. The most economical material that will adequately do the job is usually the best choice of material. On the other hand, many manufacturing operations are controlled by material specifications; therefore, there is no choice of material. Some materials contain flaws of unacceptable magnitude; consequently, in many operations a thorough search is made to verify the soundness of the material before processing commences (Fig. 1–18). Then, at some point along the production line

FIGURE 1–18   A steel ingot at 2200 °F (3992 °C) is being removed from the soaking pit for subsequent rolling into a predetermined shape. Prior to additional working the large mass of metal may be inspected by X radiation or by ultrasonics to verify the absence of discontinuities. (Courtesy of Bethlehem Steel Corporation)

another inspection is often made to verify the continued soundness of the material. In these instances, when metals are the materials, nondestructive testing methods and techniques are applied as needed. At the end of the process the finished part is again inspected to verify compliance with specifications.

Somewhere a compromise is often made in regard to the use of the material.

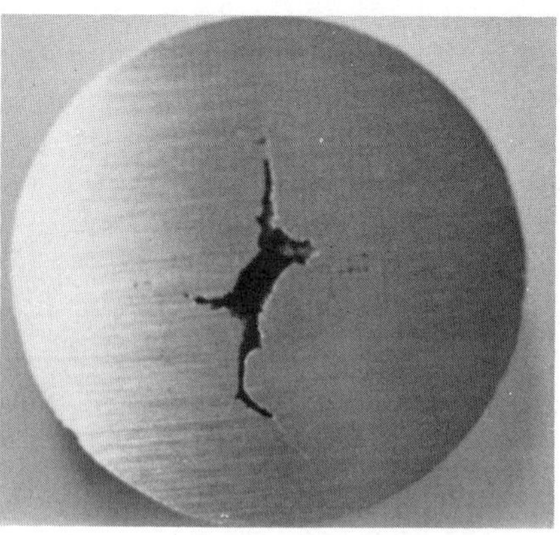

FIGURE 1–19   An internal burst in a large metal bar.

Large sections of metal which are rejected because of internal bursts (Fig. 1–19) subtract from profits. Yet the use of this raw material will allow a lifelong flaw. Failure of this part in service could then cause a catastrophe. The huge area of materials is consequently relying more and more on the assistance provided in the field of nondestructive testing to help assure validity of materials. The part shown in Figure 1–19 was inspected by ultrasonics and was immediately bisected. As a matter of information, the completed part showed no external clues related to its destroyed interior. Then a large section of steel was measured for a highly stressed welded structure. As the torch cut through the plate the lamination was evident (Fig. 1–20). There is absolutely no reason to fabricate a part containing a large

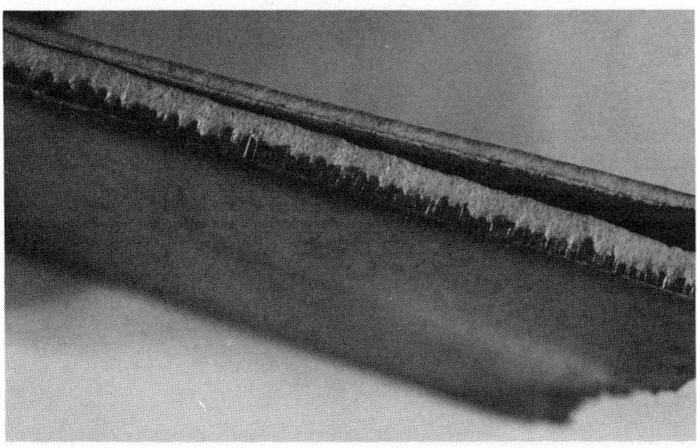

FIGURE 1–20   An edge view of a thick metal plate which has split due to large quantities of trapped oxide.

slag inclusion if the finished part is to be subjected to high levels of stress. In fact, such an act forecasts possible disaster.

RAW MATERIALS BECOME FINISHED PRODUCTS

Iron ore from the earth eventually becomes a kitchen knife or an automobile steering column. First there is the material, then there is the process, and ultimately there is the finished product. How, then, are the millions of items produced and what is the constant relationship between material and process?

Geometry of the finished part, its chemistry, and its resulting physical and mechanical properties combine to form a concept which ultimately results in a planned manufacturing procedure. Certain processing steps must be chronological while others may be variable. The manufacturing plan provides details from receipt of the raw material to final inspection and stocking. Again, geometry, chemistry, and expected properties form the foundation for the processing plan. Because the field of engineering materials is extremely diverse, the first thoughts are focused on the finished product and its material. The various stages of processing are arranged from this point on. In this regard, the use of metal requires entirely different machines than those used for plastics. In the area of plastics, ceramics, glass, concrete, rubber, and wood, specific machines and support equipment are arranged in the production sequence.

Metals Processing Varies

In the area of metals, the largest manufacturing area, different methods of processing are also arranged in the production line (Fig. 1–21). Some of the methods used

FIGURE 1–21   An aircraft production line which manufactures and assembles parts for these tail sections of large aircraft. (Courtesy of Vought Corporation)

to shape a metal include melting and pouring; powdering, pressing, and sintering; and forging, rolling, extruding, bending, pressing, or cutting. In many of these forming operations heat treating of the metal is usually required. Also, additional processes such as welding, plating, anodizing, or painting are often performed on the part. Many of the forming operations are accomplished by hand controlling of the equipment, but many operations are completely automatic or semiautomatic. Economics is always present in a manufacturing operation due mainly to quality of the product, costs, and the profit motive, while the manufacture of spacecraft demands quality first (Fig. 1–22).

Knowledge of Material Capabilities Is Mandatory

From the moment of initial processing to the instant that the part is finished, responsible personnel are cognizant of processing capabilities as they relate to the chemical, physical, and mechanical properties of the material. Consequently, manufacturing personnel must not only understand the operations of forming, treating, and finishing, they must also understand the capabilities of the raw material. Persons who are proficient in metallurgy and materials science are great assets to the manufacturing team and the company. These engineers and technicians realize the need for heat treatment of shaped parts to insure that proper strength will be imparted to the parts. They realize that the chemistry of the material provides the potential properties of the material, and they realize the service life hazards that the

FIGURE 1–22   Artist's concept of United States' Apollo command module and service module docking with Soviet Union's Soyuz spacecraft. (Courtesy of Rockwell International)

part must endure. Technicians must be cognizant of details such as the relationship between microstructure and strength of the material. Engineers must be able to control the levels of stress in the part, and the team must be aware of fatigue. The ship shown in Figure 1–23, for example, is a container of balanced stresses among

FIGURE 1–23   The *Alaskan Mail* makes a turn while its carefully constructed structure safely resists the heavy loads and stresses. (Courtesy of Newport News Shipbuilding)

steel plates, beams, columns, rods, and special shapes. Subsequent chapters of this text provide much of the information in materials science needed for effective support to manufacturing processes.

Questions

1. Explain why manufacturing processes produce uneven levels of production throughout a society.

2. Define a cultural lag and describe one which is related to manufacturing technology.

3. Why must the selling price of an article be established at a level high enough to include a fair return for factory maintenance?

4. List some goals of manufacturing.

5. Describe some consumer products that are available as the result of aerospace technology.

6. List some advantages for materials improvement that have occurred as the result of space explorations.

7. Explain why manufacturing is the backbone of progress.

8. List some raw materials which are extracted from the earth for use in manufacturing.

9. What has caused the rising need for materials that can withstand extremely cold and extremely hot environments?

10. Why must many materials be tested before being processed into parts?

11. List several manufacturing processes.

12. What is the relationship between strength of material and size of material when heat treating and weight are factors?

13. Name some very hard nonmetallic materials.

14. When do you believe it becomes essential to install automatic machines in a manufacturing plant?

15. Describe the makeup of the manufacturing team.

16. Why must there be an integration of people, machines, and materials for effective manufacturing?

17. Why is choice of material a critical decision?

18. Why must raw materials be challenged for internal soundness prior to manufacturing when stressed parts are to be produced?

19. What is nondestructive testing, and why should it be used throughout the manufacturing process when stressed parts are being made?

20. The study of manufacturing materials is concerned with the chemical, physical, and mechanical properties of materials. Why?

An Introduction to Manufacturing Materials

MATERIALS OF INDUSTRY

A *manufacturing material,* for purposes of this text, is a material in the solid state having engineering properties that can be calculated prior to the processing of the material into an object. Such a definition excludes liquids at room temperature, some gases, and many organic materials such as cloth, leather, and some petroleum products. Included in the definition are groups of materials such as metals, polymers, ceramics, glasses, concrete, rubber, and wood. By far, metals constitute the largest percentage of manufacturing materials.

Solid Materials

Many materials in the solid state are easily adaptable to forming operations through one or more of the manufacturing processes. During manufacture the solid may be changed into a liquid or mushy state, but the resultant product is solid. In the solid state these materials are divided into crystalline and amorphous types, depending on how their atoms are structurally arranged and the kinds of atoms constituting their identity. When in the crystalline state the material consists of grains or crystals and is often called granular. (Metals and most minerals are crystalline.) In this arrangement the metal's atoms are patterned into larger lattice-type structures, and, in turn, these are organized into grains which constitute the basic material, as in Figure 2–1. Note the individual grain that has been extracted from the bar. Surfaces of the metal bar are smooth due to manufacturing processes, but the material's interior, steel in this example, is crystalline and remains crystalline as long as it is solid. When a material is in the amorphous state, however, the interior of the material is noncrystalline and often resembles the surface characteristics. A common amorphous material is glass. It has been noted that common glasses fracture in a

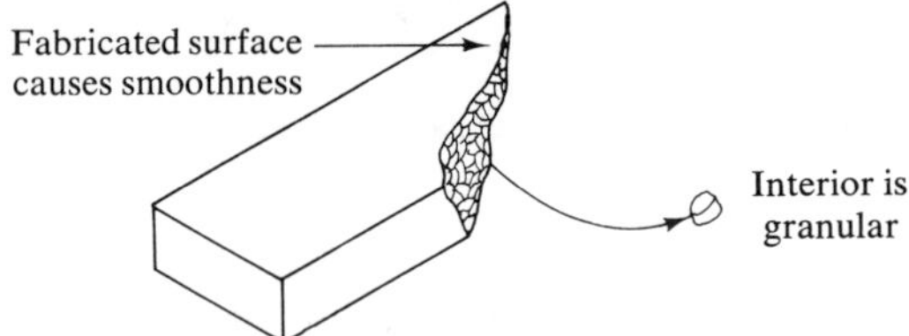

FIGURE 2–1   A coarse-grained fractured steel bar showing the removal of an individual grain.

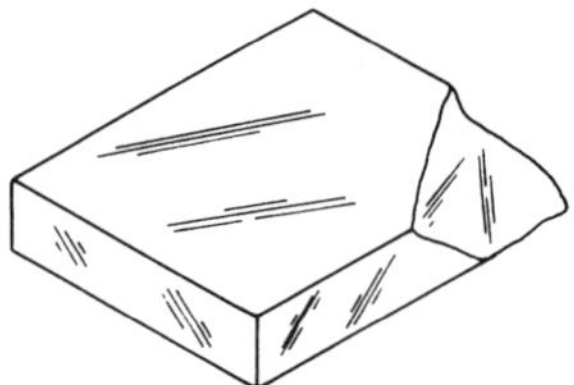

FIGURE 2–2   A fractured glass plate indicating the amorphous nature of glass. Internal and external appearances are similar.

noncrystalline manner, having no structural pattern resembling the granular structure of metal. Figure 2–2 illustrates the amorphous nature of glass. Atoms or molecules of amorphous materials are less organized and have different patterns than crystalline materials. A listing of some common manufacturing materials is shown in Table 2–1. As the table points out, manufacturing materials range from

TABLE 2–1   MANUFACTURING MATERIALS

| Material | Metallic | | Nonmetallic | | Characteristics |
	Element	Alloy	Organic	Inorganic	
Brass		X			Soft and ductile
Lead	X				Soft and weak
Concrete				X	Hard and brittle
Nylon			X		Ductile plastic
Hard tool steel		X			Hard, strong, and brittle
Wood, oak			X		Tough
Wood, white pine			X		Soft and weak
Aluminum	X				Soft and weak
Ceramic				X	Hard and brittle
Gold	X				Soft and corrosion resistant
Magnesium	X				Lightweight and will burn
Bronze		X			Good bearing qualities
Acetate			X		Transparent plastic
Brick				X	Hard and brittle
Titanium	X				Lightweight and heat resistant
Copper	X				Soft and heat conductor
Cast iron, gray		X			Soft and brittle
Glass, common				X	Hard and brittle
Wrought iron		X			Soft but tough
Mild steel		X			Structural strength

crystalline to amorphous and from organic to inorganic. Each material has its particular characteristics, and these characteristics help the engineer and technician to decide the material to be used.

Capability Predictions of Materials

Manufacturing materials have engineering properties which identify the material so that capability predictions can be made prior to manufacturing. Capability is incorporated within chemical, physical, and mechanical properties of the material. Together, these three properties provide scientific data for use by the engineer and technician. Chemical properties refer to how much of each constituent is present. This analysis ultimately originates with the material's atoms and their lattices, or molecules, which are the atomic building blocks of nature. Physical properties based on the material's chemistry refer to constants such as weight and density, melting point, and electrical factors. Mechanical properties, on the other hand, pertain to the large mass of engineering factors surrounding the material's capability, other than physical or chemical. Several physical and mechanical properties are listed in Table 2–2.

TABLE 2–2 MATERIAL CAPABILITY FACTORS

Physical Properties	Mechanical Properties
Density	Hardness
Melting point	Softness
Boiling point	Elasticity
Corrosion factor	Ductility
Magnetic property	Malleability
Electrical property	Elongation
Coefficient of expansion	Reduction in area
Thermal conductivity	Rigidity
Atomic weight	Brittleness
Atomic number	Plasticity
Specific heat	Tensile strength
Crystal structure	Yield strength
	Elastic limit
	Modulus of elasticity
	Toughness
	Shear strength
	Compression strength

The mechanical properties are often the chief concern of the engineer and technician. Examples of mechanical properties include hardness, toughness, strength, and malleability. Hardness is the resistance to penetration or deformation from an external source. Strength, on the other hand, is the ability of a material to resist being pulled apart as in a tensile force, or the ability to resist being crushed as in a compression force, or the ability to resist being cut as in a shear force. The relationship between hardness and strength is approximate. Toughness, however, is the ability of a material to resist a combination of forces at the same time, such as the impact of a falling object. Malleability of a material refers to the ease with which it is permanently deformed under compression forces, while ductility, very closely related, refers to the ease with which a material is permanently deformed under tension forces. In metals, these properties often change, whereas chemical and physical properties are fixed, unless altered by a special treatment.

A material's chemistry, the summation of the atomic arrangements, lays the foundation for what a material can do with respect to resisting external forces or with regard to manufacturing qualities. Mechanical properties of a material originate from the chemical factor. From the engineer's point of view, three specific properties—*tensile, compression,* and *shear*—have primary engineering concern. Figure 2–3 illustrates these three main engineering strengths of materials.

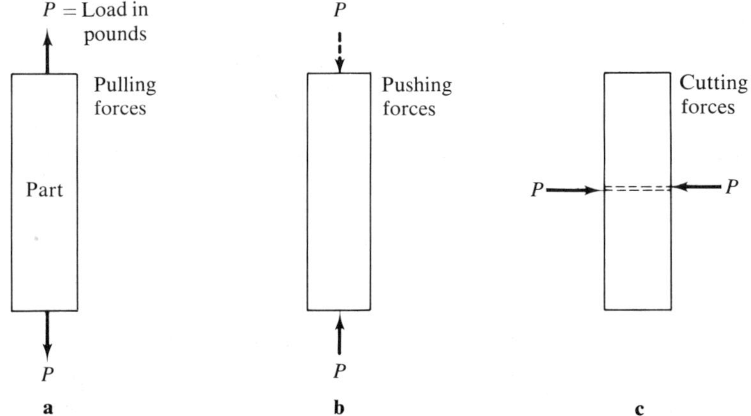

FIGURE 2–3   Three metal specimens being exposed to tensile loading (*a*), compression loading (*b*), and shear loading (*c*). Note that the tensile load attempts to stretch the specimen, the compression load attempts to crush the specimen, and the shear load attempts to cut the specimen.

Availability of Materials

Manufacturing materials are available to manufacturers in sufficient quantities to satisfy consumer demands in most situations. This quick source has not always existed however. Only during the past few decades have sufficient quantities of most materials been available. During the 1940s shortages existed in some materials because of wartime conditions and because some metals were not yet available in production quantities. Titanium was used primarily as an alloying element in steel during World War II, whereas today it is produced as a pure metal or alloy in shapes sufficient to meet most design requirements. Economics, geographical sources, and know-how have provided the technology with an abundance of industrial materials for use in manufacturing. But this great abundance does not signal the limits in material availability. Demands of design cause the continuation of the search for new materials.

Engineers, architects, designers, technicians, and specialists continually search for materials to fulfill their production requirements. Many materials available today were nonexistent in the past decade. The coming decade will introduce materials that are unheard of today. Many of today's materials will, in the near future, be cleaner and freer from discontinuities so more reliance can be placed on their capability. There is a vast spectrum of available metals. Just a partial listing would fill a small book; however, the alloys of steel, cast iron, aluminum, magnesium, nickel, copper, zinc, tin, and titanium provide the bulk of manufactured objects from metal. The nonmetals also provide a sizeable percentage of manufactured parts, especially the plastics and ceramics, but the metals provide the foundation for highly advanced cultures.

Diversity of Materials

The array of manufacturing materials is huge. (Using melting points of metals as a measure, for example, metals range from one used in fire extinguisher systems which melts at 150 °F (66 °C) to tungsten, which is used in high temperature applications and will not melt at 6000 °F (3316 °C).) Even though metals continue to be the manufacturer's number one material in fabrication processes, plastics are rapidly becoming more important in manufacturing. The ability of some plastics to be laminated, which greatly increases stiffness, has provided increased capabilities in design configurations, especially in the airfoil industry. The speed of aircraft (Fig. 2–4) and missiles has increased to such proportions that heat generated by

FIGURE 2–4 The A-7E combat aircraft includes materials and designs not available a decade ago. (Courtesy of Vought Corporation)

air friction has limited the high temperature capabilities of most metals in these vehicles. Therefore, more capable materials such as titanium, special alloys, austenitic steels, the graphites, and compounds are being used where materials are exposed to higher temperatures. A great advantage of titanium, for example, is its light weight along with its ability to structurally function, in some intsances, at 1000 °F (538 °C). Austenitic steel has even greater heat resisting capabilities.

Experiments with materials at elevated temperatures continue to provide more capable materials. Most engineering materials cannot be used at temperatures in excess of 1000 °F (538 °C). Presently, however, there is a material consisting of carbides or oxides which functions adequately in specified areas of structures at temperatures in excess of 3000 °F (1650 °C).

But heat is at one end of the temperature utilization spectrum for materials. At the other end is cold—temperatures approaching absolute zero. Iron or ferrite, for example, becomes brittle in the presence of subzero temperatures. Here again, cryogenic temperatures challenge and defeat most engineering materials which function satisfactorily in seasonal temperatures from winter through summer.

FORMABILITY OF MATERIALS

Engineering designs pertaining to metals and some plastics increasingly demand new shapes in all kinds of structural parts and consumer objects. These demands cause materials producers to increase their research in production shapes. Consequently, many manufacturing materials, the metals and plastics especially, are available in numerous standard shapes. Forming includes such operations as bending, compressing through plastic deformation, cutting, plastically stretching to shape, powdering the material and compression molding, and various melting techniques. Formability is then often a requisite of materials such as metals.

Some metals are more easily shaped than others; low carbon steel, for example, can be bent cold or hot into many shapes (Fig. 2–5) and has good machining properties. On the other hand, magnesium will often crack if bent cold, as illustrated in figure 2–6, but when heated will deform to the desired shape. The ability

FIGURE 2–5 Magnesium and its alloys fracture when bending loads are applied while the metal is cold. A low carbon steel bar bends easily while hot or cold (70 °F or 21 °C).

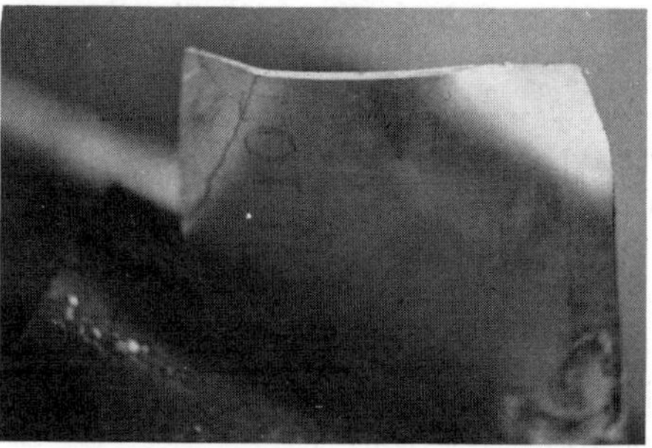

FIGURE 2–6 When magnesium is heated to approximately 400 °F (204 °C) bending is easily accomplished, but the metal fractures when a bending load is applied at room temperature.

of a metal to be shaped, either hot or cold, gives it high acceptance in the industrial market. Sometimes, however, metals are often cast from the liquid state because of poor forming qualities in the solid state. For example, alloys containing large amounts of chromium and molybdenum are frequently melted and cast because of their inherent stiffness, even in the white-hot condition.

Manufacturing materials can be formed into desired shapes with the use of modern manufacturing methods. Melting a metal (Fig. 2–7) and then pouring the

FIGURE 2–7 Molten steel is poured into ingot molds for subsequent rolling operations. (Courtesy of Bethlehem Steel Corporation)

liquid into a hole is one of the oldest methods of shaping, and hammering a hot metal was a technique used several centuries before the time of Christ. Today these two methods constitute a large portion of forming operations. On the other hand, many metals and nonmetals lend themselves favorably to powdering for subsequent pressing processes (Fig. 2–8). Shaping a material is accomplished by molding and casting, pressure forming such as hammering or rolling (Fig. 2–9), and by the use of powdered materials which are pressed into desired shapes and sintered.

STRENGTHS OF MATERIALS

All materials have certain strengths; that is, they are able to withstand specific loads and pressures without failure. The load may be a single load as in tension, or multiple loads may exist concurrently. A material's resistance to being broken or deformed while under load relates to its strength. The specific strength may be in tension (pulling), compression (pushing), or shear (cutting), as illustrated in Figure 2–3. As an example, a one-inch square section of mild carbon steel is pulled

FIGURE 2–8 Powdered metal is placed into the die and the ram descends and compacts the powders into a briquette. The pressed sprocket is subsequently sintered and is then ready for use. (Courtesy of Martin Sprocket and Gear)

FIGURE 2–9 Steel plate is rolled on the 160-inch plate mill. A white-hot slab enters the mill's roughing stand while another already through the stand moves to the modern four-high finishing stand. (Courtesy of United States Steel Corporation)

apart under a load of 72,000 pounds. For comparison purposes, the same size bar of a specific alloy steel requires 144,000 pounds to pull it apart or to crush it in compression. (For most ductile steels the tension and compression strengths are usually approximately equal.) It is then stated that some materials have low strengths and others have high strengths. Another example relates to a nylon plastic failing in tension at 11,500 pounds per square inch of cross-sectional area while 300,000 pounds fails to separate the same size alloy steel specimen containing nickel, chromium, and molybdenum. The term *pounds per square inch* (psi) relates to the number of pounds that is applied to the cross section of the material which is measured in square inches. In the International System of Units (SI) the term is the pascal (Pa), which can also be expressed as the newton per square meter, $N m^2$.

Metals Are the Strongest Materials

Information in Table 2–3 points out some common strengths of various materials. Values are pertinent to the specific specimens tested; therefore, values indicated are approximate. Note that the metals excel in the various types of strength; the plastics and other nonmetals are weaker in resisting forces tending to pull them apart or to crush them. The nonmetals vary considerably: the hardwoods are twice as strong as the softwoods; yet both types of wood are much weaker than the metals. Another example is the inorganic nonmetal, such as concrete and clay products, which are both hard and brittle. This brittleness in the nonmetals is demonstrated when tensile forces are applied onto the materials. The hard and brittle nonmetals have their strongest strengths in compression loading. For comparison purposes, a brittle metal, gray cast iron, is very similar to concrete in its mechanical properties and is also mostly used in compression stress. This brittleness in gray iron is pointed out in its microstructure (Fig. 2–10) where the dark stringers of graphite provide the brittleness. In most instances, metals are stronger than nonmetals, due mainly to the absence of a high brittleness factor. On the other hand, most of the structural metals have equal tensile and compression stresses while their shear values are about 60% of their tensile.

The strength of a material is exhibited in many ways. The material often resists static or steady loads, but in many instances the load may be applied suddenly or dynamically. In the latter example, failure in the material occurs more readily than

TABLE 2–3 STRENGTHS OF VARIOUS MATERIALS

Material	Condition	Strength (psi)		
		Tensile	Compression	Shear
Low carbon steel	As rolled	55,000	55,000	39,000
Structural steel	As rolled	72,000	72,000	42,000
Nickel-chromium steel	Heat treated	220,000	220,000	160,000
Ni-Cr-Mo steel	Heat treated	240,000		
Wrought iron	As produced	49,000	48,000	40,000
Gray cast iron	As cast	20,000	80,000	18,500
Stainless steel	Heat treated	270,000		
Nickel-copper	As rolled	87,000		51,000
Yellow brass	As rolled	55,000		
Aluminum-copper	Heat treated	69,000		41,000
Magnesium	Heat treated	48,000		
Nylon	Cast	11,500		
Concrete	Cured		6,000	
Hard maple	Dried		8,400	
White pine	Dried		4,600	
Brick, clay	Dried		3,500	

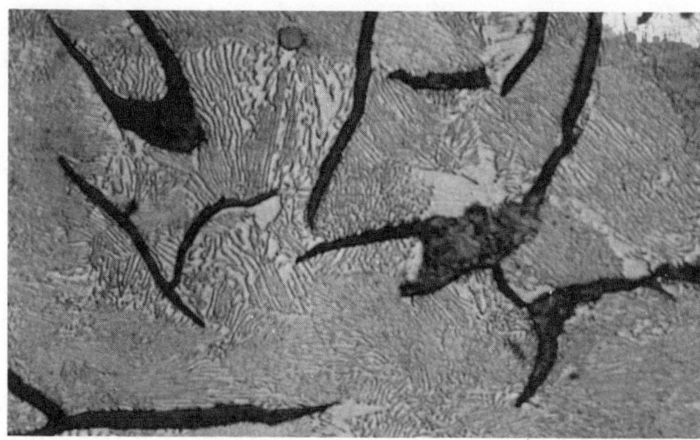

FIGURE 2–10   Concrete and gray cast iron are brittle materials.

in static loading when all other factors are equal. Certain materials resist shock loading better than others. Molybdenum steel, for example, is tougher in impact than carbon steel of equal hardness; therefore, it is used in aircraft landing gears where shock loading occurs.

IMPORTANCE OF COSTS OF MATERIALS

The cost of a material often influences its design. Basically, the least expensive material that will adequately do the job is frequently considered the best buy. For instance, in twist drill manufacture carbon steels are less expensive than special alloys and, in some cases, the carbon steel tool will adequately drill the hole to specifications. Should the need exist to drill the hole faster or deeper, however, the carbon steel drill fails at the cutting edges due to heat generation and microstructural change. The special alloy drill, on the other hand, continues for many hours under adverse conditions without sharpening. In this example the least expensive drill over an extended period of time is the one which initially costs the most—the special alloy drill. The carbon steel tool is the least expensive for other types of manufacture; for example, it drills effectively in soft woods and some metals.

Raw materials are available in many shapes and sizes, the needed shape depending mainly on the design of the part to be made. Metals are available in bars, rods, tubes, sheets, plates, powders, and wires, while plastics and glass are also available in most of these shapes. On the other hand, most other nonmetals are available in powder and organic forms, such as cement, wood, and rubber. Choice of raw material is then also influenced by cost.

A design requiring a safe operating tensile load of 45,000 psi of cross-sectional area could adequately and economically, in specified circumstances, be satisfied by the use of an aluminum alloy. The purchase of nickel-chromium steel having five times the needed strength in addition to a higher cost factor is wasteful. In this illustration, the extra strengths in the alloy steel would never be called upon. On the other hand, a plastic or even wood may suffice in lower strength applications. Cost consideration is therefore a compromise, a compromise among the capability factors in the material and the purchase price. Again, the lowest priced material that will effectively accomplish the task is usually the best purchase.

Choice of Material

Choice of material influences processing all the way to the finished part. Processing includes the total forming, treating, and finishing operations. A material which responds to heat treatment, for example, may sometimes be less expensive due to the need for less material for a given strength purpose. Heat treatment can increase the strength in a given volume of receptive material. Heat treatment, in turn, allows smaller dimensions of the material and less cost and weight. Engineers, designers, and technicians work as a team in material selection and processing procedures because proper choice of materials is usually critical with reference to safety purposes. Naturally, the proper choice influences manufacturing profits.

WHAT IS A MANUFACTURING MATERIAL?

As previously stated, manufacturing materials include many of the metals and nonmetals. In order to better describe a manufacturing material, we must present certain fundamental data. Basically, all materials are produced from the elements, the earthly things from which all matter is made. Table 2–4 is a list of some of the common elements included in industrial manufacturing; most of them are metals.

Pure Metals and Alloys

Many metals are often used individually in the manufacture of a part and many are combined to form alloys. A pure metal, aluminum, for example, is an element. When two or more elements, such as nickel and chromium, are melted together, an alloy is formed. Alloys are common among the metals. Further, when two or more elements combine chemically, a compound is formed, such as the combination of iron and carbon to form iron carbide, $3Fe + C \rightarrow Fe_3C$. The compound is quite unlike the parent materials. (This is an example of the combination of a metal and a nonmetal.) Three basic types of materials are therefore available for manufacturing use—the elements, the alloys or combinations of constituents, and the compounds. Frequently, the metallic alloy also includes compounds with the pure metal and sometimes contains compounds as impurities. In any situation, the metals solidify in the crystalline state, the elements or the alloy and compound constituting the material of the many grains.

Nonmetals

Some inorganic and nonmetallic materials such as clay and lime products also have crystalline structures, sometimes similar to metals, but the molecular arrangements of nonmetals are more complex and their atoms in molecular arrangements vary in size when compared to metallic atoms in lattice arrangement. An orderly molecular arrangement of a silicate structure, quartz, for example, will form crystalline material, whereas a disorderly structural arrangement of silicate will form glass and is amorphous. In clays the principle matrix or base materials include silicon, aluminum, oxygen, and hydrogen in compound forms. In addition, there are often other metallic compounds mixed in the clay matrix. In a common glass, for example, the constituents include limestone, soda ash, and silica sand.

TABLE 2–4 COMMON ELEMENTS USED IN MANUFACTURING

Element	Symbol	Typical Manufacturing Use
Aluminum*	Al	Housewares, aircraft structures
Antimony*	Sb	Tin alloys, battery plates, type metal
Beryllium*	Be	High strength and heat resistant; brakes
Bismuth*	Bi	Used in fusible alloys melting at 160°F (71°C)
Boron	B	Hardening element in steels; fibers stiffen plastics
Cadmium*	Cd	Plating on steel tools; corrosion resistant
Carbon	C	Chief hardener in steels
Chromium*	Cr	Hardener in steels and for plating
Cobalt*	Co	High temperature applications and as alloy in steel
Copper*	Cu	Electrical apparatus; excellent conductor
Gold*	Au	Extremely corrosion resistant; instrumentation
Iron*	Fe	Base of all steels as ferrite
Lead*	Pb	Radiation shielding, solders, cable sheathing
Magnesium*	Mg	High strength-weight ratio; aircraft parts; will burn
Manganese*	Mn	Hardener in steel; constituent in all steels
Mercury*	Hg	Very toxic; silent electrical switches
Molybdenum*	Mo	Alloying element in steel; toughener and hardener
Nickel*	Ni	Alloying element in steel; toughener and is stainless
Niobium*	Nb	In stainless and high temperature applications
Nitrogen	N	A gas; when combined with iron forms iron nitride which is very hard
Oxygen	O	A gas used in a type of steel manufacture; forms film on metal
Phosphorus	P	Used in some castings such as bronze for deoxidation
Platinum*	Pt	Highly resistant to corrosion and heat
Potassium*	K	Highly reactive; used as compounds such as in cyanide for case hardening steel
Rhodium*	Rh	High temperature use; thermocouples
Selenium*	Se	Electrical apparatus, photo cells
Silicon	Si	Rectifier of electricity; alloying element
Silver*	Ag	Corrosion resistant; heat and electrical conductor
Sodium*	Na	Coolant for exhaust valves; a reduction agent in metals manufacture
Sulfur	S	Used in making acid; a constituent in many chemical processes
Tantalum*	Ta	Combines to form hard carbides; used in heat and electrical applications
Thorium*	Th	Forms a high temperature resistant oxide
Tin*	Sn	Plating and in solder; an alloying element; used in food canning
Titanium*	Ti	As an alloy, subzero usage; high strength-weight ratio for structures
Tungsten*	W	As electrodes; an alloying element for high temperature use. Has highest melting point
Uranium*	U	Nuclear fission
Vanadium*	V	Control of grain size in steel; forms hard carbides
Zinc*	Zn	Corrosion resistant applications as in galvanizing
Zirconium*	Zr	Toxic and will burn; used in electrical apparatus

*Indicates a metal.

Being inorganic and thermoplastic, common glass is brittle when cold. Failure occurs more often in tension loading, but when heated above 1300 °F (704 °C), glass becomes deformable. In the cold or solid state a noncrystalline structure exists, this condition being similar to a liquid.

Organic Materials

Organic materials such as polymers are structures of large molecules having carbon, hydrogen, nitrogen, and oxygen in their analyses. The large molecule is a grouping of atoms clinging together to form its shape. This physical arrangement allows quick deformation through sliding action of the molecules in certain polymers. Their chemistry sometimes becomes very complex, however. These materials are mostly synthetic and are recognized as plastics and resins. During processing of certain chemicals, polymerization produces a molecular structure of long chains of atoms, some chains being connected along the several chains. These molecular arrangements in the polymers are the macromolecules. A linear type of structural arrangement in these materials can produce nylon, which is ductile up to a certain point and fairly strong for a plastic. When linking of molecules occurs along the chains, however, a different structural arrangement exists and the product is cross-linked; a typical thermosetting plastic, such as bakelite which is not ductile, is formed. Polymers are frequently amorphous because of their usually disorganized molecular arrangement. Because of differences in production processes and molecular arrangements, however, some crystalline areas may form in the matrix of amorphous material. The common plastics which are produced through polymerization are available in the thermoplastic or thermosetting types. The thermoplastic can be reheated repeatedly to reshape it, whereas the thermosetting type cannot. In order that manufacturing personnel may better understand the capabilities of materials, a short explanation of the building blocks of nature, atoms, will be provided.

ATOMIC STRUCTURE

An examination of Figure 2–1 shows a granular mass of material, all grains or crystals approximately the same size but each many sided in three dimensions. Each grain contains a quantity of matter or energy. To the eye the crystals are small, actually smaller than pin heads. However, when a prepared group of grains of gold, for example, is magnified (Fig. 2–11), interior structural arrangement

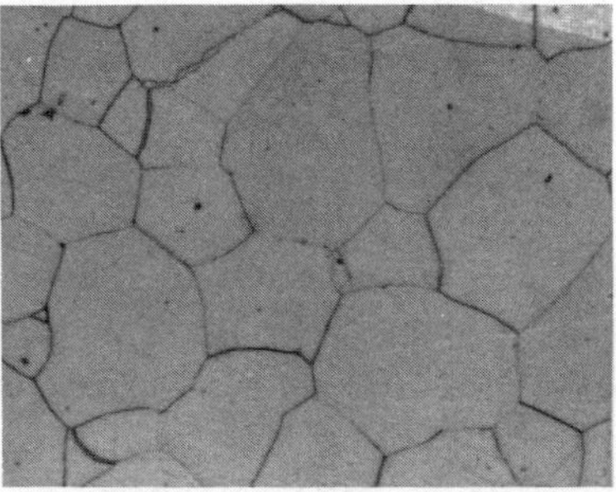

FIGURE 2–11 Magnified view of gold, 200×. Dark dots are polishing pits. Granular makeup is equiaxed.

is noted as only a homogeneous mass of parts of several grains. Upon further magnification, no clues as to atomic arrangement are found. Should the magnification process continue for many millions of times, however, a final atomic arrangement or structural pattern would be noted and this pattern would be the atoms in lattice form. No one has seen an atom, but its effects have been noted.

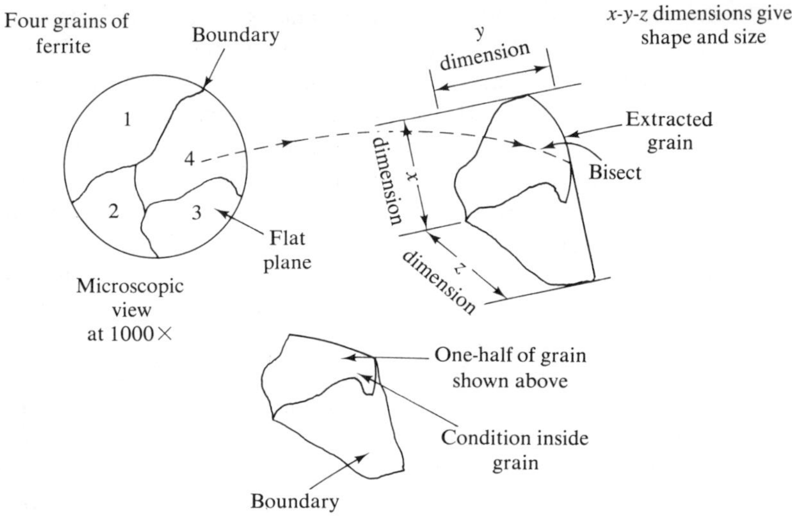

FIGURE 2–12   Parts of four grains are shown, one grain being extracted
and bisected to show the condition in the grain.

The Atom

An investigator who fractures or bisects a single grain of pure metal (Fig. 2–12)
finds the same basic material in the resulting parts. When a single part of the
bisected grain is also bisected, the same basic material is found again. In this
regard, a material can be bisected only a limited number of times; therefore, an
attempt to visualize such minute particles of matter by this method is impractical
and eventually becomes impossible. Nevertheless, should the halving process be
possible and then be continued, a size would be reached which demonstrates the
last bisection; the next attempt of material separation would cause immediate
material loss of identity. This last particle size which identifies the material is the
atom. No further reduction in size is possible without losing the atom's identity.
The little particle would be observed as a vibrating mass of energy, maintaining
its identity according to its place in the large family of atoms. It makes little
difference whether the atom is considered a particle or a mass of energy; however,
in these discussions it will be thought of as a particle due to the nature of solids.

Scientists have determined that the atom exists in organized or disorganized
company with other atoms forming a larger particle of the same material, a
molecule in some materials or a lattice network in metals(Fig. 2–13). The
mclecule may be a build-up of similar atoms or it may consist of several different
atoms resulting in a new material such as the compound sodium chloride (Fig.
2–14). As the molecules in plastics or the lattices in metals grow in number to
form larger quantities of material, a size is reached which is visible to the naked
eye. This mass of material is many millions of time larger than the atom. The
organized pattern of atoms in metals (Fig. 2–13) is unlike the loosely oriented
pattern having random positions, such as in amorphous materials like glass
(Fig. 2–2).

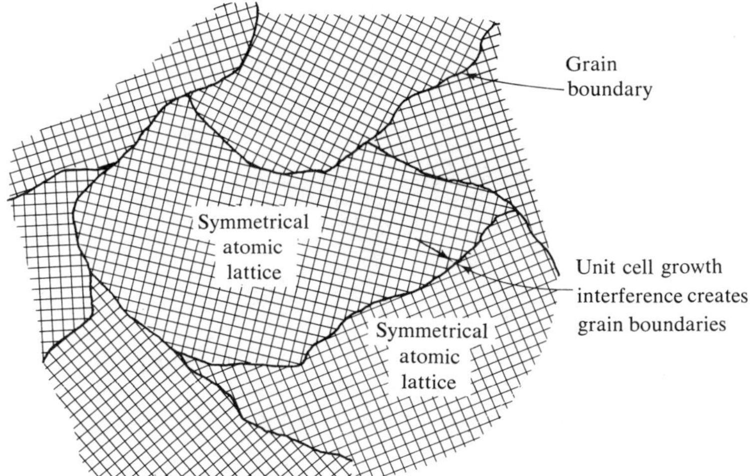

FIGURE 2–13   Grains of metal consist of lattice formations of atoms organized into rows of unit cells.

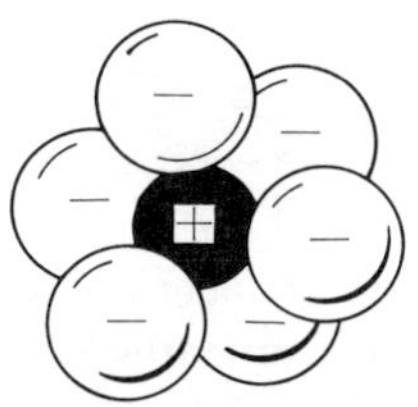

Sodium chloride, NaCl

FIGURE 2–14   Sodium is positive and chlorine is negative. The negative ions are attracted to the positive ion.

What Is An Atom?   An atom consumes space; therefore, its mass is measurable. It is predominantly spherical in shape, and its core consists of particles known as *protons,* which have positive electrical charges. Mixed with these protons and in close proximity in the nucleus are particles known as *neutrons,* which have approximately the same weight as protons but no electrical charge. The atom's weight is based on the weights of protons and neutrons, which is about 99% of the total weight. Spinning in elliptical orbits around the atom's nucleus at exceptionally high speeds are much smaller, negatively charged particles called *electrons.* The number of electrons in an atom equals the number of protons, giving the atom its identity. Such an electrical balance promotes stability in the atom, but not permanence in energy arrangement. In recent times, both theories about the nucleus and about the electron shells of the atom have been altered, as evidenced by the use of atomic power from the nucleus and the concept of electricity through electron mobility. Figure 2–15 illustrates an atom's probable arrangement in schematic form. Only the electrons in the first shell are shown; the three dimensional arrangement of the atom is illustrated in Figure 2–16.

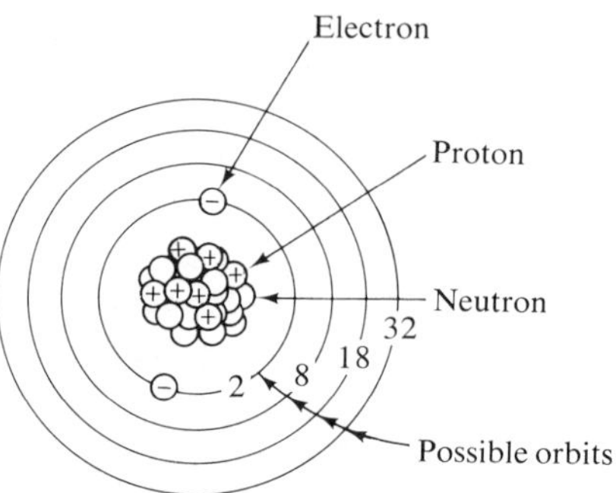

FIGURE 2–15 An atom's probable arrangement schematically illustrated in two dimensions, but the atom has a third dimension. It is spherical because of the rapidly spinning electrons around the cluster of positive protons and neutral neutrons.

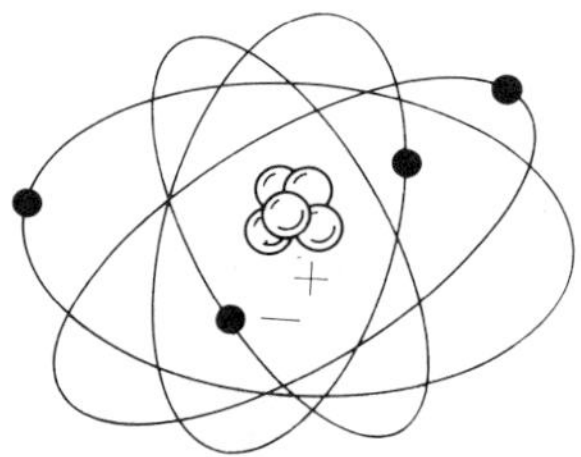

FIGURE 2–16 Orbital shells of electrons surrounding the proton-neutron nucleus. Three-dimensional electron paths act as outer limits of the atom.

Electron Shells   Electrons, spinning on their axes as they revolve, occupy paths around the atom; electrons orbit the nucleus in circuits at such tremendous speeds that a hard shell seems to surround the nucleus. Several electron orbits may exist around an atom and these orbits are called *shells*. The outer shell establishes a mean radius from the nucleus that provides a dimension to the atom. The number of electrons in the shells (paths) and the number of shells are variable and these different combinations identify the atom. The inner orbit of the atom, the lowest energy shell, has a capacity for only two electrons, the second shell can accommodate eight, the third will allow eighteen, the fourth up to thirty-two, and so on. This arrangement is schematically illustrated in Figure 2–15.

The energy level of the atom increases as the radius of the orbit increases. As the internal energy of the atom increases, electrons may shift to the next outer orbit. Orbits, or shells, which are not filled to their natural capacities with electrons are primarily responsible for chemical actions of the atom. Electrons in unfilled shells are called *valence electrons*. These outer shell electrons account for most of the engineering properties of the material because they establish the chemical properties of the material. In turn, the chemistry determines the atomic bonding and provides for atom capability. Consequently, the physical and potential mechanical properties of the material develop from the atoms. In metals valence electrons provide high mobility potential within the metal, accounting for excellent electrical and temperature conductivity. This shell concept of electron activity does not imply, however, that all electrons have the same energy levels or that all electrons are equivalent.

An Atom Is Mostly Space   The structure of an atom is such that most of the spherically shaped mass of energy is void of material. Distances between the nucleus and orbital paths of electrons are tremendous in proportion to the sizes of the particles involved. To illustrate this phenomenon the most common example of atom arrangement described by authorities is the solar system arrangement, and this is shown in Figures 2–15 and 2–16. In this respect, volumes of space between planets and the sun of this solar system and between electrons and the nucleus of an atom are mostly void of material; this absence of material is *space*. An atom is mostly space then, because space exists between the particles of energy, protons and electrons. Orbital radii of electrons are maintained by the resultant forces of attraction and repulsion. Electrons repel each other while protons maintain the atom's radius through electron attraction. When an atom's shell radius is changed, energy is absorbed or liberated. The loss of outer shells decreases the atom's size. Interatomic activity results from individual atom activity; that is, atoms cluster together in accordance with the powerful forces of equilibrium and maintain interatomic distances according to their natural arrangement.

ATOMIC BONDING

A solid material exists as a solid due to the exceptionally strong forces of attraction or bonding among the atoms. Engineering properties, both existing and potential, are basically established by atomic bonding. The simplest interatomic bond, used here for the purpose of clarity, is the *ionic bond* whereby equal positive and negative charges are mutually attracted in matter. Figure 2–17 shows this bonding arrangement for sodium chloride. Electron stability is established in the outer shells when an electron from the sodium outer shell moves to the chlorine outer shell. The resulting positively and negatively charged *ions* have a strong attraction, and a strong ionic bond produces this new three-dimensional material from two completely different materials, a solid and a gas. An ion is an atom with a positive or negative charge.

Covalent Bonding

Another common bond is illustrated in the example of two chlorine atoms (Fig. 2–18) sharing two electrons in their outer shells, thus forming a stable chlorine molecule. In this example two like materials join to produce a larger mass of the material. This *covalent bonding* produces very strong bonds in different materials. Many materials, especially the brittle, have covalent and ionic bonds. It is the final assembly of atoms, beginning from a minute starting point of mass, that provides the material which is observed and used. An analysis of a piece of material ultimately divides it into its smallest components, the atoms. Synthesis and continued build-up of these millions of energy particles establish the observable piece of material.

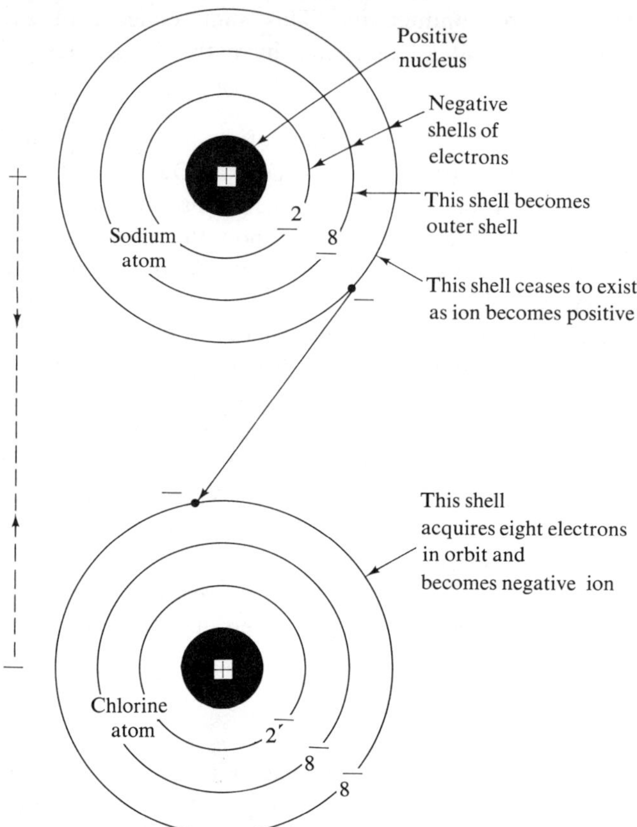

FIGURE 2–17 Electron transfer from sodium atom to chlorine atom provides stability in outer shells. This ionic bonding is between unlike charged ions and produces sodium chloride.

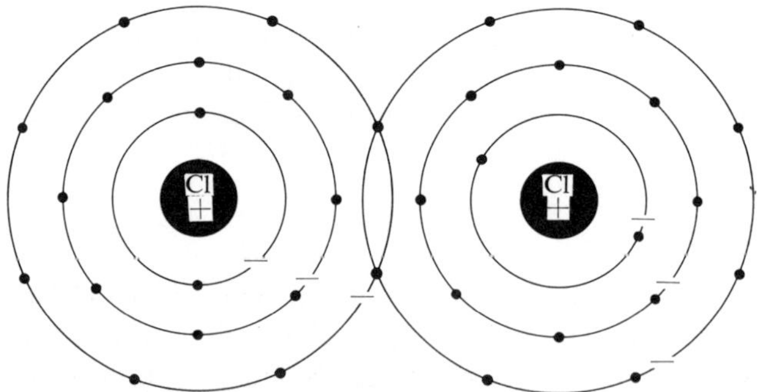

FIGURE 2–18 Covalent bonding in a chlorine molecule. Note the sharing of electrons in the outer shells of both atoms.

Metallic Bonding

In metals a different bonding arrangement, *metallic bonding,* exists due to extreme flexibility and mobility of the electrons. This mobility establishes engineering factors such as electrical current, heat dissipation, opaqueness, and metallic plasticity. Metallic bonding occurs among the atoms (ions) and their valence electrons roam at great relative distances from the nuclei, this movement depending on environmental conditions. Positive cores of the proton-neutron masses, together with the nonvalence electrons, are oriented and attracted to the negative cloud of highly mobile electrons. Electron mobility in the metals is so great and flexible that the gaseous or liquid description appears to be the best for this phenomenal activity. The attraction is strong and variable between the positive ions, the atoms with lost electrons, and the negative mass of electrons. The strength of the bond is like a super powerful glue. In fact, as an example, attempts to pull metal apart demand very strong forces; often thousands of pounds is required to separate the binding force within a very small section of metal. This strength, along with other mechanical and chemical properties, is mainly reflected from the activities of the valence electrons.

TEMPERATURE EFFECTS ON THE ATOM

Gases, liquids, and solids behave in certain patterns of activity. As the temperature of a solid increases, energy vibrations increase until the atoms lose their attachment in the lattice and form a liquid. In the gaseous stage the atoms are more extended from each other. Because most elements are metals, the physics of the metallic bond is very important to the engineer and technician.

The silent activity in the atoms' submicroscopic world persists in maintaining its inherent order and balance even though continuing compromise is part of this phenomenal activity. Thermal agitation and pressure cause variable or increased vibrations among the atoms. Solid stability in the material ceases as temperature increases to a point where shear strength of the bond reduces toward zero. Atoms are very close together, but in liquids they are loosely bonded as compared to solids. The closeness is evidenced by the near incompressibility of liquids. As temperature increases, however, atoms become more randomly arranged and move farther from their neighbors, accounting for the compressibility of gaseous matter. On the other hand, metals in the liquid stage become solids as temperature is reduced. Reduction in liquid metal temperature causes solidification, and solidification in the metals is crystalline. The crystalline structure is produced naturally from the millions of well-organized atomic building blocks known as lattice (Fig. 2–13). With respect to future failure of a metal, the failure in no way is caused by crystallization because crystals constitute the metal from time of creation in the solid state.

In metals the lattice arrangement of assembly exists whereby atoms share positions with adjacent atoms in the build-up of the metal. Because of its complexity in arrangement and phenomenal activity, the lattice network will be explained in subsequent chapters. In the meantime, the summation of the particular types of lattice networks of atoms should be accepted as providing the particular granular material. Consequently, all materials are large masses of atoms bonded together in three dimensions.

MATERIAL CONSTITUENTS

Manufacturing materials are produced according to the need for these materials. As Table 2–1 shows, a material may be an element such as gold, a combination of elements such as steel, a biological product such as wood, or a combination of compounds or elements such as ceramic and plastic. Consequently, the term *manufacturing material* is very general and gives no hint to its identity other than it is a solid that has definite engineering properties. These properties are derived from the inherent nature of the material. Atomic patterns and physical arrangements of the material within the solid set the capability performance of the material. Four main physical arrangements are subsequently discussed: the element, the mechanical mixture of the elements and/or compounds, the solid solution of the elements and/or compounds, and a combination of any of these arrangements.

The Element

Most of the elements, according to the definition of a manufacturing material, are metals (see Table 2–4). In this respect, a large percentage of metal production is devoted to pure metals, for example, magnesium (Fig. 2–19). Any pure metal

FIGURE 2–19   The Dow cells operate at about 1292 °F (700 °C) using more than 100,000 amperes of current. Graphite bars are anodes which are used in the production of liquid magnesium within these cells. (Courtesy of Dow Chemical U.S.A.)

therefore, being an element and being granular, exhibits a homogeneous structure within the granular arrangement. Figure 2–1 depicts the granular structure of the alloy called steel, and the same granular arrangement exists within pure metals, as shown in Figure 2–11. Of chief concern to the engineer and technician is the structural pattern or arrangement of the material within the individual grains. This pattern of chemicals reflects most of the material's mechanical properties.

Because grains or crystals of material are small in size, the metallurgical microscope is used to magnify the structure in order to comprehend the arrangement and predict the material capabilities. Figure 2–11 is a photomicrograph of a pure metal at 200 magnifications. A homogeneous mass is observed. The flat surface is actually the section of metal which was cut, polished, and etched. (Specimen preparation is similar to slicing an apple to reveal the seed arrangement.) Accidentally, and unavoidably, several grains were penetrated and partly removed during the cutting and polishing operations (indicated by the irregular shaped lines). Some grains will appear smaller than others, therefore, because they were parted near an edge of their boundaries. The lines on the photomicrograph are the grain or crystal boundaries and these grains have three dimensions, depth not being shown in the illustration. The material within the lines of this example, the volume, shows only one constituent and this is the element. Aluminum, gold, and other pure metals have similar appearances when viewed under the microscope. Microscopic magnification enables the observer to study the magnified microstructure so that material arrangement can be analyzed and mechanical properties predicted.

The Mechanical Mixture of Constituents

The next example of material arrangement is illustrated in Figure 2–20. The sketch shows more than one constituent. This structure consists of several

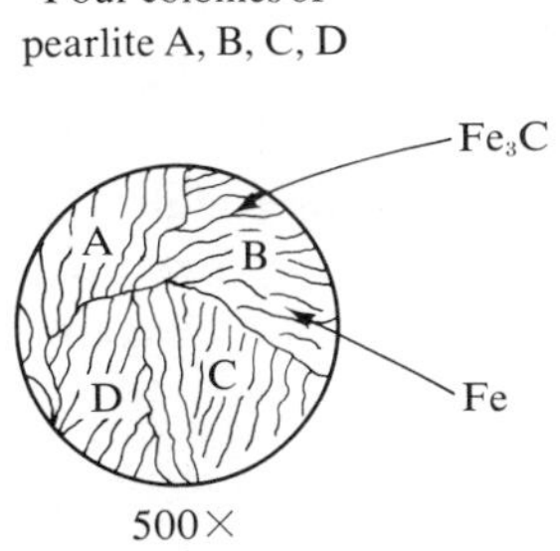

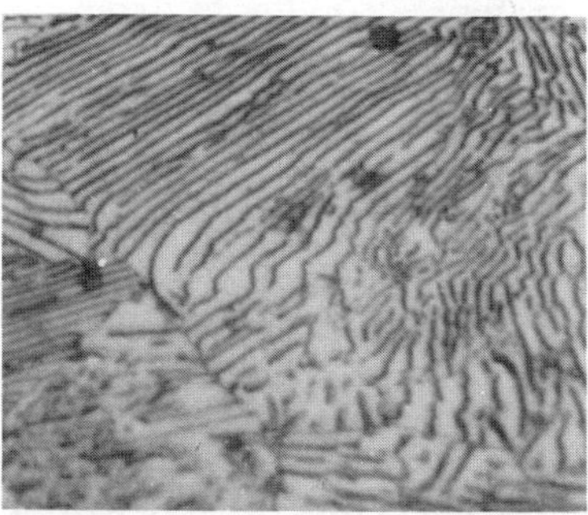

FIGURE 2–20   The microstructure of annealed eutectoid steel consists only of pearlite

elements. The base material, ferrite (iron), is the lightly colored constituent. The dark constituent is the chemical compound, iron carbide, and is mixed within the matrix of ferrite and other elements. Two constituents are shown; therefore, this material must be an alloy rather than a one-constituent metal, an element. This particular alloy is annealed steel, and the mixture arrangement among the two constituents shown is called *pearlite* and is only one of several possible arrangements. *Annealing* is a heat treatment that induces softness in a metal.

Each arrangement of constituents, called a *microstructure,* provides specific mechanical properties. A steel's strength variability due to microstructure within a given chemical analysis of 0.8% carbon, for example, is from approximately 90,000 psi tensile in a mixture (Fig. 2–20) to over 300,000 psi in a solid solution (Fig. 2–21). A solid solution exists when one material is absorbed into another. The continuous and curved lines shown in Figure 2–20 are grain boundaries where individual crystals rest against each other.

The structural arrangement of a material's contents within the particular type of irregularly shaped grains predominately provides the mechanical properties of

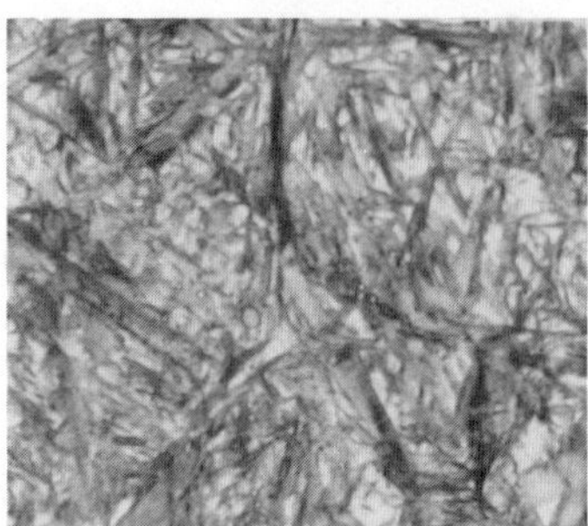

FIGURE 2–21   Martensite in a eutectoid steel at 1000×, etch is 2% Nital.

the material. Atoms of an element assemble into larger masses of the same atoms. Atoms of different materials assemble according to their specific bonding arrangements and ultimately form the observable mass of material.

The Solid Solution of Constituents

The third example of material arrangement is reflected in Figure 2–22. Only one constituent is seen, even though several elements are present. The lines are grain boundaries. Basically, the appearance of this structure resembles the pure metal shown in Figure 2–11. The resemblance to the pure metal is caused by the elements copper and zinc being in solution and resulting in one matrix, or practically one constituent for observation and comparison purposes. The solid solution is made possible by flexible atomic spacing, whereby one element diffuses into another while in the solid state.

Interstitial and Substitutional Atoms   Atoms may space themselves between other atoms or substitute for one another. In an interstitial solid solution atoms of the additional materials randomly find spaces between atoms of the base material.

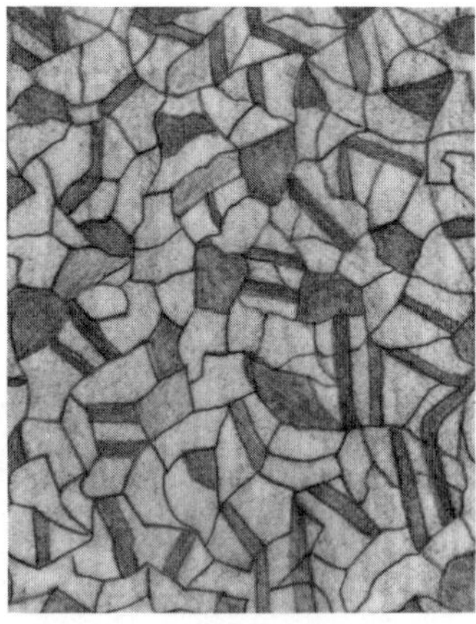

FIGURE 2–22   A solid solution of zinc and copper called brass. The etchant attacks each grain differently, causing light and dark surfaces on the many grains. Some twins are also observable. Pure metals and solutions have common characteristics.

In the substitutional solid solution some atoms of the added material, the alloying element or elements, randomly replace some of the atoms of the base material. In other words, two or more certain elements are soluble in each other in the solid condition. The result is a homogeneous solid solution. Sugar dissolved in coffee and salt dissolved in water are examples of solutions. The latter is no longer water, but is brine. The sugar or salt can be precipitated from the solution, however, resulting in a mixture of the two constituents which can again be observed. In the same way that salt is dissolved in water, one solid can diffuse into another solid. In the example shown in Figure 2–22, common yellow brass is used to illustrate the phenomenon of a solid solution, the diffusion of zinc and copper.

Another matter of interest with reference to metallic solid solutions is that the solution may or may not change back to a mixture as the metal's environment changes. As a solution the material then has several of the qualities of a pure metal; it is observed as one constituent even though several materials are present in the solution. Many of the solution's mechanical properties are so reflected. Pure metals and solid solutions have many very close relationships; however, properties of solutions and elements do differ tremendously, in some instances. As a mixture, resulting from a solution, mechanical properties are quite different from the solution. In fact, a mixture formed into a solution and then back to a mixture may have exceptionally different properties from those of the original mixture. Such is the case when red-hot carbon steel is cooled in oil, for example, to change the originally coarse pearlite into fine pearlite or possibly martensite. The term *pearlite* refers to a mixture of iron and iron carbide in layer form throughout the grain. These layers may be close together or separated into a coarse appearance. Martensite is a solid solution of carbon in iron and appears as one material.

Complex Structures

The fourth possibility is nothing more than one of several combinations of the above examples. This means that the material is structurally arranged whereby part of the material is in solution or is in compound form and part exists as a mixture. The total structure is therefore a mixture. For example, when annealed, the wrought alloy of aluminum with 4% copper produces a structure of a solid solution of copper in aluminum and a compound of aluminum and copper, $CuAl_2$, copper-aluminide. At room temperature the two constituents, the aluminum-copper solution and the copper-aluminide compound, are mixed as illustrated in Figure 2–23. There are three material arrangements present in this illustration: the solid solution of copper in aluminum, the lightly colored constituent; the compound $CuAl_2$, the dark constituent; and the total arrangement which shows the mixture of the solution and the compound. On the other hand, when this metal is heated to approximately 925 °F, the total mixture becomes one solution in the solid state. When rapidly quenched from 925 °F, the solid solution is retained and a super-saturated solid solution exists as the solution is unstable at room temperature. Then after a time lapse of several hours at room temperature, the solid solution of the two constituents begins to precipitate the $CuAl_2$ throughout the solid. The resultant material (Fig. 2–24) is quite different from the condition shown in Figure 2–23.

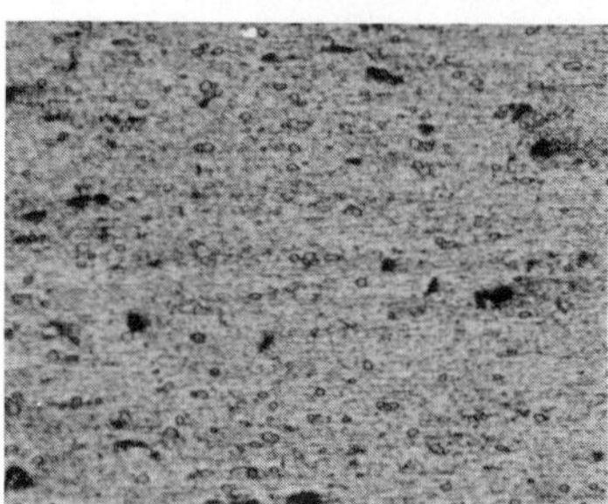

FIGURE 2–23 Annealed 2024 aluminum alloy. The dark particles are $CuAl_2$ (copper-aluminide) and the gray constituent is the aluminum-copper solution. Keller's etch at 200×.

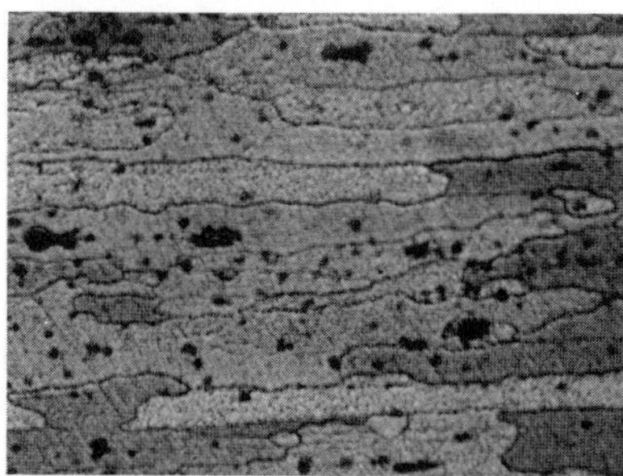

FIGURE 2–24 Aluminum alloy 2024 that has been solution treated and naturally aged. Keller's etch at 200×.

Another example of a complex structure is illustrated in the alloy of copper and zinc known as brass (Fig. 2–22). Copper and zinc are elements, and brass is an alloy of the two elements. When approximately 41% zinc is present, the alloy exhibits two distinct solutions when polished and etched, one being mixed with the other. Differences observed between the two solutions, light and dark regions on the surface of the metal, are due to one solution being richer in zinc than the other.

When intermetallic compounds, Fe_3C, for example, are formed in the alloyed material, atoms of the alloys replace atoms in the base metal according to certain proportions. Such an arrangement induces more atomic rigidity in the total material and, in turn, greater hardness and strength are available in the new material. With regard to atomic build-up in the alloys, the same general lattice pattern is established as in the pure metals. The different atoms are spaced in the lattice network in an organized pattern (Fig. 2–13), resulting in greater strength of the granular pattern in alloys than that in the pure metals. The other mechanical property potentials of the alloys are usually far greater than the potentials of pure metals because of the presence and spacing of unlike atoms in the lattice of the alloy.

NONMETALLIC STRUCTURE

In nonmetals—inorganic materials, for example—the atomic structural bonding is more complex and more fixed than in the metals. Also, atoms constituting the nonmetals have great size differences, some being fairly large. Flexibility of atomic arrangement exists within the metals, notably in electron relationships, whereas electron activity in the inorganic nonmetals provides a more rigid structural arrangement. Physical arrangement of material constituents described mainly for metals becomes increasingly complex as different compounds are formed in the nonmetals. Each kind of basic chemical constituting the non-metallic material provides its own combination in the organization of its ultimate structure. Man changes the structural patterns, existing or potential, by adding or deleting constituents or by causing chemical changes brought about by heating or other means during the material's processing. Nonmetals such as concrete, ceramics, wood, and thermosetting plastics are basically fixed in mechanical properties at the time of production, with some exception such as aging, while most metals are subjected to great changes in these properties long after production.

THE CHEMISTRY OF A METAL

The manufacturing process is based on the presence of a material. To the responsible person, the engineer or technician, a material is exactly what its analysis describes. Simply stated, a quantity of specific materials constitute the analysis. The analysis may include only one element, or it may contain at least six as in carbon-type steels, or it may involve only two metals, such as nickel and chromium. Further, the analysis may pertain to the plastics or other nonmetals. In numerous manufacturing situations, however, foreign materials are present but frequently not identified in the analysis. As an example, oxides may develop in the material somewhere along the manufacturing process because of mechanical handling. When these foreign particles work their way into the material, several undesirable situations may result in the end product such as cracks, other discontinuities, or potential areas for cracks. Consequently, the more the engineer knows about the material's constituents, the better will be the finished product. The manufacturing technician and engineer must understand the effects of total material chemistry if quality is to be effectively controlled during processing.

Questions

1. What is the main difference between a granular and an amorphous material?
2. Explain the lattice structure of atomic arrangement in metals.
3. What factors in metals provide for mechanical capability predictions?
4. Explain the three main engineering forces.
5. What elements help to resist fatigue failure in steels?
6. What are some of the differences between a pure metal and an alloy?
7. Define *iron carbide* and explain its difference from carbon.
8. Name two main classes of plastics.
9. Describe the total structure of the atom.

Mechanical Properties of Manufacturing Materials

The ability of a material to adequately perform in service demonstrates several mechanical properties of the material. For example, the network of bridges and roadways shown in Figure 3–1 typically illustrates the capabilities of materials.

FIGURE 3–1 A network of bridges and roads that safely holds the many vehicles that move along daily.

While in service, a material internally resists the forces which tend to deform it. Many of these forces are negligible, but frequently many are measured in thousands of pounds.

Resistance to deformation is illustrated in Figure 3–2. A weight of 10,000 pounds is applied to the beam and columns. Beam and column materials must be

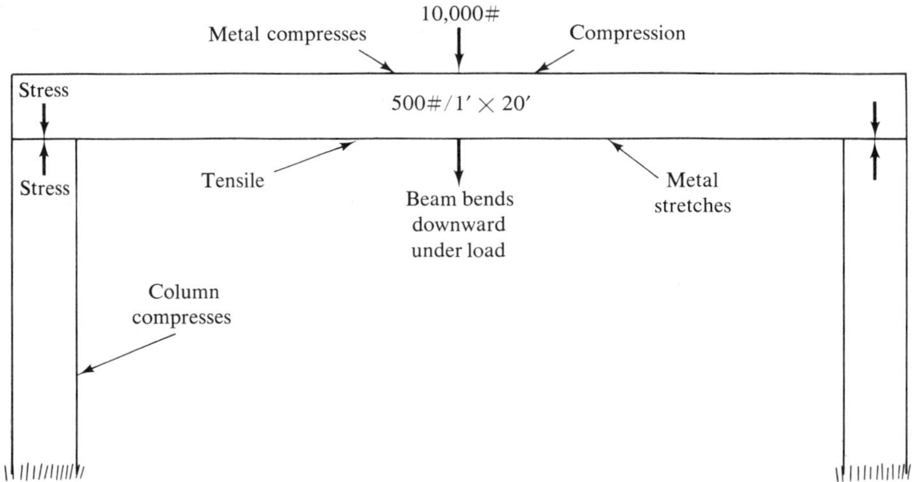

FIGURE 3–2 A schematic arrangement of a beam and two columns under load. The beam weighs 500 lb per linear foot and is 20 ft in length.

strong enough in their several strengths to safely resist the load. The columns are in compression and the beam bends downward elastically under the load. Deflection of the beam is controlled by the beam's shape and its modulus of elasticity (a stiffness factor), the tensile and compression strengths being in balance.

Compression and tensile strengths are only two of several mechanical properties existing in a loaded beam. Other properties such as shear strength and toughness are also present. Concurrent with mechanical properties, physical properties also react to service conditions, but it is usually the mechanical properties (Fig. 3–2) which mainly enable a material to perform day after day.

MECHANICAL PROPERTIES BECOME COMPROMISES

Time in service is sometimes considered to be a material factor because the engineer and technician want to know how long the material will satisfactorily function in its assigned capacity. Some materials, such as many household items of hardware, are used in the absence of loading conditions and, in this situation, the time factor is infinite. Even so, the material must still be able to hold itself together in its intended use. On the other hand, a metal bolt must hold parts together, parts which often attempt to cut the bolt (shear) or pull it apart (tension). These two forces, shear and tension, as illustrated in Figure 3–3, very often include loads of several thousand pounds; therefore, caution must be used to assure that proper mechanical properties exist in the associated materials so than no failure occurs at the joint. When a steel ax is suddenly applied to a block of wood, for example, the tendency of motion is to temporarily stop at the instant the ax and wood are in contact. The metal ax must absorb the shock loading and, at the same time, be able to penetrate the softer wood. In this situation, if

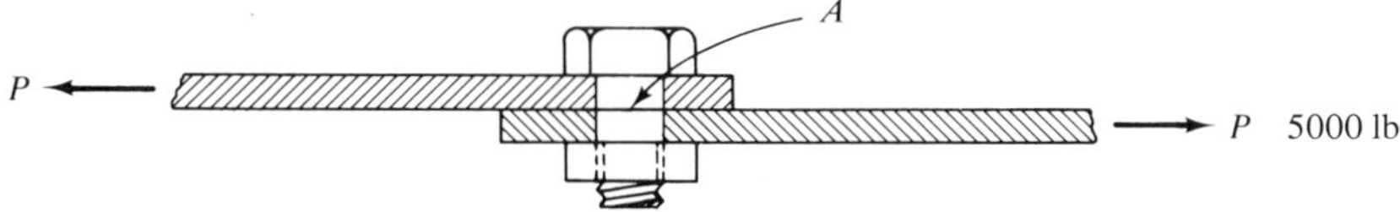

FIGURE 3–3   A bolt in shear. The tensile load P acts on the bolt at area A causing shear stresses in the bolt and tensile stresses in the two plates of metal.

the combination of structural conditions, such as hardness and softness, is not correct for such usage, the ax will fly apart at the moment of contact with the wood or the cutting edge will roll over and lose its cutting ability.

A manufacturing material usually exists as a planned combination of mechanical properties, the combination always being a compromise of one property with another. Table 3–1 illustrates some of these compromises. As pointed out in

TABLE 3–1   MECHANICAL PROPERTIES BECOME COMPROMISES

Material	Condition	Very Hard	Elastic	Soft	Ductile	Malleable	Brittle	High Strength*	Medium Strength*	Low Strength*	Plastic	High Toughness	Low Toughness
Lead sheet	Rolled			X	X	X				X	X		
Ceramic tool**	Sintered	X					X	X					
Mild carbon steel bolt	Rolled			X	X	X			X		X		X
Concrete column	Aged	X					X			X			
High carbon steel spring	Heat treated		X					X				X	
Alloy steel drill	Heat treated	X						X					X
Cr-Mo steel bolt	Heat treated		X					X				X	
Low carbon steel plate	Rolled			X	X	X				X	X		
Common glass plate	Rolled	X					X			X			
Nylon	Cast			X	X					X	X		
Clay brick	Fired	X					X			X			
Cr-Ni-Mo aircraft beam	Heat treated		X					X				X	
Aluminum sheet	Rolled			X	X	X				X	X		
Gray cast iron	Cast			X			X			X			

* Strengths relate to material strength potentials.
** Tested in compression.

the table, mild carbon steel formed into a bolt and furnished in the "as produced" or rolled condition is soft, malleable, and ductile with a medium strength factor, but also with a low toughness factor. The compromise in mechanical properties is aimed at resisting total service conditions so that the material's service time factor will hopefully be infinite. In the example of the ax, metal behind the hardened cutting edge is softer than the edge, and the softer metal acts as a cushion when impact is applied, allowing the fast moving weight to effectively penetrate deeply into the wood without shattering or deforming the metal.

For purposes of the following discussion, the metal *steel* will be used to illustrate several mechanical properties. Steel is an alloy of iron, carbon, silicon, manganese, sulfur, phosphorus, and often other elements.

MECHANICAL PROPERTIES ARE VARIABLE IN METALS

It has been pointed out that chemical and physical properties of a material are fixed to a high degree; that is, at the time of material production the constituents are known and are measurable, resulting in a given set of physical capabilities in the material, such as an expansion factor in the presence of heat. But the ability of this given chemistry and physical condition to mechanically react to stress or loading conditions in metals is variable. Because of complexity of atomic bonding in clay and other compound products, mechanical properties of nonmetals are not as variable, but remain in a near fixed condition after material aging. A clay brick or section of concrete, for example, remains hard and brittle throughout its lifetime. On the other hand, most metals respond to various treatments which change their mechanical properties in accordance with the metallurgical treatment. A tool steel, for example, can be made very soft for purposes of being cut and shaped and then be made very hard so it can perform the cutting or shaping operations. Varying degrees of hardness are therefore mechanical properties and are changeable in many metals, but not changeable in many nonmetals. The effects of mechanical properties in metals and nonmetals are compared in Figure 3–4.

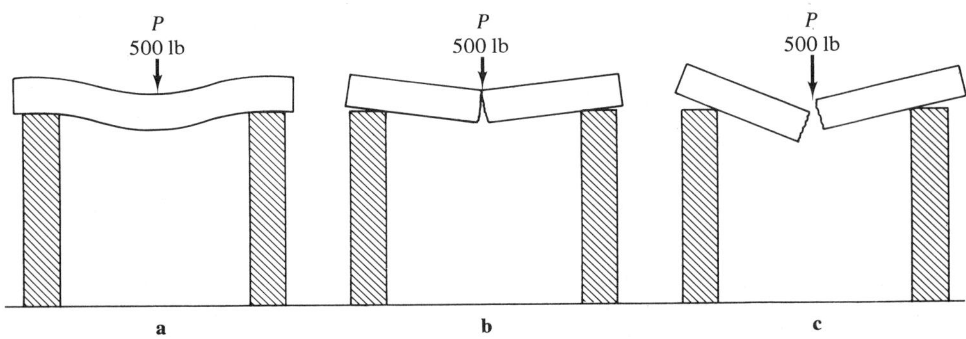

FIGURE 3–4 *a,* The load *P* permanently deflects the ductile steel bar; *b,* the load fractures the hard and brittle steel bar; *c,* the load fractures the brittle clay brick.

MECHANICAL PROPERTIES ARE GROUPED

Hardness is a manifestation of strength, another mechanical property. This hardness-strength correlation for steel is pointed out in Table 3–2. For example, the RB 92 hardness value (92 on the Rockwell B scale of hardness) shown in Figure 3–5 is approximately equal to 93,000 psi in tensile or pulling strength. These values indicate a soft steel, whereas the RC 54 hardness value (54 on the Rockwell C scale of hardness) is approximately equal to 291,000 psi in tensile

TABLE 3–2 HARDNESS-TENSILE STRENGTH CONVERSION FOR STEEL

C	A	15-N	30-N	Knoop	Br'l	TENSILE STRENGTH
150 kg Brale	60 kg Brale	15 kg N Brale	30 kg N Brale	500 gr & Over	3000 kg 10mm Ball	
ROCKWELL	ROCKWELL	ROCKWELL Superficial	ROCKWELL Superficial	KNOOP	BRINELL (Hultgren Ball)	Thousand psi
70	86.5	94.0	86.0	972	—	
69	86.0	93.5	85.0	946	—	
68	85.5	—	84.5	920	—	
67	85.0	93.0	83.5	895	—	
66	84.5	92.5	83.0	870	—	
65	84.0	92.0	82.0	846	—	
64	83.5	—	81.0	822	—	
63	83.0	91.5	80.0	799	—	
62	82.5	91.0	79.0	776	—	Inexact and only for steel
61	81.5	90.5	78.5	754	—	—
60	81.0	90.0	77.5	732	614	—
59	80.5	89.5	76.5	710	600	—
58	80.0	—	75.5	690	587	—
57	79.5	89.0	75.0	670	573	—
56	79.0	88.5	74.0	650	560	—
55	78.5	88.0	73.0	630	547	301
54	78.0	87.5	72.0	612	534	291
53	77.5	87.0	71.0	594	522	282
52	77.0	86.5	70.5	576	509	273
51	76.5	86.0	69.5	558	496	264
50	76.0	85.5	68.5	542	484	255
49	75.5	85.0	67.5	526	472	246
48	74.5	84.5	66.5	510	460	237
47	74.0	84.0	66.0	495	448	229
46	73.5	83.5	65.0	480	437	221
45	73.0	83.0	64.0	466	426	214
44	72.5	82.5	63.0	452	415	207
42	71.5	81.5	61.5	426	393	194
40	70.5	80.5	59.5	402	372	182
38	69.5	79.5	57.5	380	352	171
36	68.5	78.5	56.0	360	332	162
34	67.5	77.0	54.0	342	313	153
32	66.5	76.0	52.0	326	297	144
30	65.5	75.0	50.5	311	283	136
28	64.5	74.0	48.5	297	270	129
26	63.5	72.5	47.0	284	260	123
24	62.5	71.5	45.0	272	250	117
22	61.5	70.5	43.0	261	240	112
20	60.5	69.5	41.5	251	230	108

TABLE 3–2 Continued

| B | F | 30-T | E | Knoop | Br'l | TENSILE STRENGTH |
| 100 kg 1/16″ Ball | 60 kg 1/16″ Ball | 30 kg 1/16″ Ball | 100 kg 1/8″ Ball | 500 gr & Over | 3000 kg / D. P. H. 10 kg | |
ROCKWELL	ROCKWELL	ROCKWELL Superficial	ROCKWELL	KNOOP	BRINELL	Thousand psi
100	—	82.0	—	251	240	116
99	—	81.5	—	246	234	112
98	—	81.0	—	241	228	109
97	—	80.5	—	236	222	106
96	—	80.0	—	231	216	103
95	—	79.0	—	226	210	101
94	—	78.5	—	221	205	98
93	—	78.0	—	216	200	96
92	—	77.5	—	211	195	93
91	—	77.0	—	206	190	91
90	—	76.0	—	201	185	89
89	—	75.5	—	196	180	87
88	—	75.0	—	192	176	85
87	—	74.5	—	188	172	83
86	—	74.0	—	184	169	81
85	—	73.5	—	180	165	80
84	—	73.0	—	176	162	78
83	—	72.0	—	173	159	77
82	—	71.5	—	170	156	75
81	—	71.0	—	167	153	74
80	—	70.0	—	164	150	72
79	—	69.5	—	161	147	
78	—	69.0	—	158	144	
77	—	68.0	—	155	141	
76	—	67.5	—	152	139	
75	99.5	67.0	—	150	137	
74	99.0	66.0	—	147	135	
72	98.0	65.0	—	143	130	
70	97.0	63.5	99.5	139	125	
68	95.5	62.0	98.0	135	121	
66	94.5	60.5	97.0	131	117	
64	93.5	59.5	95.5	127	114	
62	92.0	58.0	94.5	124	110	
60	91.0	56.5	93.0	120	107	
58	90.0	55.0	92.0	117	104	
56	89.0	54.0	90.5	114	101	
54	87.5	52.5	89.5	111	87	
52	86.5	51.0	88.0	109	85	
50	85.5	49.5	87.0	107	83	
48	84.5	48.5	85.5	105	81	
46	83.0	47.0	84.5	103	79	
44	82.0	45.5	83.5	101	78	

Even for steel, tensile strength relation to hardness is inexact, unless determined for specific material.

B	F	30-T	E	Knoop	Br'l	TENSILE STRENGTH
100 kg 1/16″ Ball	60 kg 1/16″ Ball	30 kg 1/16″ Ball	100 kg 1/8″ Ball	500 gr & Over	3000 kg / D. P. H. 10 kg	
ROCKWELL	ROCKWELL	ROCKWELL Superficial	ROCKWELL	KNOOP	BRINELL	Thousand psi
42	81.0	44.0	82.0	99	76	
40	79.5	43.0	81.0	97	74	
38	78.5	41.5	79.5	95	73	
36	77.5	40.0	78.5	93	71	
34	76.5	38.5	77.0	91	70	
32	75.0	37.5	76.0	89	68	
30	74.0	36.0	75.0	87	67	
28	73.0	34.5	73.5	85	66	
24	70.5	32.0	71.0	82	64	
20	68.5	29.0	68.5	79	62	
16	66.0	26.0	66.5	76	60	
12	64.0	23.5	64.0	73	58	
8	61.5	20.5	61.5	71	56	
4	59.5	18.0	59.0	69	55	
0	57.0	15.0	57.0	67	53	

Even for steel, tensile strength relation to hardness is inexact, unless determined for specific material.

Source: Courtesy Wilson Instrument Division, ACCO

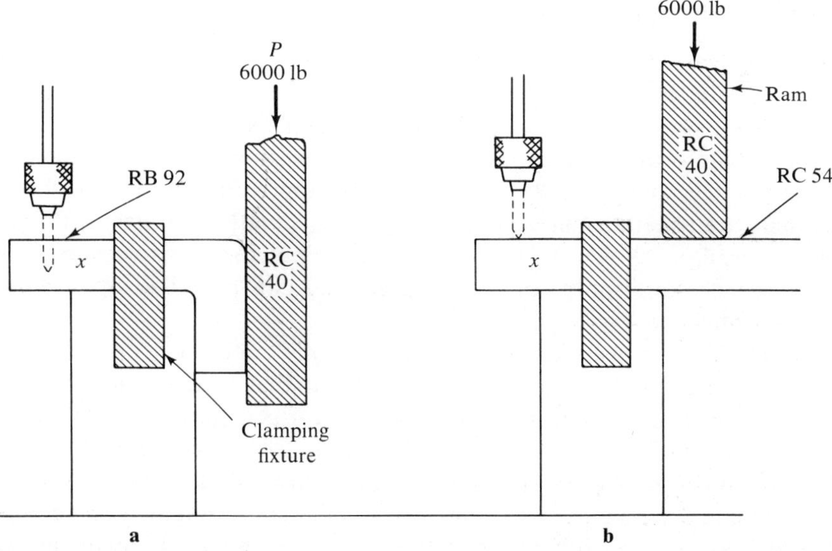

FIGURE 3–5 The steel bar x shown in a is soft and enables drilling and bending under the load P. In b the metal is very hard and strong in tensile; therefore, it cannot be drilled or bent permanently.

strength and indicates a very hard steel. The Rockwell hardness-tensile conversion system will be used throughout this text because of the continuing need for hardness and tensile strength comparisons. The Rockwell is a hardness testing machine. As a metal's hardness increases, its several strengths also increase, but the metal's ability to deform is decreased. In other words, a certain soft metal can easily be plastically or permanently deformed by means of pressure, but as this softness moves toward hardness the ability to deform decreases due to permanent deformation and work hardness. Work hardness results when most metals are permanently deformed. Therefore, the engineer or technician must determine where along the hardness spectrum the desired hardness of the material must exist.

Diversity of Mechanical Properties

A given metal, AISI (American Iron and Steel Institute) 1095 (high carbon steel), for example, or an AISI W1, will respond to mechanical property changes within a large strength or hardness variability area. This steel can be reduced in hardness to a low value in order for it to be cut, as illustrated in Figure 3–5a. Also, it can be increased in hardness to a value near maximum whereby it cannot be cut with metal tools under ordinary conditions, as illustrated in part b. In the soft condition it can be easily bent to shape, but in the hard condition it will fracture before bending. Tensile strength correlation with hardness values is shown in Figure 3–5; basically, hardness increases as strength increases. Yet, the total chemistry of the metal remains constant during different treatments as the metal's hardness and strength factors are changed back and forth at will throughout the life of the metal.

Hardness Is Strength

Experiments have demonstrated that there is a positive correlation between hardness and tensile strength. As hardness increases, tensile strength increases. It has also been demonstrated that a metal can be twice as strong as another metal. The stronger metal is then much harder than the weaker metal. In this regard, it must be pointed out that all conversions are approximate, and as the material varies there are also unequal variations in certain situations. Further, the hardness-tensile conversions shown in Table 3–2 are mainly for steels; however, they are being used more and more for other metals and can be relied upon with a high degree of confidence.

Information previously cited in Table 3–1 illustrates a scope of mechanical properties in certain situations. It is pointed out that lead is soft; therefore, this element is ductile and malleable and can be deformed plastically without tearing, but it has very low strength. (Plastic deformation allows permanent shape change without fracturing.) Again, it must be emphasized that a material's chemistry fixes its potential strengths. Lead has a low inherent strength when compared to steels which have higher inherent strengths, and these higher strengths are variable. However, softness in a material does not always produce plasticity. Common gray iron has a soft reading on a hardness scale but when pulled apart in tension, it exhibits brittleness and a low tensile strength, 22,500 psi, for example. Gray iron is not malleable or ductile at any temperature and when bending is attempted, fracture results. Nevertheless, most other soft metals are ductile, malleable, and plastic, have low strengths, and have a very low toughness factor in impact.

Should a higher tensile strength be desirable in a material, 290,000 psi, for example, an alloy or high carbon steel would be selected and then heat treated in a specific manner to attain the desired strength (Fig. 3–6). Before heat treatment, the tensile strength of the specimen in Figure 3–6 was estimated to be

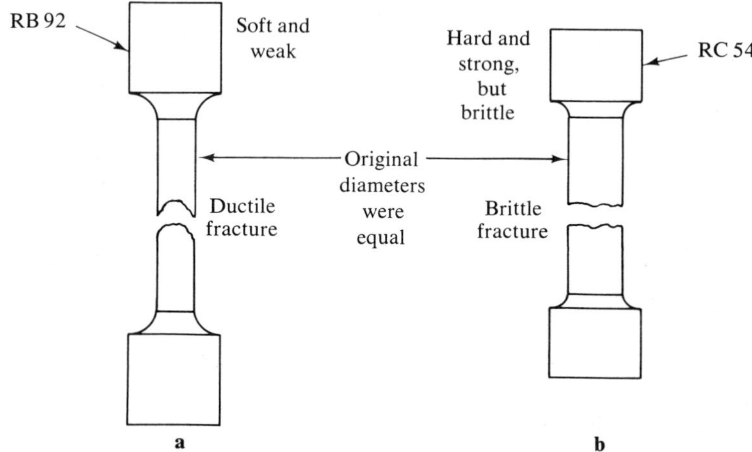

FIGURE 3–6 The AISI 1095 tensile specimen shown in *a* was pulled apart under a load of 91,500 psi. A specimen of the same material and size was pulled apart at 288,000 psi, as shown at *b,* after a different metallurgical treatment which rearranged the constituents.

91,500 psi. Mechanical properties in the heat-treated specimen would then reflect a fairly high strength and hardness factor but only a very small ability to deform plastically before fracturing under higher loads.

A study of Table 3–1 reveals the combination of circumstances which normally produces certain mechanical properties in a given material. An increase in one property often demands a decrease in another. The term heat treatment refers to the heating and cooling of a metal to obtain specific strengths. Steels are often heated to red temperatures varying from a dull red at 1300 °F (704 °C) to bright red at 1600 °F (871 °C).

CHEMISTRY SETS THE POTENTIAL MECHANICAL PROPERTIES

Materials possess potentials which can be obtained only when the material is structurally arranged by special treatment in specific patterns. These patterns are variable and numerous in metals. Microstructural arrangements in a steel, for example, the internal arrangement of the materials, are illustrated in Figure 3–7. The microstructural arrangement shown in *a* reveals only a single constituent, iron or ferrite, with a tensile strength of about 42,000 psi. Grain boundaries are

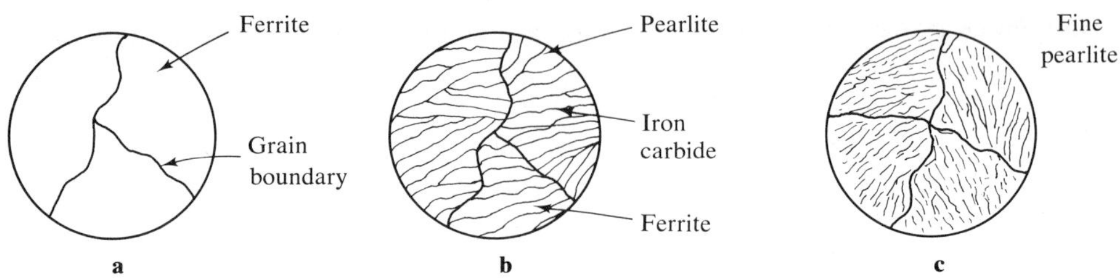

FIGURE 3–7 A simulation of ferrite and pearlite at 400×. Parts of three grains are shown after their surfaces have been polished and etched with acid. Granular materials are three-dimensional. *a,* One constituent is shown, ferrite; *b,* two constituents are observable, ferrite and carbide, which form pearlite; *c,* pearlite is again shown, having two constituents. The latter microscopic view presents a finer mixture of the ferrite and carbide which results in an increase in hardness and strength.

the lines which intersect near the center of the view. The arrangement shown in *b* reveals two constituents, ferrite and iron carbide, a metal and a chemical compound. This chemical mixture provides an increase in strength over the single constituent material in *a*; in fact, tensile strength doubles (87,000 psi) as an increase in hardness occurs. When the material in *b* is rearranged through heat treatment the material in *c* results. The material in *c* provides approximately twice the strength (176,000 psi) of the material shown in *b,* yet the chemistry of *b* and *c* are equal. Further structural arrangements are possible whereby many different hardnesses and strengths can be obtained. Should the material shown in *b* or *c* be water quenched from a bright red temperature, another strength increase will occur, and the resulting material will be three times stronger that it is at *b*. Close observation shows that the particles of the intermetallic compound, Fe_3C, are closer together in *c* than in *b*. Water quenching from austenite either places the carbide particles in solution with iron or arranges them at a minimum size. This type of carbide dispersion increases hardness by a large factor. Austenite is a solid solution of constituents in steel, more often existing in the red heat temperature range.

Even though the chemistry provides the capability potential of materials, it is the microstructural arrangement that decides the degree of hardness and strengths. Without the necessary chemistry, a desired microstructure will never be obtained. For instance, if a one-inch square block of steel must effectively withstand 200,000 pounds in compression, many structural steels and other materials cannot be used. No amount of treatment can induce this mechanical capability in a low carbon steel, for example, because this steel does not have the inherent capability to resist this load. However, should a higher carbon content or alloy steel be specified, which requires a chemical change, the 200,000-psi potential would exist, and through a specified treatment or certain arrangement of the microstructure, this compression property would then be real. It is the arrangement, shape, quantity, and kind of chemical constituents in the metal that provide designated mechanical properties, and these factors provide the microstructure.

Chemistry and the Microstructure

A knowledge of the chemistry of a metal provides information for possible development of mechanical property changes by means of heating and cooling and through hot or cold working processes. Because two types of metals exist, pure metals and alloys, each is discussed separately.

An element is a single material having no other constituent to combine with itself for production of various structural arrangements by heat treatment. The element will respond only to a microstructural change through a grain distortion process known as *cold working* and through a softening process known as *stress relief* or *annealing* after cold working. Cold working the element or reshaping the material at room or warm temperatures (below recrystallization) induces plastic flow of the metal, and a permanent distortion of the granular structure results. This distortion in metal is illustrated in Figure 3–8. The equiaxed or normal grain arrangement is illustrated in part *a*. Cold rolling compresses and elongates the grains, as shown in part *b*. In effect, hardness of the element increases along with

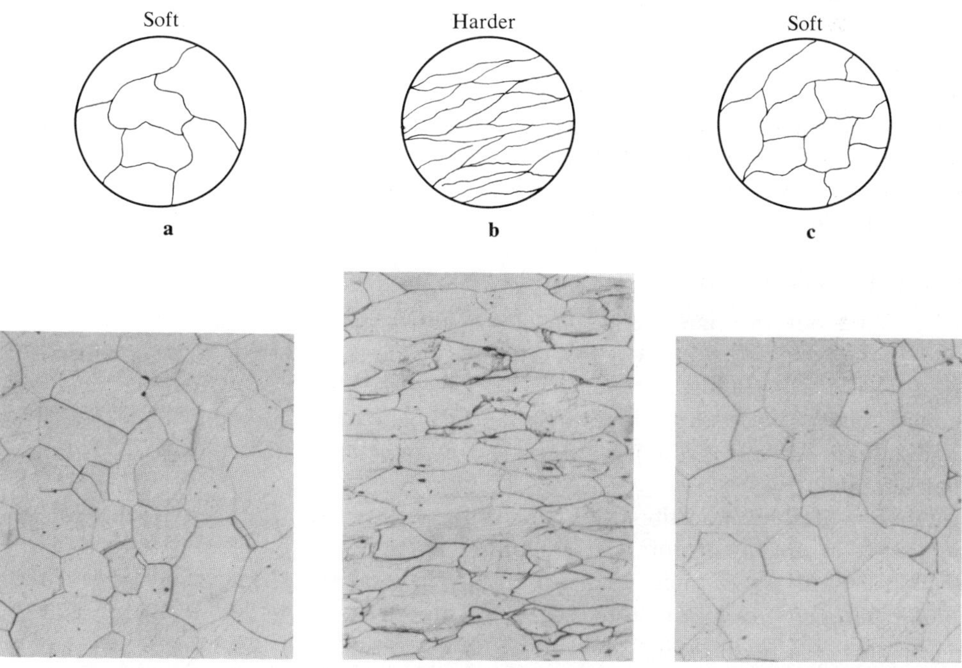

FIGURE 3–8 Effects of cold work on a pure metal. In *a,* the element is equiaxed, the grains being stress free. In *b,* the same material was deformed at room temperature, resulting in equiaxed grains being compressed into longer and thinner dimensions. Hardness and strength increased as a result. In *c* the deformed grains were heated to an elevated temperature which caused recrystallization, or formation of entirely new grains, and these grains are stress free. Drawings are shown at top and photomicrographs are shown at bottom.

strength during cold-working processes. But the ability of the element to continue an increase in hardness as working progresses is limited because the brittleness factor approaches as material malleability or softness is decreased. Somewhere, just prior to cracking, the hardness-softness compromise is noted. This increase in strength factor, induced by cold work, is then pertinent to the specific element and is different from cold-worked properties in other elements. In this regard, cold-worked solid solutions, such as brass, have mechanical characteristics similar to the pure metals. Also, other mechanical properties of pure metals are often similar to solid solutions.

Temperature vs Cold Work

Heat removes effects of cold work, as pointed out in Figure 3–8. When heated to certain temperatures above recrystallization, any cold-worked and distorted metal returns to its stress-free condition and again resembles the soft and weak condition shown in *a*. Recrystallization is a temperature factor where grains of metal change in size. As distortion is removed through heat, hardness and strength decrease, allowing the metal to return to its natural hardness. Grains rearrange themselves in stress-free equiaxed shapes. The mechanical properties of hardness and strength, from minimum to maximum, for the pure metals (elements) are then set in terms of potential capabilities, and these capabilities are due to the chemical properties of the pure metal. Cold-worked copper, for example, is harder and stronger than cold-worked aluminum. The element then, but not all elements, is capable of being cold worked for increased degrees of hardness and strength and then softened, if necessary, through means of heat which also removes the increased strength.

Microstructure and Strength

As pointed out in Figure 3–7, the extremely hard constituent iron carbide is mixed in a matrix of soft and weak iron, or ferrite. Iron carbide is inherently hard while ferrite is naturally soft. Figures 3–7*a* and 3–8*a* appear similar, but both materials have different mechanical properties, as each is a different metal, ferrite in Figure 3–7*a* and aluminum in Figure 3–8*a*. Ferrite is stronger and harder than aluminum.

When a second constituent mixes with another, as shown in Figure 3–7, numerous arrangements of the structural constituents are often possible. Each microstructural arrangement then provides a general set of mechanical properties for the particular metal or other material. This set of properties will be in general consonance with data presented in Table 3–1. The microstructural properties of a material, when understood, are used to determine existing or potential mechanical properties. Figure 3–9, for example, is a photomicrograph of an annealed hypereutectoid steel (a steel having more than 0.8% carbon) which has a potential tensile strength capability in excess of 270,000 psi after hardening. This structural pattern or material arrangement is similar to Figure 3–7*b*.

Constituent Arrangements

A microscopic examination of the ferrite illustrated in Figure 3–10 and Figure 3–7*a* reveals a very malleable material which is soft and weak in tensile strength with

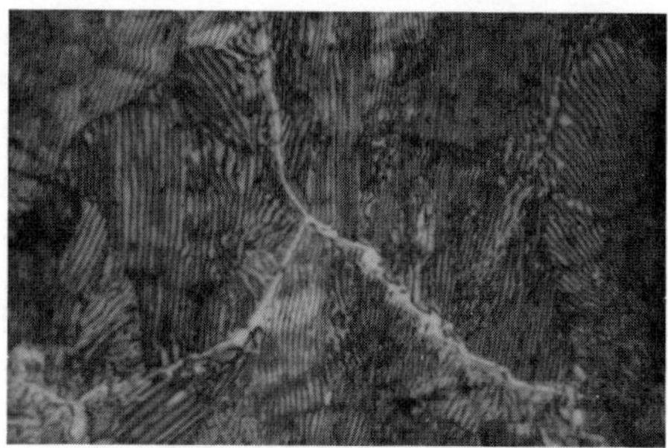

FIGURE 3–9 Microstructures of annealed hypereutectoid steel showing excess cementite (white) at the pearlitic grain boundaries.

regard to steels. These mechanical properties of softness, malleability, and weakness infer plasticity and bendability without tearing. Also, no brittleness or toughness is present in such a combination. When a specimen of ferrite is placed in increasing tension beyond its yield strength (a load causing permanent change in shape), for example, ductility provides elongation (stretch) in the material. As elongation proceeds, work hardness is induced along with increased stress or internal resistance; therefore, the metal becomes harder and stronger as it plastically elongates and is stressed to a higher degree. When the ultimate or tensile strength is reached, fracture of the metal may follow or the metal may suddenly begin to elongate rapidly to rupture.

The pearlitic structure found in a high carbon steel, as shown in Figure 3–7b, is less malleable and ductile than ferrite, but its tensile strength and hardness are much greater due to the presence of particles of hard carbides. The pearlitic structure will not bend as far as ferrite, but it will resist a combination of deformations better than ferrite. The arrangement of the microstructure of a material establishes a set of mechanical properties for the specific material. As materials vary, so do the mechanical properties.

Microstructures Reflect Mechanical Properties

Microstructures appear to be infinite in numbers of constituent arrangements in metals. When the pearlitic steel structure presented in Figures 3–7b and 3–9 is

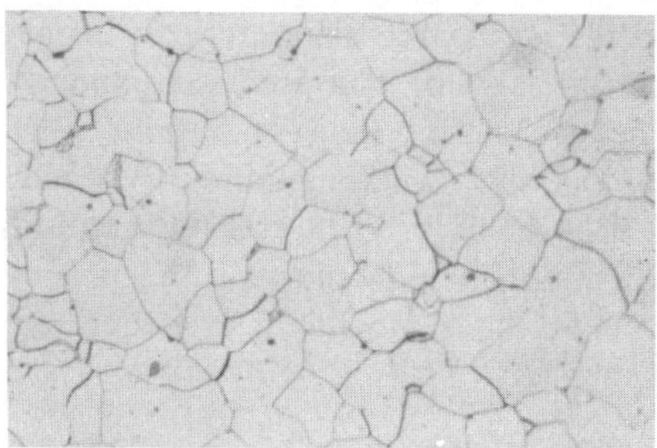

FIGURE 3–10 Annealed ferrite. This solid solution has characteristics of a pure metal because only one constituent is observable.

quenched in water from a bright red temperature and prepared for microscopic examination, the martensitic structure in Figure 3–11 results. The appearance of pearlite and martensite are contrasting because the arrangement of the total quantity of material in each specimen is different. The rearrangement of ferrite and carbides observed in Figure 3–9 produces the following tremendous changes in the metal's mechanical properties, as reflected in Figure 3–11: a hardness which prohibits cutting with a sharp file; an ability to safely hold more than 300,000 pounds to the square inch of area in tension, but with a high degree of brittleness and little toughness in impact loading; and no capability of being deformed because brittle fracture will occur if loads overcome the material's strength. When ground into a cutting edge, the metal shown in Figure 3–11 will quickly cut into

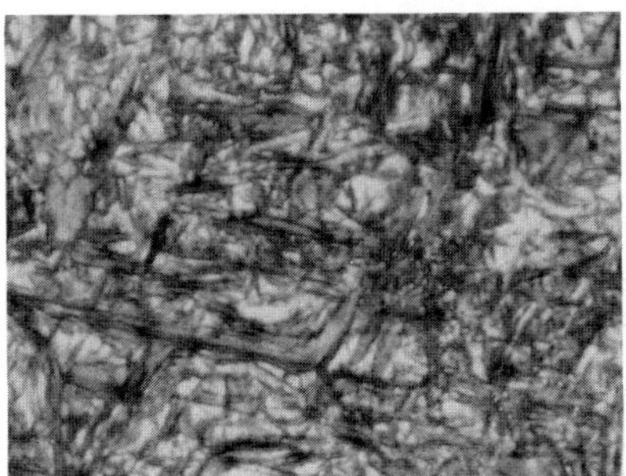

FIGURE 3–11   Martensite in a high carbon steel. Nital etch at 1000×.

the materials shown in Figure 3–9. Consequently, manufacturing personnel require as much information as possible pertaining to mechanical properties of materials in order that manufacturing processes can be expedited and materials can be relied upon to a high degree to safely hold their loads during lifetime services.

As has been pointed out, mechanical properties in most metals are variable, depending on the structural arrangement of the metal's constituents. The alloys provide the greatest capabilities in combination of mechanical properties because of the several microstructural possibilities for each alloy and the presence of inherently different and often stronger materials. There are hundreds of alloys, each providing a known mechanical capability or series of capabilities. However, some of the elements, tungsten, for example, are hard in whatever situation they are used and do not provide a variable characteristic in mechanical properties as most of the other common metals do. Tungsten is frequently used for hard facing on tools, and it responds to cold-working operations but does not respond to softening by heating and cooling, unlike aluminum which can be hardened by cold working and softened by heating. With a few exceptions, the metals mechanically respond to heating, cooling, and cold-working processes. Sometimes all of these treatments are given to the same metal in order to produce a designated mechanical property and shape.

Effects of the Final Heat Treatment   A given set of mechanical properties in a specified metal often yields to a new set of properties when different treatments are performed on the metal. Treatments can be performed over and over on the same metal, the resulting mechanical properties being those mainly developed through the final treatment or final series of treatments. The foregoing statement is fundamental in the study of metallurgy because, in most instances, the final treatment establishes a new group of properties which will remain with the metal until it is exposed to another metallurgical treatment.

In the production of a higher tensile strength in a steel, for example, the final treatment destroys most of the previously existing properties as it induces new properties. For instance, a steel having a tensile strength of 150,000 psi exhibits a specified hardness and a specified percentage factor of ductility. When heat treating this steel to a new tensile strength of 200,000 psi, previous mechanical properties, including the 150,000 psi with related properties, are eliminated from the metal and the new tensile strength of 200,000 psi with its related properties is established. This final treatment will remain throughout the life of the metal until a situation occurs that changes this last mechanical condition.

The presence of elevated temperatures removes existing strengths in most metals and, according to the specific situation, new mechanical properties appear on cooling. Some metals respond to subzero temperature conditions, and mechanical properties change accordingly. Time in service also may alter a material's properties because of submicroscopic shifting of the material arrangement. In other words, most steels and many other metals can be hardened and softened or strengthened and weakened repeatedly. The final treatment, deliberate or accidental, induces the final group of mechanical properties in the metals; however, a previous treatment is often essential to establish a specified property during the final treatment.

GRAIN SIZE IN METALS AND MECHANICAL PROPERTY RELATIONSHIPS

The crystalline structure of metals is inherent in the metal's solid state, but the size and shape of these crystals or grains are not fixed. Mechanical properties in metals are somewhat affected by the size and shape of the grains. In fact, the grain size of a metal can be changed from fine to very coarse and then back to fine over and over. In steels, as it is in all metals, grain size is an important factor with regard to impact resistance and toughness.

Grain Sizes Are Numbered

Grain size is often measured microscopically, and a number is assigned to the size. As an example in steel, an ASTM (American Society for Testing and Materials) grain size 3 is considered coarse or large when compared to a 7 which is fine. This system of measuring grain size depends on a magnification of 100 (Fig. 3–12). Individual grains within a square inch of observable space are counted under magnification. Each grain size number is the average number of grains within the magnified and measuring area. Obviously, when the specimen is prepared for examination, many of the grains are penetrated at different cross sections; therefore, some grains appear smaller than others because all grains are three-dimensional and many-sided. Often, many grains are longer in one dimension than in

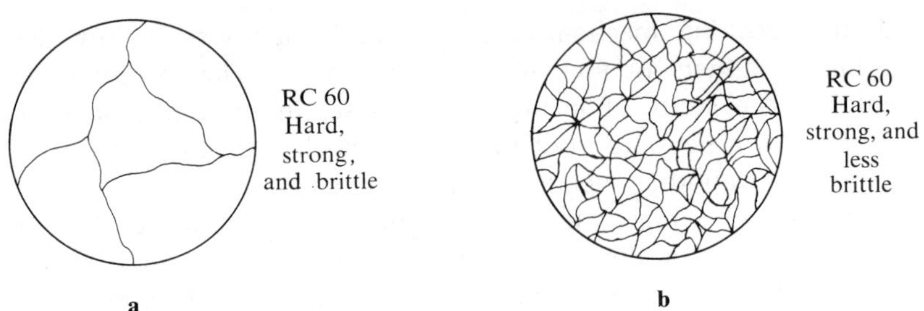

FIGURE 3–12   Comparison of a coarse-grained and a fine-grained steel at 100×. In *a* the grains are very coarse and produce only a small amount of grain boundary material. In *b* the grains are fine and produce much more boundary material.

another; that is, a cross section of one end of an elogated grain will naturally be smaller than a cross section of the widest part of the grain. Grains are randomly arranged by nature at a certain temperature; consequently, all observable grains will not be the same size. However, the average grain size of a fine-grained structure is exceedingly smaller than the average size of a coarse-grained structure. Figure 3–12 illustrates a comparison between two different size grains in a steel. Grain size reflects no fixed hardness factor, but it drastically influences the ability of the material to resist impact (suddenly applied load). Figure 3–13 illustrates a fractured view of a coarse-grained and a fine-grained steel, both steels having the same hardness. Rolled out or plastically elongated grains increase the hardness of metal. In this respect, the hardness results from atomic rearrangement and not just from the granular deformation.

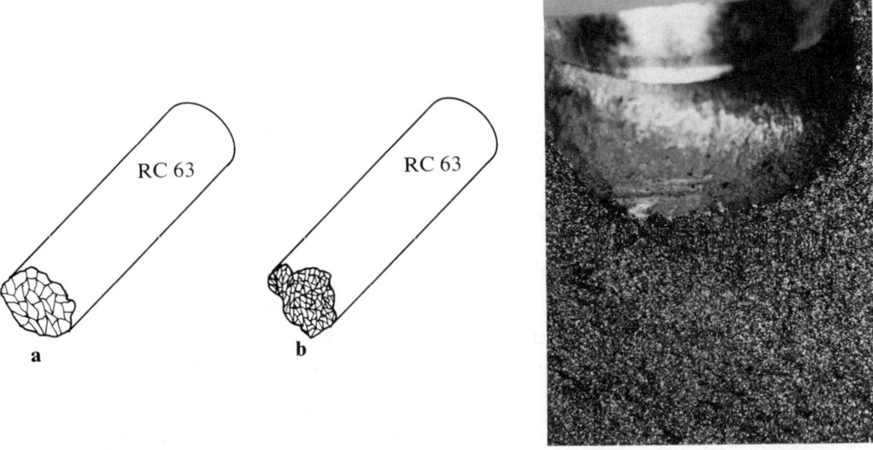

FIGURE 3–13   Fractured view of coarse- and fine-grained steel. The bar in *b* required twice the impact load to break as compared to *a*. In *c,* the ram fractured the coarse-grained steel under compression loading.

Effects of Grain Size

Grain size in metals is a product of temperature and forming methods and, in turn, the sizes and shapes of grains help determine certain mechanical capabilities of the material. Specifically, fine grains promote a higher all-around toughness in a metal at temperatures below recrystallization, and coarse grains tend to decrease the toughness factor, especially when observed during impact loading. Large grains have fewer holding surfaces among the other grains and are consequently more susceptible to fracture during impact when the metal is very hard. Grain size does not impart a specific hardness or a certain strength in a metal, but it does alter the metal's shock resistance and depth of hardness capabilities. Grain size is basically independent of hardness when temperature has established the size.

Distorted grains resulting from cold-forming processes alter hardness and strength properties of the metal because the material slips along the atomic planes within the grains. In simplified terms, two metals of the same material can have the same hardness, but the one with a fine grain or distorted grain structure will be much tougher than the one with a coarse grain or equiaxed grain structure. Figure 3–14 illustrates these two mechanical property relationships, the toughness

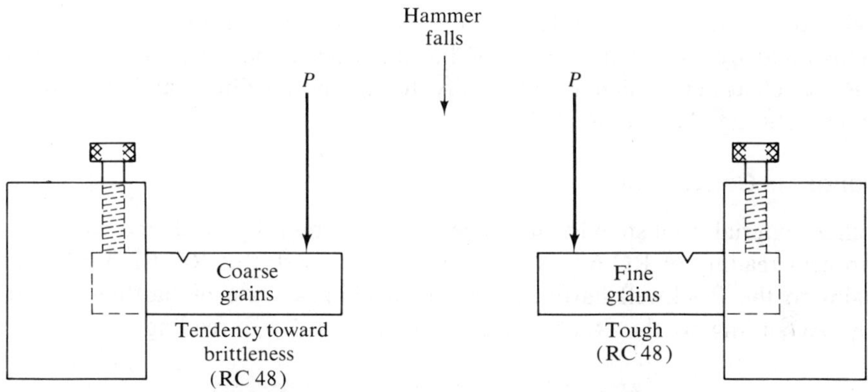

FIGURE 3–14 Fine-grained metal is tougher than coarse-grained metal when tested in impact, even though both metals are equal in hardness.

factor being mainly determined by grain size and grain shape differences with a given hardness. Most mechanical properties of materials do not evolve from the grain size or shape, however, but result from the microstructure existing inside each grain. Two factors then portray a metal's microstructure, grain size and arrangement of material within the grain.

PROPERTIES RELATED TO HARDNESS

Hardness is a relative term; one material's hardness is compared to another or hardness is compared to another property. Hardness is defined as the ability of a material to resist deformation or indentation. Different materials have different resistance to penetration. This is illustrated in a comparison of mineral hardness. The mineralologist Mohs prepared a scale of hardness in which talc was rated as 1, being the softest possible mineral, and the diamond was rated as 10, indicating the hardest material. Different minerals, ranging in hardness values from 1 to 10 (Table 3–3) reflect hardness increases from talc to the diamond. This hardness

TABLE 3–3 Mohs Scale of Hardness

Mineral	Numerical Value
Talc	1
Gypsum	2
Calcite	3
Fluorite	4
Apatite	5
Feldspar	6
Quartz	7
Topaz	8
Corundum	9
Diamond	10

test is demonstrated by scratching one material with another. The diamond still remains the hardest material (boron nitride is a competitor) and its hardness is demonstrated by its use in the tip of hardness tester penetrators, such as are in the Rockwell tester, which is frequently found in metallurgical laboratories and in manufacturing plants (Fig. 3–15).

Hardness is Convertible

Another material, tool steel for instance, is said to be fully hardened when it shows a hardness reading of RC 67. As previously explained, this RC 67 hardness value pertains to the Rockwell hardness testing machine's scale of hardness values for steels. According to the Rockwell scale (Table 3–2) a reading of C 70 shows

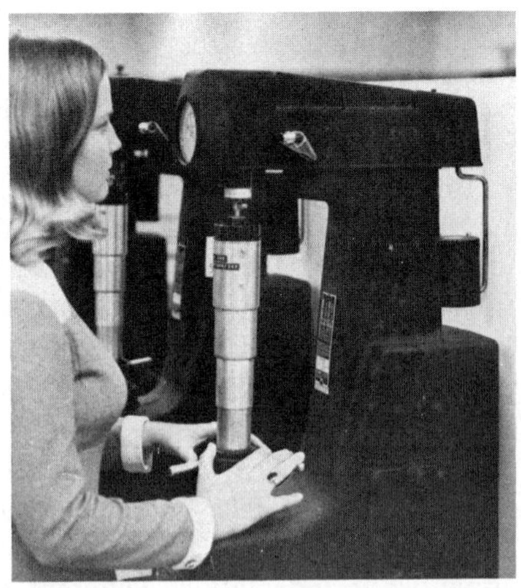

FIGURE 3–15 A technician prepares to make a Rockwell hardness test.

maximum hardness and a reading of B 72 indicates a soft seel. Therefore, numerical values between maximum hardness and soft steel are indicated along a constantly changing surface hardness value from maximum hardness to near maximum softness in steels. Even though the Rockwell Hardness-Tensile Conversion chart is prepared for steels with conversions from one testing method to another and from approximate hardness values and tensile strengths, there is enough evidence of reliability of the tests for these values to be safely used in quality assurance processes. Hardness conversions for other metals should be made on other scales or charts; however, many nonferrous metals are properly inspected on the Rockwell testers, using modified strength conversions.

Softness Is a Degree of Hardness

Softness is opposite to hardness. A file used to remove small particles of metal from a part's surface illustrates a relationship between a very hard metal and a soft metal. The steel file often has a hardness of RC 64 which is enough hardness to allow its fine cutting edges to remove small particles from the surface of most metals that are not fully hardened. When a file skids off the surface of a metal, the metal is said to be file hard. On Mohs' scale the hardened file is somewhat below the diamond in hardness, even though it is near maximum in hardness for any metal. A very few materials exceed RC 70 in hardness and these are found in a few minerals and some carbides, nitrides, and oxides. The opposite of the hardened condition is softness, which is relatively compared to hardness.

Types of Hardness Testers

Hardness values of materials, especially of the metals, are relevent and are convertible to several other scales of hardness. Table 3–2 includes relative hardness values for scales of hardness testing machines other than the Rockwell, such as the Brinell. Some hardness conversion charts also include conversion values for the Vickers, Tukon, and Scleroscope hardness testing machines. When a hardness value is obtained on one scale, it can then be approximately correlated or converted to the scales of other hardness testers. As an example, a Rockwell reading of RC 46 is converted to a Brinell 437, a Knoop 480, and a Rockwell A 73.5. All of these hardness values have an approximate converted tensile strength in steels of approximately 221,000 psi. Because hardness is reasonably related to tensile strength in metals, the tensile strength value is therefore included in the Hardness-Conversion table. Again, it must be pointed out that all hardness values and conversions are approximate but are normally valid enough to be used in engineering designs.

Knowing the tensile strength of a steel, for example, can lead to finding its hardness value through conversion. A tensile strength of 117,000 psi is equivalent to a Rockwell C 24. Should a tensile strength of 123,000 psi be needed in a steel bolt, for example, an RC 26 would be used to verify the hardness of the bolt and, in turn, this hardness value is correlated to the needed tensile strength. Obviously, to test the bolt in tension would destroy it; therefore, hardness-tensile correlations are necessary. Potential mechanical properties of hardness and tensile strength can be estimated in a steel when only its chemical analysis and treatments are known. This is a common procedure among engineers and technicians.

Hardness and Stress

There is a close relationship between hardness and stress. A hardened steel will not bend because it is rigid and brittle. Rigidity and brittleness are mechanical properties. When a load, as illustrated in Figure 3–5, is applied to a steel bar having a hardness value of RC 54, it attempts to penetrate the bar's surface or push it downward, and this force increases resistance in the bar to bending. This resistance is called *stress*. When the external load becomes greater than the metal's internal ability to contain the stress in equilibrium, the metal in this highly stressed and hardened condition suddenly shatters. This illustration shows that a high hardness includes: a high stress, a high tensile strength, a high rigidity and brittleness factor, and a very low ductility factor because little elongation in the metal occurs at fracture. In other words, failure occurs without warning.

TENSILE STRENGTH

Possibly the most used mechanical properties are tensile strength and hardness. Tensile strength is the ultimate strength of the material during tension loading. Hardness conversions to tensile values are common occurrences. While these conversions are only approximate because of the large number of variables associated with microstructure, they are within the safe limits of operating conditions. During the tensile test a material elastically stretches as illustrated in Figure 3–16*b,* the difference between dimensions x and x'. At a loading point where plastic deformation begins, elastic deformation ceases. This *elastic limit,* pointed out in *b,* is an important mechanical property because engineering design rests within this elastic dimension. The terms *elastic limit* and *proportional limit* are sometimes used interchangeably because of the equal ratio between stress (load) and elongation (strain). However, the elastic limit appears most frequently in the literature. As the tensile test continues the elastic stretch suddenly changes to a permanent or plastic stretch, and this point, measured in psi, is the *yield strength.*

Effects of Yield Strength

A material's yield point, or yield strength, verifies the load which causes permanent change in the material's dimensions (Fig. 3–17). As the pulling load continues to yield to the tensile, there exists an area of permanent dimensional and microstructural change in the metal. Soft and ductile metals plastically deform in this range, while hard materials fracture. Sometimes, the fracture point of a metal is beyond the tensile strength. This is demonstrated in the example in Figure 3–16 of an annealed low carbon steel which separates at a load somewhat below the tensile. Notice the severe necking of the soft metal prior to fracture. A summary of the test reveals that the tensile strength indicates the highest load held during the test. When the tensile strength is known, safe loads can be designed.

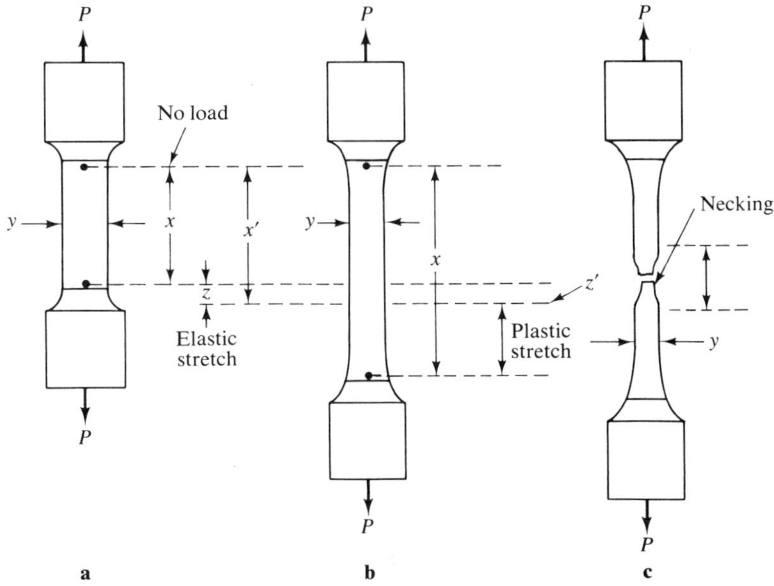

FIGURE 3–16 A tensile speciment *b* shows dimensional changes resulting from tensile loading *P*. Dimensions shown in *a* are original dimensions. In *c*, the specimen has fractured. In *d*, the tensile specimen is placed in the stamping fixture for producing the punch marks prior to testing.

COMPRESSION STRENGTH

Another mechanical property is compression strength. This strength is approximately equal to the tensile strength in many, but not all, instances. Figure 3–18 illustrates two steel specimens exposed to tensile and compression loads. Both specimens have the same chemistry and the same cross-sectional areas, and both have been metallurgically treated identically. The specimen in part *a* was pulled apart at approximately the same load that crushed specimen *b* to fracture (part *c*). The tensile-compression correlation factors relate more to ductile materials, however, because many very hard materials will withstand higher compression loads than tensile. In this regard, some materials are used primarily in compression. Often, all three forces, including shear, act on the material concurrently.

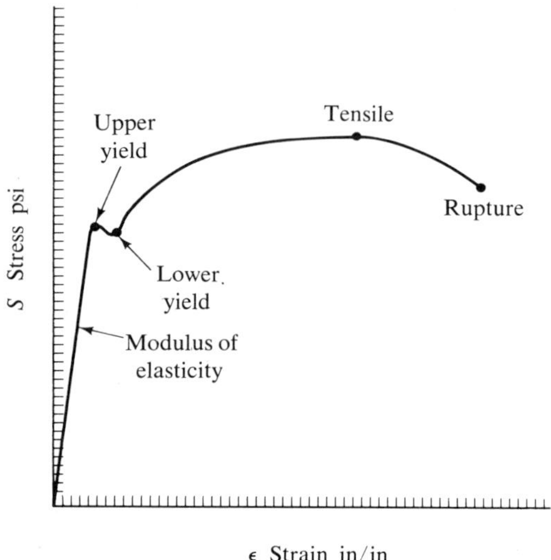

FIGURE 3–17   A stress-strain diagram for annealed steel.

MECHANICAL PROPERTIES INFER MATERIAL CAPABILITY

Mechanical properties for engineering purposes refer mainly to stressed or loading conditions and infer material capability. A column, for example, is increased in compression loading when a load is applied to the beam. This situation is illustrated in Figure 3–2. The top side of the beam is in compression, but the bottom side is in tension. In the beam tension and compression forces must not be too incompatible. When these two forces, or loads, are not properly placed in static

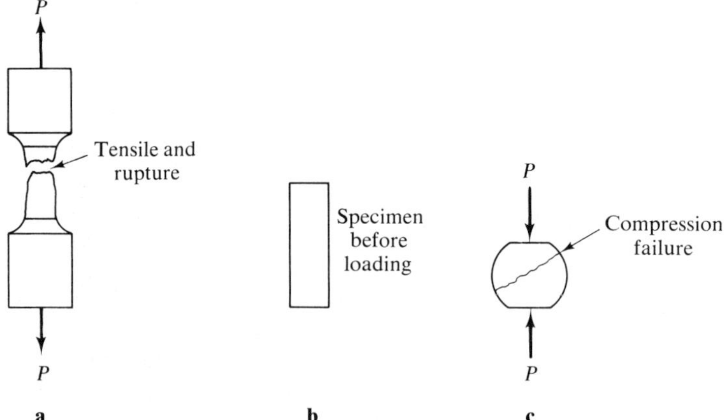

FIGURE 3–18   Two steel specimens (*a* and *b*) exposed to tensile, *a*, and compression loads, *c*. Both specimens have approximately the same tensile and compression strengths. Specimen *b* is before loading, while specimen *c* is after loading. Note the shear failure crack running diagonally across specimen *c*, indicating it had deformed to a point where compression stress reached its ultimate.

equilibrium, or when the deflected beam is overstressed, the beam fails, usually in tension. Another example, pointed out in Figure 3–19, pertains to a threaded rod supporting a platform from a suspension cable. The rod is in tension while the threads at the end of the rod are in shear. This is an example of one force causing a multistressed condition in a material as tension and shear act simultaneously; tension attempts to pull the material apart, while the shear force attempts to slide the threads from the rod or the nut.

A stressed condition similar to that indicated in Figure 3–3 is further detailed in Figure 3–20. This example illustrates what happens in a specific part of an airplane structure in flight. The two sheets of 2024 aluminum alloy are riveted together along a lap seam. Bending of the structure causes one side of the sheet

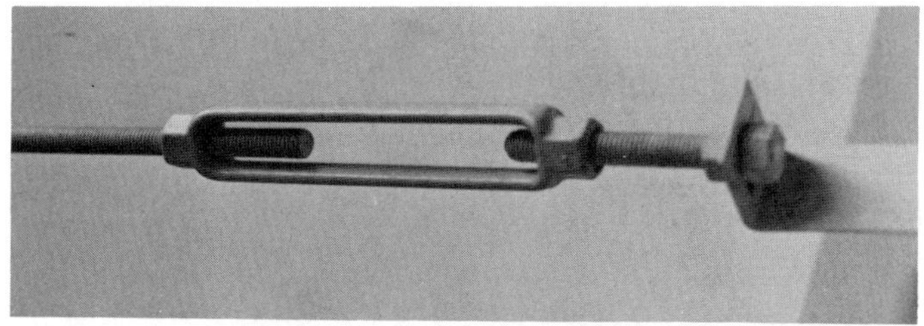

FIGURE 3–19 A threaded suspension rod under tensile load. The threads are placed in shear.

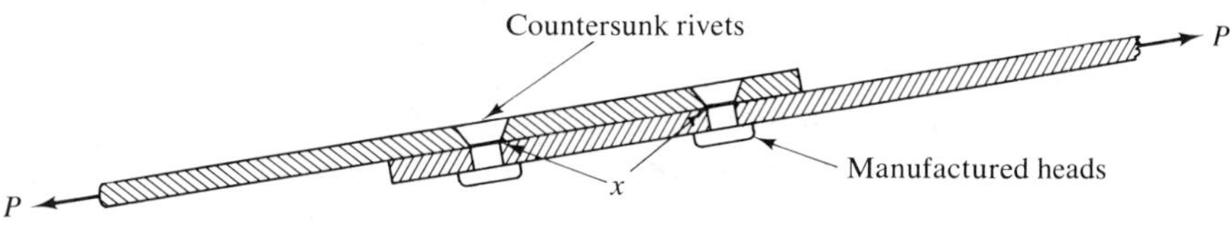

FIGURE 3–20 The tensile load P on the two aluminum alloy sheets causes a shear load on the two rivets at points x. Pulling forces of the tensile load attempt to permanently stretch the sheets and cut the rivets. In b, the lap joint provides a strong connection between the two aluminum alloy sheets of this aircraft's fuselage.

metal to elastically stretch in tension while the other side is in compression. But the rivets transfer these tensile and compressive stresses, even though the stresses are trying to shear, or cut, the rivets. Shear strength is then a mechanical property and often exists concurrently with tensile and compression stresses.

ELASTICITY

Many persons have witnessed the phenomenal action of a powerful steel compression spring as it stretches and compresses under loads (Fig. 3–21). This elastic

FIGURE 3–21   A coil spring used to resist compression loading must function in adverse conditions of static and dynamic loading.

property does not result because of the spaced coils of wire being in their particular shape. A metallurgical explanation portrays the presence of a specific type of microstructure in the metal which assures original coil spacings after release of the load. The mechanical property known as elasticity is, in reality, a microstructural compromise between hardness and softness. It is a condition in the steel, for example, that reduces brittleness to elasticity, which is approximately midway between RC 67 and RB 100. It is a combination of hardness and strength values far below maximum hardness and far above the softness of ferrite. Basically, the arrangement of carbide in the ferrite matrix of the steel is so patterned by heat treatment that the coil is capable of flexing like a rubber band.

Elastic capability is unlike ductility and rigidity. In fact, it is a compromise between these two properties, also. As rigidity decreases, hardness reduces, softness increases, which, in turn, promotes elasticity. As long as the loads on the spring are safe (below the yield strength), the spring's material will elastically function satisfactorily. Overloading of an elastic material however, causes failure. According to a Rockwell hardness reading, in the example of the spring, the surface hardness reflects an RC 45. According to Table 3–2 the tensile strength of the metal is approximately 214,000 psi. Experiments have proven that tensile strengths in the general area of 214,000 psi in steels allow wide limits of elastic movement of the metal, such as in spring operations, without permanent deformation or facture. Elasticity pertains equally as well to the flat leaf type springs as to coil springs.

Variations of Elasticity

An elastic property which fails because of overloading follows a certain pattern. Overload causes the elastic property to give in to other deformation patterns. An observation of the RC 45 value in a steel subjected to increasing loading conditions reveals that the steel continues to elastically deform until it reaches the end of its elastic capability. Deformation changes to a very small plastic deformation, and then, without warning, fracture occurs. The RS 45 hardness value has an excellent elastic mechanical property. However, when elasticity in a material is compromised, either the brittleness factor enters the material or the ductility factor is imparted. If the hardness value of a spring was RC 54, little elasticity would be demonstrated because of rigidity. A value of RC 65, on the other hand, would provide immediate fracture of a brittle nature when the load attempted to deform the spring. A miscalculated value of RC 30 would not provide elasticity, but would result instead in some plasticity or permanent deformation when the operating load was applied. A badly miscalculated value of RC 20 would cause the spring's coils to immediately and permanently collapse under load. Elasticity, then, is a particular type of mechanical property and must not be equated with the modulus of elasticity, which remains approximately constant during any of the above situations in the steel spring.

IMPACT RESISTANCE

Another mechanical property of great concern to the designer is impact resistance. As example of impact is illustrated in the action of the landing gear system of an aircraft (Fig. 3–22). When the wheels of an aircraft suddenly contact the runway, a force known as *impact* occurs as the result of two opposing forces quickly colliding. Impact is basically an accelerated compression loading. The ability of the

FIGURE 3–22   The landing gear system on this F-16 aircraft must sustain severe impact loading on landing. (Courtesy of General Dynamics, Fort Worth Division)

concrete runway to flex downward without cracking and the ability of the metallic structure and shock-absorbing system of the aircraft to flex upward provide a safe landing. Impact resistance is a mechanical property that privides a high degree of capability to a material to retain its permanent dimensions after two or more opposing forces have quickly collided. Tremendous quantities of energy are often absorbed in impact situations as the elastic property functions and the design maintains its shape.

Impact Resistance Is Toughness

It is pointed out that impact resistance, or a kind of toughness measurement, is a microstructural arrangement within a specified metal that provides the best arrangement to withstand a combination of loads. Aircraft hardware, such as a bolt, is frequently subjected to shock loadings. The microstructure (carbides and ferrite) of the material is arranged so that a slightly softer condition will exist than is in the steel spring in order to eliminate most of the fracturing factor that occurs beyond the elastic factor should overload occur. Yet, the metal must be high enough in a hardness value to eliminate most of the softness factor. Therefore, a metallurgical compromise in hardness is established. Hardness values between RC 26 and RC 38 are used where toughness is needed. At approximately RC 38, for example, the toughness factor in a metal has excellent resistance to loads applied suddenly; the designed material effectively absorbs the load and stands ready to receive another. The RC 38 value has an approximate tensile strength of 171,000 psi, meaning high strength with a high shock-absorbing ability.

What condition exists in a steel with an RC 38 hardness value for it to function in such an adverse situation? The answer is a proper blending of several mechanical properties. The blending process is the compromising process whereby hardness, brittleness, and strength are reduced, and toughness is increased. Toughness in a material includes a small material capability to plastically elongate or reduce its cross-sectional area under load when overloaded rather than fracturing first. Therefore, the softness property is changed toward hardness and includes a high degree of elasticity. Toughness in a material is exhibited in the blades of road scrapers and in selected areas of armor plate.

PROPERTIES RELATED TO SOFTNESS

Relative to a material's mechanical property, softness exists in the absence of hardness. The ability of a material to resist permanent deformation is hardness, whereas the ability of the material to be permanently deformed without tearing or fracturing is a degree of softness. Softness is a compromise in hardness. Softness is also a compromise in a material's strength, because as hardness is reduced, strength is proportionally reduced. A soft material then has a low strength when compared to a hard material having a high strength, but the kind of material is also a governing factor in accounting for hardness or softness.

Properties related to softness include ductility, malleability, and bendability, but all soft metals are not ductile. Gray cast iron is soft and brittle, the brittleness resulting from the free graphite in its microstructure. A ductile material, such as ferrite, aluminum, or lead, can be stretched or compressed into many shapes without the metal tearing or rupturing. Ductility indicates elongation ability, while malleability denotes the ability of the material to be rolled, stretched, and pressed into numerous configurations. If a material is ductile, then it is malleable, and it can also be bent into a small radius. Ductility and malleability are mechanical properties which infer an ability to be plastically formed into shape under pressure.

The Region Between the Yield and Tensile Strengths

An engineer who notes that a particular metal has an elongation factor of 41% in two inches is pleased because he knows that the metal can be bent to a 90° angle with a very small radius without fracturing or tearing. When elongation is high in a metal, the reduction in area factor is also high because these two properties are closely related, as illustrated in Figure 3–16. Under a tensile load a metal beyond its yield elongates permanently, the plastic stretch being illustrated in part b. This stretching reduces the cross-sectional area (y) in relation to the elongation (x). The elastic property known as the *modulus of elasticity,* indicated as the area z, exists below the yield. The *elastic limit* exists at a point just below the yield, indicated as z'. A material that is ductile is also soft, its hardness value is low, its tensile strength is low, its brittleness value is nonexistent, and it is malleable, but its toughness factor is also very low. In other words, whether a load is applied in a static situation or in impact, metal failure will occur in a soft and ductile material through some plastic separation as opposed to complete brittle fracture in hard materials.

FATIGUE AND FATIGUE RESISTANCE

Fatigue in a material is associated with time and loading conditions. Fatigue has many definitions. Fatigue in a material refers to changing loads on the material or to cyclic loading over a period of time with one of the loads being tensile. One student explained, "The metal becomes tired." The student's definition has some merit. How many times can a lever be loaded before it moves for the last time due to fatigue failure? This type of failure in a material usually occurs without warning, even though the part has been satisfactorily functioning over an extended period of time. Fatigue, then, is cyclic tensile loading such as loading and unloading or changing loads on a material over an extended period of service. Even though the material is safely loaded, that is, loaded below its yield strength, the time factor often causes the material to begin to change its structural arrangement. This subject will be discussed in a subsequent chapter.

THE LOAD-STRESS RELATIONSHIP

Many industrial materials undergo service loading from time to time. A material which is designed for a working load of 60,000 psi, for example, will hold 60,000 pounds safely in each square inch of stressed area. In this respect, the external load is measured in pounds, while the internal reaction of the material to the load is measured in pounds per square inch of stressed area across the line of stress.

This internal loading is called *stress* and it is equal to the load. Stress acts as an internal resistance to the deformation or fracture of the material from an external source; stress is the internal reaction to an external force on the surface of the material. Stress is a reaction to all loading situations. As long as the load remains below the elastic limit of the material, satisfactory material performance is expected. However, engineers and technicians must be on the alert for fatigue indications in a material, even though the material is operating under a safe load. The term *safe load* means that only a fraction of the yield strength is called upon to sustain the load. The remaining strength is then available for unknown overloading.

THE MODULUS OF ELASTICITY

The modulus of elasticity is the elastic relationship between stress and strain in a material. When a material is tensile loaded below the permanent deformation point or yield, the material stretches elastically and not plastically. The elastic stretch is proportional to the stress or load. This equal ratio of stress to strain, being below the yield point, is known as the modulus of elasticity, E. This factor or measurement of stiffness in a material varies with different materials, but remains reasonably constant in a given material under differing conditions. Table 3–4 indicates the modulus of elasticity for several materials.

The Modulus of Elasticity Is Fixed

The modulus of elasticity may be thought of as a constant because this value is fixed in the material at a given temperature. All other mechanical properties in metals are variable, but the ratio of stress to strain remains fairly equal to a point just below the yield. As an example, steel has an average E value of approximately 30,000,000 psi, while wrought iron has an E value of 27,000,000 psi. According to the table the metals have high E values and the nonmetals have low values.

TABLE 3–4 MODULUS OF ELASTICITY OF VARIOUS MATERIALS

Material	E Value (psi) Tension
Gray cast iron	12,000,000–18,000,000
Brass	13,500,000
Bronze	15,000,000
Copper	16,000,000
Aluminum alloy	10,000,000
Steel	29,000,000–30,000,000
Wood, hard	2,100,000
Wood, soft	800,000
Concrete	1,500,000–4,500,000
Gold	11,000,000
Magnesium	6,000,000
Silver	11,000,000
Titanium	16,000,000
Tungsten	50,100,000
Nylon, cast	500,000
Nylon, asbestos reinforced	1,100,000
Acrylic	450,000
Copper-nickel alloy	26,000,000
Wrought iron	27,000,000

The Modulus of Elasticity Is a Stiffness Factor

The modulus of elasticity can be better explained by an illustration, for example, loading a bolt in tension below its yield. When a 1000-lb load is placed on a structural steel bolt, the bolt elastically elongates a given amount. The bolt's extended length will diminish and return to its original length when the load is removed. Reapplication of the load will again elongate the bolt. Elongation and load or strain and stress are proportional in the material through the elastic limit, which is just below the yield. The release of any load below the material's yield point or, specifically, its elastic limit, will cause the material to spring back to its original length with a stress reduction to zero. In other words, solid materials are elastic, and this elasticity varies according to the material. As stress increases in a material, energy also increases in the material. As stress is reduced, energy is reduced. A typical example of the E factor is the relationship of stiffness between aluminum and steel; aluminum has an E value of 10,000,000 psi and steel has an E value of 30,000,000 psi, so steel is three times stiffer than aluminum. It is important to know the modulus of elasticity of a material when design calls for this stiffness factor, such as in a bending situation.

Questions

1. Why does the yield point of a material follow the elastic limits?
2. A stressed material contains energy. Why?
3. Differentiate between elasticity and toughness.
4. What is hardness?
5. Why is brittleness sometimes associated with stress?
6. Why are most mechanical properties variable?
7. What mechanical property is not variable? Why?
8. Why are mechanical properties grouped?
9. How does chemistry set the potential for a material?
10. Explain psi.
11. What determines microstructure in a material?
12. Grain size affects what mechanical property the most? Why?
13. How does microstructure affect mechanical properties?
14. How do distorted grains affect the mechanical properties of a material?
15. List five properties related to hardness.
16. List five properties related to softness.
17. Why is the modulus of elasticity related to stiffness of a material?
18. Describe fatigue failure. How can it be prevented?
19. In ductile materials why are the tensile and compression strengths approximately equal?
20. Differentiate between elastic and plastic deformation.

Hardness and Tensile Testing

Manufacturing processes produce millions of finely finished parts daily. Many of these parts, such as bridge and building beams, railroad rails and wheels, ship plate, and aircraft components, are for stressed applications and require minimum strengths. Failure of the metals in any of these items could cause injury or death and loss of money. In essence, the manufacturing process is incomplete without valid certification that the material and design are what they are supposed to be and that the finished parts are safe for use.

THE NEED FOR DESTRUCTIVE TESTING

There is no such thing as a perfect material. However, knowledge of the imperfections of materials and of the relationships of materials to stress flow can help establish safe loading on the material. In turn, long and safe operating conditions can be projected. But a safe operating load cannot be established unless the failing load is known. Consequently, several mechanical types of destructive tests must be made to produce a profile of the material's capabilities. The designer uses the profile to choose a safety factor and safely design the part. For example, if forty high strength bolts are to be manufactured, one or two must be destroyed under controlled conditions, such as tensile or shear testing, to infer the capabilities of the others. These tests are valid provided the metal comes from the same batch of steel at the initial processing and that all the bolts have been heat treated alike. If forty bolts are needed, then forty-two are produced. The remaining forty can be assumed to have the tensile and shear properties of the destroyed bolts. However, these remaining bolts must then be tested in other ways, such as hardness testing, to help assure mechanical property validity and removal of the assumption. If hardness testing cannot be performed on a part because of the slight depression caused by the testing, nondestructive testing must follow.

Testing Machines Are Essential

Often, it is said that mechanical properties for specified materials are indicated in the several engineering handbooks and that testing equipment is not necessary. This assumption is false because theoretical values must be translated to practical and operational values. Frequently, too much reliance is placed on material identification, and often the wrong material is used because of ignorance or accidental misidentification. Improper processing can cause different properties, also. Such situations usually end in material failure.

If the design of a part is to be accomplished based only on printed data, then the part should subsequently be tested by machines to validate both the material and the process if the design is to be reasonably stressed. To validate the design, both destructive and nondestructive testing may be essential. Too often, an automobile trailer hitch, for example, has broken at the weld because of insufficient knowledge by the welder of strength of materials or because of lack of welding proficiency. The finished hitch is only visually observed for acceptability. Guesswork and "eyeballing" simply do not have a place in designs where human life is involved. On the other hand, when materials, designs, and welding proficiencies have been approved, and when a qualified and certified welder is welding according to a standard or specification, then the finished work can be assumed to be acceptable. Keep in mind, however, that unless some form of testing is done on the part, such as a nondestructive test, safe operation of the part is still only an assumption. To help validate the mechanical properties of a material, several other tests are available, the most common being the hardness test.

Hardness Testing

The mechanical testing of a material, especially metal, for hardness is the most common way of determining one of the mechanical properties. This type of test is considered destructive only because of the small depression resulting from the test; no other destruction of the material occurs. However, some work hardness results around the depression. If the hardness value of a metal is known or is even approximated, several conclusions can be drawn. It is well known that hardness values vary, and as these values vary other mechanical properties such as tensile strength also vary. For example, if a metal is hard, what does *hard* mean? If a metal is soft, what does *soft* mean? These terms are meaningless until they are correlated with tangible values; the only practical way is to equate *hard* with a range of meaningful measurements such as numbers. If these numbers are derived from a scientific experiment which has been repeated hundreds of times, such as hardness testing, values of acceptable meaning and logic can be attached to a number when it expresses hardness. For instance, the letter-number symbol for hardness RC 67 connotes a series of properties. As will be explained, the hardness value of RC 67 means that a standard hardness testing machine, the Rockwell Hardness Tester (Fig. 4–1), was used to test a section of the metal (steel in this example), and the conclusion from the test is that near maximum hardness exists in the metal at the point of the test. Only a very small percentage of metals are harder than RC 67, such as an RC 70 existing in the surface of a nitrided steel. Then, for comparison purposes, what does the RC 67 mean? Such an explanation

FIGURE 4–1   The Rockwell Hardness Tester. (Courtesy of Wilson Instrument Division ACCO, Bridgeport, Connecticut)

requires correlation with other mechanical properties, along with simple illustrations.

Range of Hardness Values

The spectrum of hardness exists within a continuum from maximum to minimum. One hardness scale uses the diamond's hardness as maximum because no other material is harder than the diamond. For example, Mohs' scale of hardness equates the diamond to 10. Subsequent hardness values include the ruby, 9; topaz, 8; quartz, 7; feldspar, 6; apatite, 5; fluorite, 4; calcite, 3; gypsum, 2; and talc, 1. But the hardness of minerals is not practical for use in metallurgy, even though most metals originate in minerals. Somewhere between the hardness of the diamond and talc is found the hardness spectrum of metals. This range of hardness is far below the hardness of the diamond and well above the softness of talc.

To illustrate the concept of hardness, take a sharp file and attempt to remove a small quantity of steel from the surface of an RC 67 part. Press as hard as is practical and pass the file across the surface. You will quickly see that the file is softer than the filed part because the file merely skids across the metal's surface without even scratching it. The file will possibly have an RC 63 hardness value, and this is two Rockwell hardness points greater than the minimum acceptable hardness for many wear-resistant or cutting tools such as drills.

A Concept of Hardness

In order to carry the hardness concept further, it can be stated that any metal which has a hardness value of RC 67 is extremely hard. Most all metals in any condition are not as hard as RC 67; they are, therefore, softer. But what is a soft metal, and where does hardness end and softness begin? To illustrate, a drill easily cuts into a metal and produces a hole. The drill is then harder than the metal which is cut. In order for cutting to occur one metal must be significantly harder than the other. In fact, the greater the range between hardness values of the cutter and the metal being cut the greater the cutting efficiency and ease of cutting. It can then be stated in this example that one metal is hard and the other is soft, but a hardness range exists between the two values. Accordingly, as the hardness values decrease to the hardness of the softer metal, the hardness numbers indicate softer values. In other words, a metal that is RC 61 in hardness cannot be machined with conventional cutting tools; therefore, before this steel can be machined into the shape of a drill it must be made soft by annealing. This softness value must be near the opposite end of the metal's hardness range or cutting will not occur. With reference to the Rockwell hardness scale of values, C values are harder than B, and as the scale of numbers decreases the hardness of the metal also decreases.

Hardness values of all metals can be assigned from the hardest to the softest. Table 3–2 shows the range of hardness values in a hierarchy from RC 70 at the top to RB 0 at the bottom, all B values being softer than C values. A soft steel is one which has a Rockwell hardness value less than RB 100. For instance, "as produced" steel which is to become a drill can easily be machined into that shape if it has a hardness value less than RB 100. Any value less than RB 100 is soft, and the hardness spectrum decreases to RB 0 into even softer scales. Common nails, for example, will read approximately RB 40 when annealed, but possibly RB 85 when produced and ready for use, and the steel in a tinned food can will read slightly lower. Hardness values in the lower RB hardness scale are soft because the metal, if also ductile, will always bend without breaking under load and is easily scratched and cut. Further, soft and ductile metal is weak in tensile and yield strengths and will easily deform and plastically elongate extensively when loaded beyond its yield. Soft but brittle metal, however, like gray cast iron, is not ductile even though it is soft. All soft metals are not ductile, but all soft metals are weak in tensile, yield, and shear strengths (Fig. 4–2).

Hardness Is Related to Strength

The reverse of a soft and ductile condition in a metal is hardness and high strength, but as hardness increases from softness, brittleness also increases. This hardness-

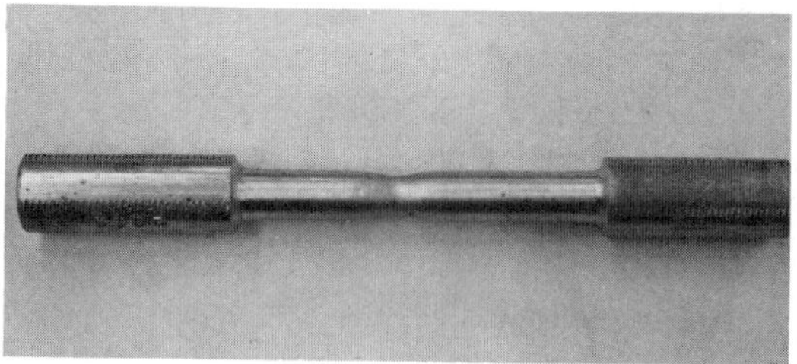

FIGURE 4–2   This specimen of aluminum is soft, is weak in strength, bends easily, and flows plastically.

strength relationship is shown in the following example. An AISI 1035 "as rolled" steel reads RB 79 in hardness, and it bends 180° without breaking when loaded in a bending stress. When pulled apart in tension loading, it reflects a tensile strength of 72,000 psi. This structural steel is easily machined because it is soft. Another example of softness is demonstrated in a section of annealed high speed tool steel which has an RB 93 hardness value and is especially suitable for cutting operations to be performed on it such as drill manufacture. The technician machines the stock into a drill by cutting flutes and edges along the length of the workpiece with hardened cutters. Machinability in this example is good because the metal being cut is soft and the metal doing the cutting is hard. After all machining is accomplished, the processed drill is then hardened and tempered and subsequently ground to size and final shape like other cutters. A hardness test is then taken, the hardness value being RC 64. In this condition the drill is much harder than when it was being shaped. But suppose the completed drill was used as a test specimen and given the tensile test with a special fixture. At fracture its tensile strength would be greater than 300,000 psi as compared to its previous 96,000 psi in the soft and machinable condition. The fracture plane would exhibit a brittle and flat type fracture as compared to a ductile and cup shaped fracture in the annealed condition. Another correlation with hardness is impact resistance. When a section of the drill is placed in a vise and struck with a hammer, it shatters into several pieces. But, if it had been exposed to impact while in the soft condition, it would have bent without fracture.

Hardness Varies According to Treatment of the Metal

Metals which have hardness values in the RC scale are hard with varying values, and metals having RB values are soft with varying values. But the RC scale extends from RC 70 to RC 20; the lower the number the less the hardness, and as hardness decreases, softness increases (Table 3–2). The RB scale begins at the bottom of the C scale and continues to 0. This hardness range then includes readings from C 70 to B 0, from hardest to softest, even though the lower B readings are not often used. Sometimes, however, during hardness testing there may be an overlap from the B to the C scale as hardness increases. This overlap in B and C scale hardness values results from the change in scales, but should present no difficulty in interpretation. With specific reference to the Rockwell Hardness Tester, any hardness value less than RC 20 (Fig. 4–3) should be retested on the B scale, and any value greater than RB 100 should be tested on the C scale. The range of hardness of metals varies with the metal and its treatments. Obviously, then, a given section of steel can be made hard or soft by exposing it to some particular kind of heat treatment.

HARDNESS

Hardness is resistance to penetration. Logically, the harder a metal, the more it will resist penetration; based on this principle, several different types of hardness testing machines are available for making the test. These testing machines are divided into the static and dynamic types. The common static testers are the Rockwell,

FIGURE 4–3   The hardness value indicated is much less than C 20, there-fore, another test should be made on the B scale.

Brinell, Vickers, and Tukon. The Scleroscope, on the other hand, is a dynamic tester. The major purpose of hardness testing is to determine a hardness value relative to a standard of some kind or to the hardness of some other object. Once the hardness value is obtained an analysis can be made of the surrounding metal-lurgical factors which brought about this value. A basic requirement for hardness testing is that the surface of the part to be tested must be fairly smooth and ap-proximately 90° to the axis of the machine's penetrator (Fig. 4–4). Basically, metal specimens can be sanded to the smoothness of a medium size grit, such as 180, for satisfactory results. However, the smoother the surface, the more accurate is the hardness value.

FIGURE 4–4   This view of the test shows the specimen resting flat against the anvil while the penetrator is 90° to the specimen's surface.

The principle of static hardness testing rests on penetration of the surface. If coarse grit has left visible scratches, the point of the penetrator will rest on the top or side of the scratch and allow a false hardness value to be obtained. Equal resistance around the periphery of the precision-shaped penetrator must occur for readings to be accurate. Also, tests must stay clear of cracks and edges of the metal being tested. One other item of importance relates to the hardness value obtained. This value indicates the hardness of the metal at that point only and not anywhere else on the surface and certainly not in the core of the metal. However, several tests along the surface will accurately infer an average hardness when processing conditions are known. On the other hand, the only way to obtain the hardness of the core is to part a section of the metal while keeping it cool and then take readings from surface to core. In other words, an RC 35 surface hardness value does not mean that a two-inch round bar of AISI 4130 has the same hardness across its cross section.

ROCKWELL HARDNESS TESTING

One of the most common and universally used hardness testers is the Rockwell, which is produced under several trade names, all using the Rockwell principle. This hardness test is based on depth of penetration into the metal being tested. Because the test relates to a very small volume of metal, several tests are essential to obtain a fairly accurate average hardness of the surface. Also, the area just a small distance away may contain hard or soft spots; therefore, three or four tests should be made. If hardness is resistance to penetration, then the harder a metal the less the penetration.

According to Figure 4–4, the specimen or part to be tested is placed on the hardened anvil and beneath the penetrator. Several options are available for testing, depending on what is known about the metallurgical conditions of the part. As indicated in the figure, the penetrator may be a precision ground diamond or a hardened steel ball of either $\frac{1}{16}$-, $\frac{1}{8}$-, $\frac{1}{4}$-, or $\frac{1}{2}$-inch diameter. As the softness of the metal increases, the diameter of the ball increases. Parts which are assumed to be hard are tested with the diamond and within a C scale range of 70–20 while using a major load of 150 kg (Fig. 4–5). Soft steel parts are tested with the $\frac{1}{16}$-inch steel ball within a B scale range of 100 to minimum while using only 100 kg major load (Fig. 4–6). Softer metals such as aluminum require the $\frac{1}{8}$-inch ball and either 100 or 60 kg major load while using the E, H, or K scale. If the approximate hardness of the part is unknown, begin with the diamond and C scale. Should the resultant reading be less than RC 20, change to the B scale or even another scale. The idea is to use the scale which causes the least penetration.

Testing Procedure

Once the setup is made and the specimen has been prepared, the test can be made. Round parts fit into V-shaped anvils and flat parts rest on flat anvils (Fig. 4–7). Slightly tapered parts fit on the ball anvil. In order to lock the specimen in place, a minor load of 10 kg is applied to the specimen by turning the capstan screw clockwise until both dial pointers are vertical and the large pointer is on C 0 (Fig. 4–8).

FIGURE 4–5　This reading on the Rockwell scale is RC 53, a valid reading, and is equivalent to approximately 282,000 psi in tensile strength.

FIGURE 4–6　This reading on the Rockwell scale is RB 79 because the 1/16-inch steel ball is used along with 100 kg major load.

Some penetration into the part occurs. By flipping the lever as shown in Figure 4–4 the major load is applied, requiring about 10 seconds and causing deeper penetration. Then, the lever is slowly pulled backwards and the hardness value is recorded. The hardness value is equal to the increment of depth, which is the depth difference between the major and minor loads. Accurate readings exist within C 70 to C 20 and any reading in the B scale from B 100 downward. Very low B readings for some nonferrous metals, however, require changing to other scales to avoid excessively deep penetrations. Three readings are usually taken, the average being recorded.

FIGURE 4–7 Rockwell hardness testing accessories: diamond penetrator, 1/16- and 1/8-inch steel ball penetrators, spot and shallow V anvils, cylindron, plane anvil, eyeball anvil, and test blocks.

FIGURE 4–8 The minor load is applied to hold the specimen in place when the major load is applied. Both pointers are vertical after the capstan screw is turned clockwise about two and one-half turns.

When very accurate values are required, a correction for rounded surface tests must be made because of a changing resistance around the cone of the penetrator, both for the diamond and ball readings to include the several scales. The correction varies with the specimen's diameter. The smaller the diameter and the softer the reading, the greater the error. For example, a Rockwell reading of 30 for the C, A, and D scales taken on a ¼-inch round steel specimen requires approximately five additional points for the reading to be accurate. The 30 value recorded on a one-inch specimen requires only one point to be added. A ¼-inch round steel part indicating 60 requires one and one-half points to be added, and the one-inch round of the same hardness requires only one-half point. With regard to the B, F, and G scales, a ¼-inch round part indicating 80 requires five points to be added, and the one-inch part requires only one and one-half points to be added. Such corrections are approximations only. It is observed that as the diameter of the specimen increases, the accuracy of the hardness value increases.

Superficial Hardness Testing

Another type of Rockwell test (Fig. 4–9) is the superficial, which is available for testing very thin parts and thin cases of case-hardened parts. Basically, if the penetrator's influence is observed on the underside of the part, the superficial test should be taken. Two common scales are available which use the 15-, 30-, or 45-kg major loads; both use 3 kg as the minor load. The diamond is used for the N scale, and the ¹⁄₁₆-inch steel ball is used for the T scale. Scale symbols are 15N, 30N, or 45N, and 15T, 30T, or 45T, respectively. Special scales are also available and include W, X, and Y with the ⅛-, ¼-, or ½-inch steel ball along with the standard 15-, or 30-, or 45-kg major load. All other operations of the machine are identical to regular Rockwell testing.

FIGURE 4–9   The model 4TT Rockwell hardness Twintester is used for testing thin sections and special tools and is known as a superficial hardness tester. (Courtesy of Wilson Instrument Division ACCO, Bridgeport, Connecticut)

According to Table 3–2, common Rockwell hardness values are indicated, and these values are in turn approximately convertible to other hardness value systems. Also, hardness values are approximately convertible to tensile strength values, but again it must be strongly pointed out that all conversions are approximate. On the other hand, experience has shown that these conversions are acceptable. A high strength steel bolt, for example, requires a heat treatment that produces a tensile strength of 125,000 psi. Certainly, an extra bolt or specimen can be heat treated along with the needed number of bolts and pulled in tensile to verify the strength of the others. If the Rockwell reading is C 27 for the destroyed bolt, it is assumed that all the others are also RC 27. Verification is made by making the tests if the small indentations are allowed. The possibility also exists that the core of the bolt is probably softer than the surface, so what is done? The destroyed bolt is sectioned and readings are taken across the cross section. Better still, a separate specimen is included for this particular test. Adjustments in heat treatments or in chemistry of the metal are then made to establish an average strength of 125,000 psi.

BRINELL HARDNESS TESTING

The Brinell is another type of hardness tester. This test is based on the area of metal affected by the load and is calculated by using the diameter of the resulting impression rather than depth. It is realized that there must be depth before there is diameter, and according to a given diameter measured in millimeters a Brinell hardness number is determined. As in other testing procedures, the same general rules apply regarding the preparation of the specimen. Because the hardened steel ball penetrator is large, 10 mm in diameter, specimens and parts to be tested must be larger than the very small parts tested by the Rockwell principle. Also, specimens must be thicker so the bottom side of the specimen is not influenced at the anvil. An advantage of the Brinell over the Rockwell is that the larger area covered in one test provides for a more reasonable average of the surface condition in one test. Several models of the Brinell, and of the Rockwell, are available. The motorized hydraulic tester is fast and efficient.

Testing Procedure

The setup for the Brinell test includes the proper placement of the specimen or part on the correct anvil, as shown in Figure 4–10, and the selection of the proper load. Depending on the hardness of the metal to be tested, one of three loads is available. The 3000-kg load is for regular testing and the other two loads are for softer metals. Choice of load is guided by the width of the impression desired, ranging between 2.5 and 6.0 mm. Also, the load should last for a minimum of time, say, 12 seconds for ferrous metals, in order that the full load be released. For nonferrous metals the load should be applied for 30 seconds. As previously pointed out, the large penetrator forbids the testing of a tool's cutting edge, but is ideal for testing castings and wrought products of sufficient size to allow for the large ball. Round objects must be at least two inches in diameter for accurate readings. Ball indentations must not be closer than about two and one-half indentations because work hardness occurs around the tested area. At least two impressions should be made at right

FIGURE 4–10   The Brinell Hardness Tester. (Courtesy of Detroit Testing
Machine Company)

angles to each other and the average indicated in the calculation of the hardness
value.

After releasing the load, the diameter of the indentation is measured with the
special microscope that has a linear scale marked in millimeters (Fig. 4–11). The
diameter of the impression is related to a conversion chart that shows the Brinell
hardness value. The conversion is a matter of stress calculation as shown in the
following formula:

$$H = \frac{P}{\pi \dfrac{D}{2}(D - \sqrt{D^2 - d^2})}$$

H equals the Brinell hardness number, P equals the load expressed in kilograms, D
equals the ball's diameter, and d equals the impression's diameter. As in Rockwell
testing, a very general conversion to other hardness scales is possible.

VICKERS HARDNESS TESTING

Specimens and parts to be tested with the Vickers penetrator are prepared so that
the surface is fairly smooth and parallel to the diamond penetrator. Also, parts to
be tested must not show any effects of the penetrator on the back side. The Vickers

FIGURE 4–11   Late models of the Brinell Tester have direct hardness values indicated on their scales. Older models use the microscope to measure the width of the ball impression (above) which is converted into hardness values.

penetrator is a square based pyramidal diamond which is forced into the metal to be tested by a predetermined load. The principle is similar to the Brinell in that the width of impression is measured across the diagonal corners and relates to the stress formula or load divided by area. Loads vary from 1 to 120 kg and the length of time under load is approximately 12 seconds. Much of the same procedure for Brinell testing is applicable to Vickers testing, including indentation spacings and measurements of impression. Also, a small correction for rounded surfaces must be made. Both thick and thin sections can be measured due to variability of loads. The microscope is mounted on the Vickers tester, whereas the Brinell impression is measured by a portable microscope. The length of each diagonal is measured with the special attached microscope and the average length is shown for calculating the Vickers hardness number. Vickers hardness numbers are approximately convertible to other hardness scales.

TUKON HARDNESS TESTING—KNOOP HARDNESS

The Tukon hardness tester (Fig. 4–12) is a microhardness testing machine that uses the Knoop pyramidal shaped diamond indentor, which has a longitudinal angle of 172° 30′ and a transverse angle of 130°. An option includes the square based diamond indentor having the 136° included angle. The indentation is applied with a light load varying from 15 to 3 kg. Again, the specimen must be prepared as described in Rockwell testing. The tester is used for finding hardness values of extremely thin parts and materials, including nonmetals. Even brittle materials that would normally fly apart under conventional loads can be tested. After the impres-

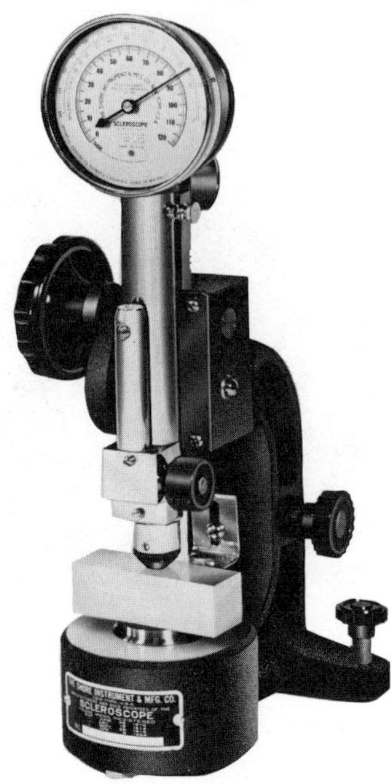

FIGURE 4–12   The model MO Tukon Microhardness Tester. (Courtesy of Wilson Instrument Division ACCO, Bridgeport, Connecticut)

sion is made with the use of a trip lever that releases the load, a microscope swings into position for measuring the length of the long diagonal. A micrometer dial allows movement of knife edges to determine the length of the long diagonal. The diagonal's length is converted to a Knoop hardness number. As in all hardness testing, the hardness number is a stress value that is found by dividing the load by the area of indentation.

SCLEROSCOPE HARDNESS TESTING

The Scleroscope (Fig. 4–13) is a dynamic hardness tester which is based on the principle of elastic indentation and height of bounce of a diamond-tipped hammer. Several models are available for testing solid materials of various shapes. The harder the metal being tested, the higher the bounce of the hammer. Therefore, hard metals will indicate high Scleroscope readings and soft metals will indicate low readings. Basically, Scleroscope testing is impact testing; both the indentor and material are momentarily elastically deformed. Immediately following the elastic deformation phase is the plastic deformation phase caused by concentrated stresses above the specimen's yield strength. This plastic deformation causes further elastic deformation. As this point the indentor's motion stops and it rests on the surface of the specimen. The highest point reached on the vertical scale during the bounce is recorded as the Scleroscope hardness value.

Operation of the Scleroscope (often called the Shore, after its manufacturer) is simple. A knob on the side of the machine is turned to allow placement of the

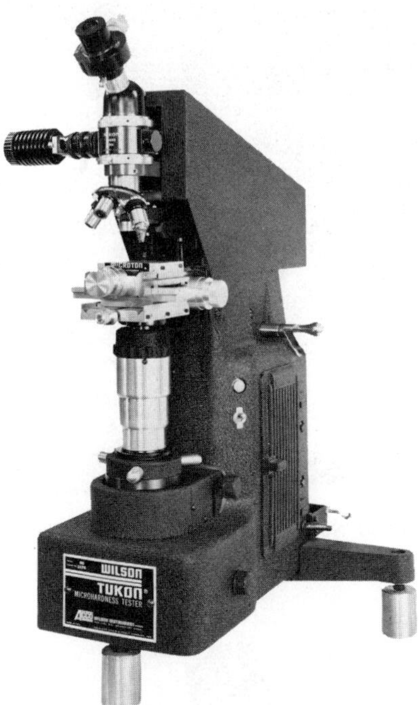

FIGURE 4–13   The Shore Scleroscope Hardness Tester Model D. (Courtesy of Shore Instrument and Manufacturing Company Incorporated)

specimen. The same knob is then turned in a reverse rotation to snugly hold the tube and working mechanisms against the smooth surface of the specimen. On the opposite side of the machine is a knurled knob which is turned until a "click" is heard, and then the knob is released. Turning of this knob raises the diamond-tipped hammer to a fixed height and quickly releases it. The harder the specimen's surface, the higher is the rebound of the hammer. The Model D (Fig. 4–14) automatically records the hardness value, while earlier models (Fig. 4–15) require quick observation to fix the number as the hammer moves upward on rebound.

CONVERSION OF HARDNESS VALUES

Conversion of values in hardness testing is possible with a reasonable degree of correlation. Thousands of tests have been made on the same kind of materials with several different testers, and their conversion characteristics have been within acceptable tolerances. In fact, enough accuracy has been demonstrated in conversions that many manufacturers use the converted values, especially tensile strength. A Rockwell C 40 may be converted to a Brinell 372, a Knoop 402, and a Scleroscope 40 with ease, and all of these hardness values show that the steel part has a tensile strength of approximately 182,000 psi. However, all conversions being approximate, the manufacturer should establish his own conversions under controlled conditions to help assure a greater degree of accuracy. Either way, the design's safety factor will absorb the possible error.

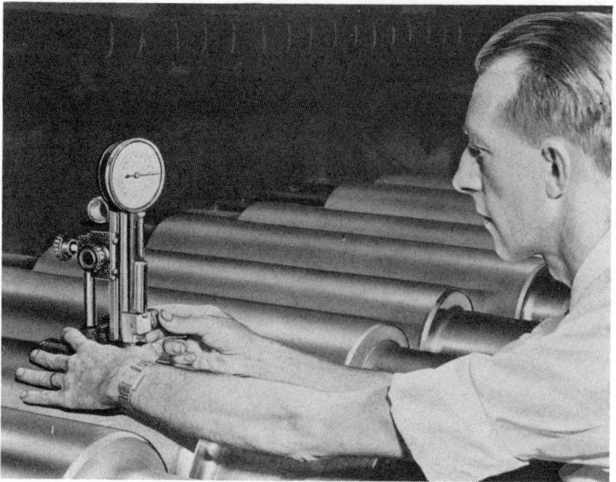

FIGURE 4–14   The model D Scleroscope is used to record readings on the roll-testing stand. (Courtesy of Shore Instrument and Manufacturing Company Incorporated)

FIGURE 4–15   The model C-2 Vertical Scale Scleroscope. (Courtesy of Shore Instrument and Manufacturing Company Incorporated)

STANDARDS AND CALIBRATIONS

Each testing machine has a set of standards (Fig. 4–16) that must be used from time to time to assure that the machines are operationally correct. Because each machine includes a copy of cleaning and calibration instructions, a few adjustments

FIGURE 4–16 Hardness testers should be calibrated from time to time to assure accuracy of scale readings. These test blocks are for the C and B scales of Rockwell testers.

often bring the indicated hardness values to acceptable tolerances. Should the readings of the machine be incorrect, only a few minutes are required to make the necessary adjustments. Each machine is calibrated differently so reference must be made to the manufacturer's literature.

TENSILE TESTING

Even though the hardness test is the most common of the mechanical tests, the tensile test produces more valuable data. Such a test requires more expensive equipment (Fig. 4–17), and the time factor for the test is greatly increased. How-

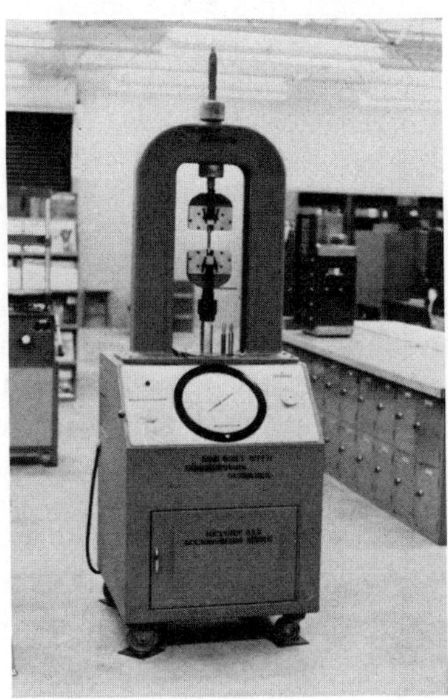

FIGURE 4–17 This tensile tester is capable of pulling standard specimens to fracture.

ever, data gathered from a combination hardness-tensile experiment furnishes a large quantity of engineering information concerning static loading and strains. Actually, the tensile test exposes a material to destruction. If the destructive load of a specific metal is known, for example, the safe operating load can be calculated. Tensile testing applies a slowly increasing load to a prepared specimen all the way to rupture (Fig. 4–18). As the external load, which is measured in pounds, is

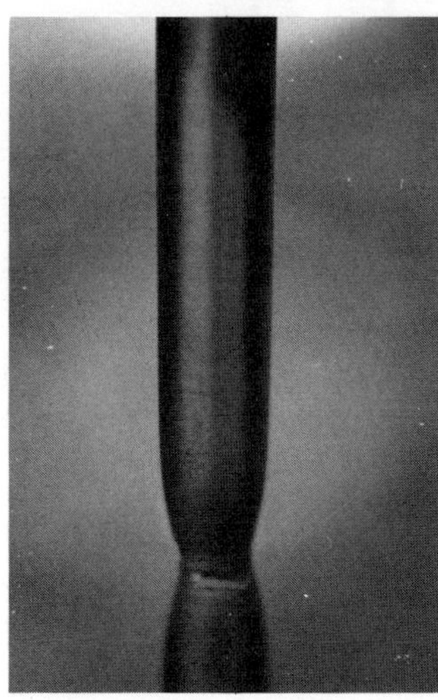

FIGURE 4–18   This standard tensile specimen has just ruptured after a large amount of plastic flow.

applied along the concentric axis of the specimen, internal stress builds up to match the load. The first group of stresses is elastic and these increase in value, being measured in pounds per square inch in a plane across the specimen's cross section. As increased loads occur, the elastic stress also increases to a point where further resistance to the load is impossible. This point is known as the elastic limit of the metal or other material. Loads beyond this value cause plastic deformation or brittle fracture.

Elastic Nature of Metals

All solid materials are elastic in nature. That is, a material under load will stretch, much like a rubber band, and if not stretched too far will return to its original length when the load is removed. Any load that stretches a material up to its elastic limit also allows the material to return to its original dimensions when the load is removed. This is because nature has provided a fixed property in solid materials known as the modulus of elasticity. The modulus of elasticity, measured in pounds per square inch, differs as the material differs. In steel, for example, and in tension, the modulus of elasticity is approximately 30,000,000 psi, while in the aluminums it is approximately 10,000,000 psi. The modulus of elasticity, being a measurement of stiffness, shows that steel is three times stiffer than aluminum alloy. Where bending loads are applied to materials and where dimensional changes occur, the modulus of elasticity becomes involved.

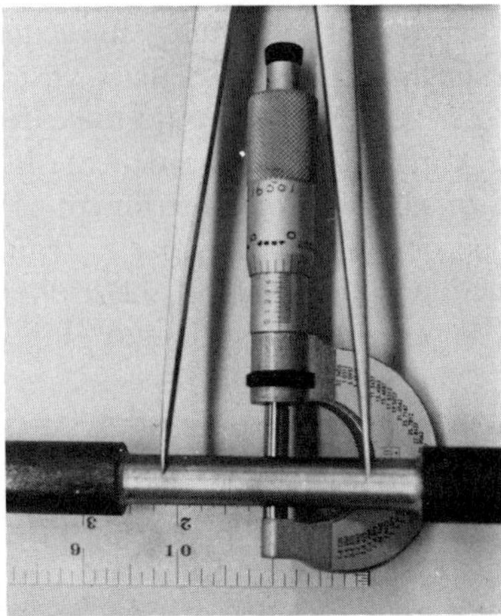

FIGURE 4–19   Dividers are used to obtain the exact distance between the punch marks, and micrometers are used to measure the diameter of the tensile specimen.

Preparation of Tensile Specimen

In order to obtain data from the tensile test, a prepared wrought steel specimen is obtained, as shown in Figure 4–19. Any diameter may be used; however, the ½-inch diameter round specimen is an ASTM standard type. On the other hand, flat specimens may also be used. Prick punch marks are lightly placed in the undercut section 2.0 inches apart (or whatever spacing is required, such as 1.0 or 6.0). Two measurements are initially taken, the specimen's diameter in the undercut portion and the original length between prick marks. Often, an extensometer is clamped onto the specimen (Fig. 4–20), its knife edges coinciding with the punch marks, or a 1.0 extensometer may be used. It stands to reason that if the specimen stretches under the pulling load, the punch marks will separate according to the load as will

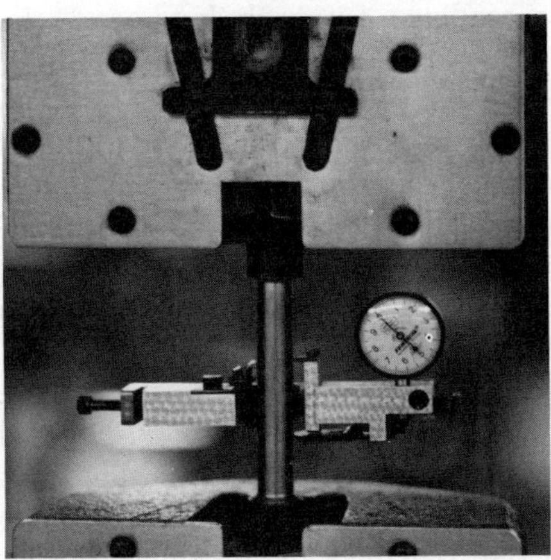

FIGURE 4–20   An extensometer is clamped onto the tensile specimen to record elongation as the loads increase.

the knife edges of the extensometer. Normally, the extensometer is removed when the load exceeds the elastic limit, therefore, it is customary to follow the whole stretch to rupture with a pair of dividers. Should the extensometer remain to tensile, however, the sudden shock may damage the delicate instrument unless it is shock resistant. The advantage of allowing the instrument to remain to fracture is that more accurate data can be obtained in the plastic regions. In order to obtain the final elongation and reduction in area values, the broken specimen is held together and the dividers are used to obtain the final length (Fig. 4–21). At the same time a

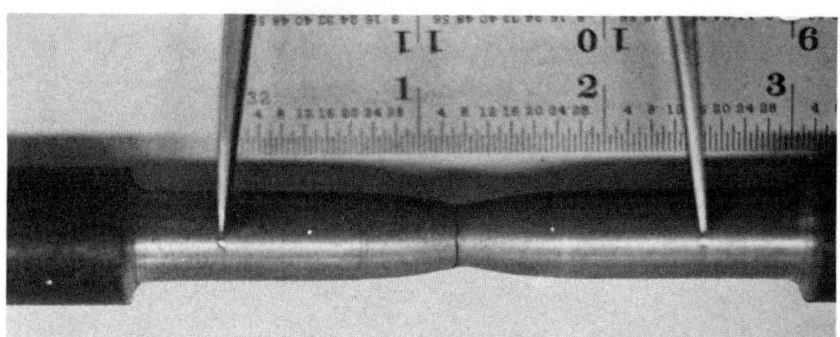

FIGURE 4–21   After fracture of the tensile specimen dividers are used to measure the greater distance between punch marks.

ballpoint type of micrometer is used to obtain the final diameter so that the final area can be calculated.

Recording the Data

When punch marks are made and when dividers are to be used, the points of the dividers are set (2.0 in) into the bottoms of the gauge mark cones and the exact length is established. Special rules and micrometers must be used for these measurements. Also, a chart similar to Table 4–1 must be made to record the following data in sequence: load in pounds in equal increments, extensometer elongation (or divider reading) for each load, and elongation rate differential between existing load and preceding load. The load is applied slowly, in 1000-pound increments in this example. (Any increment may be used.) The elongation rate is merely a quick means to show when the load has moved through the yield. For example, the elongation rates will be running fairly consistently below the yield, but when the yield is attained, the rates will be significantly greater because of plastic deformation. Table 4–1 shows the loads, elongations, and rates for a ½-inch diameter mild carbon steel specimen which was pulled to fracture. Notice that the yield occurred at a load value between 10,000 and 11,000 pounds because the rate suddenly increased. According to the table the extension rate remained at approximately 0.0003 inch through the 10,000-pound increment due to elastic stretch. However, when another 1,000 pounds were on the specimen the extension moved from elastic to plastic stretch in the amount of 0.021 inch. This substantial increase in the rate signalled attainment of the yield, but did not identify the exact load at the yield. With regard to deviations of readings, slight deviations

TABLE 4–1 TENSILE TEST DATA FOR A FRACTURED STEEL SPECIMEN

Load (lb)	e (in)	r	ϵ (in/in)	S psi	E psi
1,000	0.00034	.00000	0.00017	5,102	30011764
2,000	0.00070	.00036	0.00035	10,204	29154285
3,000	0.00104	.00034	0.00052	15,306	29434615
4,000	0.00134	.00030	0.00067	20,408	30459701
5,000	0.00172	.00038	0.00086	25,510	29662790
6,000	0.00210	.00038	0.00105	30,612	29154285
7,000	0.00236	.00026	0.00118	35,714	30266101
8,000	0.00270	.00034	0.00135	40,816	30234074
9,000	0.00300	.00030	0.00150	45,918	30612000
10,000	0.00338	.00038	0.00169	51,020	30189349
11,000	0.02100	.01762	0.01050	56,122	
12,000	0.03500	.01400	0.01750	61,224	
13,000	0.07400	.03900	0.03700	66,326	
14,000	0.25900	.18500	0.12950	71,428	
12,000	0.45700	.19800	0.22850	61,734	

Material: Mild carbon cold rolled steel Yield: 54,500 psi
Diameter: 0.500 in Tensile: 71,428 psi
Area: 0.196 in² Rupture: 61,734 psi
Original length: 2.000 in E average: 29,917,896 psi
Final length: 2.457 in
% Elongation: 22.85%

in the values below the yield are caused by human error in use of the machine and related equipment or slight errors in the machine.

Stress-Strain Diagram

Data shown in Table 4–1 are next converted to complete the stress, strain, and E values. Accordingly, the assembled information is subsequently used to plot a series of points at coordinate intersections and produce a stress-strain diagram (Fig. 4–22). First, load in pounds (P) must be converted to stress in pounds per square inch (S) by dividing the pounds by the original cross-sectional area of the specimen (A). The area in square inches is calculated by squaring the diameter, multiplying this factor by π and dividing the result by 4. This series of load-to-stress values will then be used along the vertical coordinate. Next, the strain value (ϵ) must be calculated by dividing the corresponding elongation value (e) by the original gauge length (L) of the specimen, which is usually 2.0 inches. However, some extensometers are direct reading, or 1.0 inch. These strain values will then be plotted along the horizontal coordinate. Lines are used to connect these intersecting points from no load to rupture, the lower portion being a straight line (E) to a point just below the yield. The remainder of the curve will move to the right as it smoothly moves upward to the tensile. Because this specimen exhibits a good ductility factor, rupture is plotted at a lower load value due to stress moving into the pure plastic region of the specimen. The curve is smooth from the tensile as it moves downward and to the right. This type of calculation is based on the apparent stress-strain within the specimen and is the standard method for stress-strain calculations. The true stress-strain data portrays a continually rising stress-

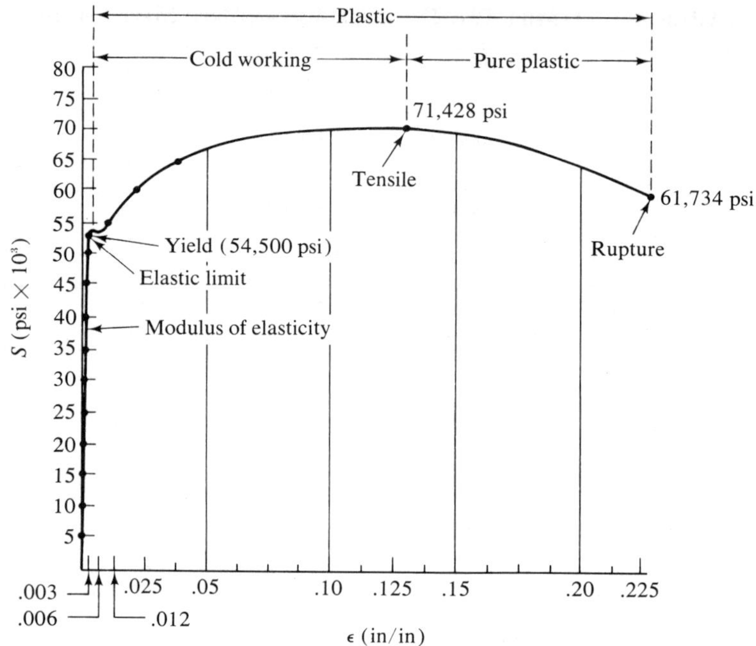

FIGURE 4–22   Stress-strain diagram for a mild carbon steel.

strain curve because of calculations of the ever changing diameter and rise of stress (Fig. 4–23).

Determining the Yield

The yield may be found by several methods. An accurate method involves the use of small load increments, such as 100 pounds or even 50 pounds. A sudden increase in the elongation rate identifies the yield, so the value will not be off more than 100 or 50 pounds. Another method is to carefully observe the load pointer

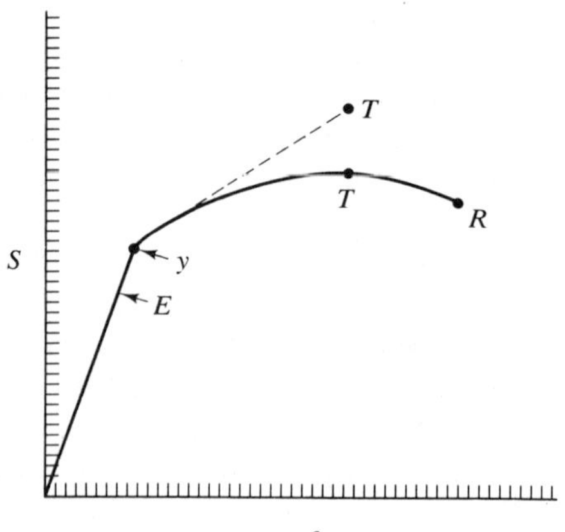

FIGURE 4–23   Apparent and true stress compared for a ductile steel.

FIGURE 4–24   A common method of finding the yield strength of a metal is to observe the load pointer. At the yield the pointer will momentarily halt or drop before increasing in load values. The pointer on the right shows highest load held, and the pointer on the left is the load pointer.

while under increasing load. The yield point (Fig. 4–24) will cause the pointer to momentarily halt or even drop and then resume the increase in loadings. When the offset method is used a line is drawn parallel to the E line (Fig. 4–25) at an offset of 0.2% or 0.002 inches to the right until it intersects the curved line in

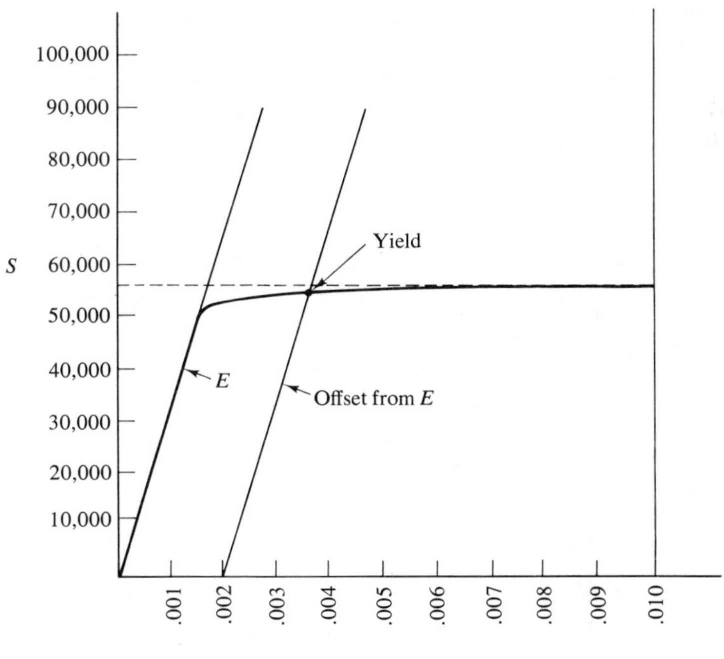

FIGURE 4–25   Offset method for determining the yield strength.

the plastic zone. The intersection is the yield, even though it is well into the plastic zone. According to the curve established in Figure 4–22 the captions of the coordinates should be made to show stress marked off in equal increments of pounds per square inch along the vertical plot and strain marked off in equal increments of inch per inch along the horizontal plot.

Analysis of the Stress-Strain Diagram

An analysis of the stress-strain diagram (Fig. 4–22) reveals numerous data of engineering value. First, it is known that the remaining metal in the bar from which the specimen was removed will react in the same manner when loaded and that all similar steels in the same condition will also react in the same way. By the way, all mechanical properties change in the particular steel as heat treatments and cold working occur, but the modulus of elasticity remain approximately the same. Reactions will reproduce the same E factor under different heat treatments. According to Figure 4–22 and Table 4–1, the following data are available: tensile, yield, and rupture strengths; modulus of elasticity; per cents of elongation and reduction in area; the elastic range; the total plastic range; the pure plastic range; the cold-working range; and the toughness factor, which is determined from the area under the curve. Also, the straight line portion of the curve from no load to the elastic limit or to a point just below the yield is given by Hooke's law, $E = S/\epsilon$, or the modulus of elasticity, which is the equal ratio between stress and strain within the elastic range. Even though this E value is not a stress value, it is still measured in pounds per square inch. Because the value of the elastic limit is shown, the amount of stored energy or recoverable energy can be calculated. Further observation shows that there is also a direct relationship between axial and lateral strains within the elastic range, and this ratio is called *Poisson's ratio*, $\mu = \epsilon_L/\epsilon$. For steel this ratio is about 0.30. With regard to the elastic range, the proportional limit or elastic limit rests slightly below the yield. In ductile steels there is usually the upper and lower yield points (Fig. 4–26), whereas

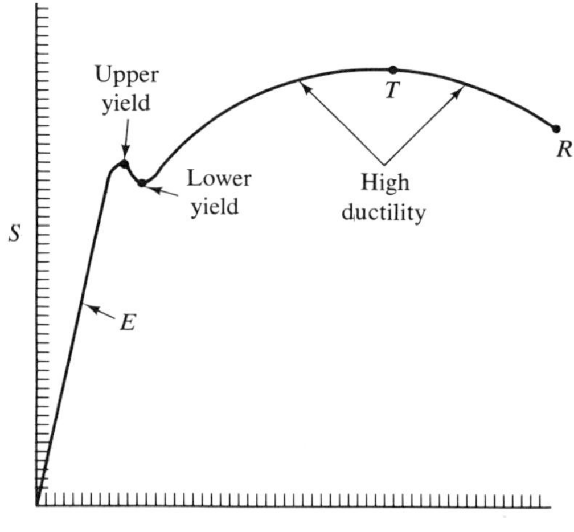

FIGURE 4–26   Upper and lower yield points occur in ductile metals.

the yield strength is midway between these two values and is the standard method of indicating the beginning of permanent deformation.

Other information is deduced from the stress-strain diagram. The percents of elongation and reduction in area indicate a high degree of ductility whereby the yield and tensile strengths are greatly extended. Such a condition allows cold forming of metal into intricate shapes without danger of tearing. Forming loads must be greater than the yield, but below the tensile. Due to the large volume of area under the stress-strain curve (Fig. 4–22), the metal has an acceptable degree of toughness with reference to freedom from brittleness. But it is much too soft and ductile to resist impact loading at higher stress levels for it to be considered very tough. The hardness value of RB 78 correlates very well with the actual tensile strength of 71,428 psi. With regard to statistics of the straight line portion of the stress-strain curve, the modulus of elasticity calculations remained within the 29–30 million psi factors as it should have for steels.

Close Observation Is Essential During the Test

During the test certain details must be followed. Unless the extensometer is engineered for shock, it should be removed prior to rupture of the specimen. In this respect, once a metal in a certain microstructural condition can be predicted with regard to its tensile and/or rupture points, the extensometer can be used up to or slightly less than a particular stress value, such as the tensile, and then removed. If the instrument is removed just prior to or at the tensile in a ductile metal test, the load and elongation values are recorded on the instruments. If dividers are used to obtain the total stretch, careful observation is essential. Accordingly, increased loading below the tensile advances the stress (pounds indication), additional elongation occurs at the same time. There is then a relationship between load and elongation. The load pointer (not the follower pointer) on the load scale of the tester must be observed closely to record the load at the instant that the load pointer begins to drop. A divider reading is then quickly taken, and this is the elongation at tensile. As the load is continued, for very ductile metal only, the load pointer continues to fall, and the follower pointer remains still and marks the tensile. The rapidly falling load pointer must be observed very closely to catch the load value in pounds as rupture suddenly occurs. When these data are plotted, the curve moves downward from the tensile and to the right. To the left of the tensile point the curve has moved upward and to the right from the yield, the tensile separating the two curves at the high point. The reason that the curve moves upward beyond the yield is that plastic flow induces higher stresses along slip planes and forces further plastic flow because of work hardening and higher strength values.

When another tensile specimen of a higher carbon-type steel is heat treated by hardening and tempering and then tested to rupture, a completely different stress-strain curve is established, as illustrated in Figure 4–27. Observation of the curve shows that the modulus of elasticity remains approximately the same, but the yield and tensile strengths are greatly increased due to heat treatment of the metal. In this respect, the load on some steels may often be doubled through heat treatment and with safety without increasing the size of the part. However, ductility is

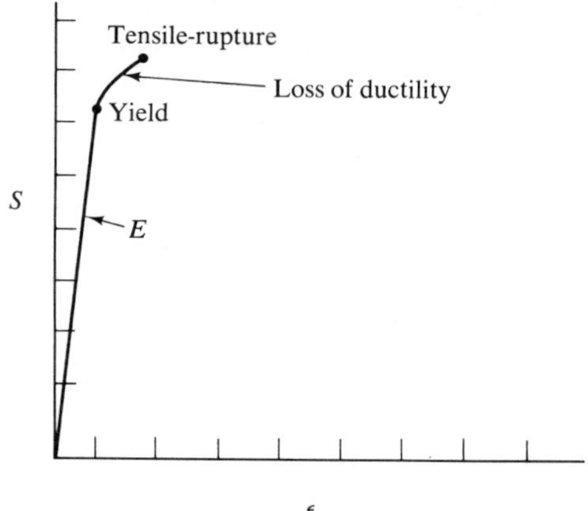

FIGURE 4–27   The stress-strain characteristics of an AISI 1080 hardened steel. The tensile and rupture occur at the same load.

greatly reduced as shown by the closeness of the yield and tensile. Little elongation occurs beyond the yield in this example. It is also pointed out that in this example the tensile and rupture occur at the same stress value because of the absence of the pure plastic region. The fracture still remains ductile to a small degree, however, indicating failure in shear.

In a further discussion of tensile loading, another specimen is pulled to rupture and the data are plotted (Fig. 4–28). This specimen is common gray cast iron. Notice that the resulting stress-strain curve lacks work hardening and ductility because the tensile quickly follows the yield. Only elastic deformation is shown; that is, only a very small amount of plastic flow occurs. Further, the stress required to fracture the specimen is much lower than either of the steel specimens because of the presence of the nonmetal graphite.

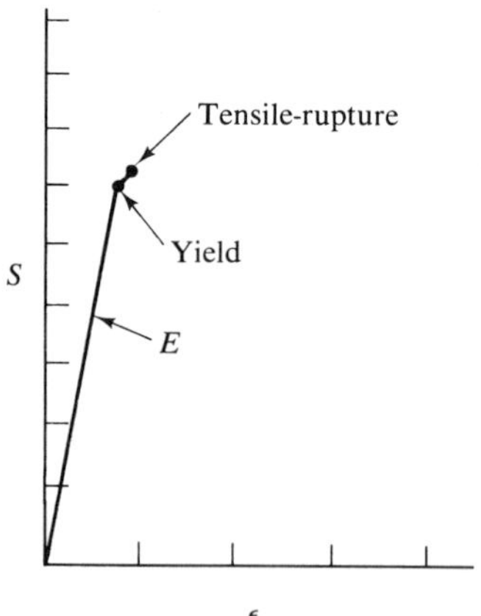

FIGURE 4–28   A stress-strain diagram for common gray cast iron in which little ductility exists, causing a soft but brittle metal.

Safe Design Requires Factual Data

Much is gained by tensile testing. Even the correlation of surface hardness and average mass hardness to tensile strength is reasonably accurate. Relationships among tensile and yield strengths are illustrated along with elastic and plastic values which enable safe operating stresses to be calculated. Once these values are known, safe designs can be drawn after the remainder of the operating environment is determined. From Hooke's law, stress is proportional to strain within the elastic limit, and from Young's Modulus (the modulus of elasticity or measurement of rigidity in tension or compression), a vast array of engineering data becomes available. And then, Poisson's ratio points out the constant narrowing of the specimen's area as its length increases under stress within the elastic range because of granular rotation of the metal's grains and more favorable slip plane alignment toward the axis of stress.

In effect, new formulas become available and, in turn, these bring additional information. When the modulus of elasticity is to be found, these following formulas are helpful:

$$E = \frac{P/A}{e/L} = \frac{PL}{Ae}$$

These calculations are derived from:

$$P = AS, \qquad S = \frac{P}{A}, \qquad A = \frac{P}{S}, \quad \text{and} \quad E = \frac{S}{\epsilon}$$

There is the constant presence of strain where there is stress, and these stress-strain relationships are as follows:

$$\epsilon = \frac{e}{L} \quad \text{and} \quad \mu = \frac{\epsilon_L}{\epsilon}$$

Other values are also needed such as the elongation and reduction in area percentages, and these are found as follows:

$$\% e = \frac{L_F - L_o}{L_o} \times 100$$

$$\% RA = \frac{A_o - A_F}{A_o} \times 100$$

Questions

1. What is the purpose of destructive testing?
2. What is hardness and how does it relate to other mechanical properties?
3. Differentiate between the yield and tensile strengths of materials.

4. Describe the importance of the yield in design.

5. Why are standards necessary when testing apparatus is being used?

6. List several factors that the tensile test identifies.

7. What is the relationship between stress and strain below the yield? above the yield?

8. Describe the difference between the pure plastic range and the range between the yield and tensile strengths.

9. Explain the difference between Rockwell and Brinell hardness testing.

10. Why is a stress-strain diagram or its reflected values essential to the design process?

11. Why does stress increase beyond the yield in a ductile steel?

12. Correlate the relationship between hardness and tensile strengths.

13. Correlate the relationship between hardness and yield strengths.

14. Why do brittle materials exhibit a cleavage type fracture?

15. Explain the significance of the region between the yield and tensile strengths.

16. How is the elastic limit related to the modulus of elasticity?

17. What condition in a metal allows an upper and lower yield strength while under tensile load?

18. What is the difference between static and impact tensile loading?

19. Describe the offset yield method of calculation.

20. What causes a metal to yield?

Other Testing Methods

Destructive testing provides essential data for design purposes. No one test provides all the needed data, however, because only limited facts about the material are accumulated from each test. For example, tensile testing relates the action of the material as it steadily elongates to constantly increasing stress levels. The compression test relates the increasing stress to elastic and collapsing deformation of the material in a manner opposite to tensile. In shear the loading forces are cutting forces and are below the tensile values. Other destructive tests include impact, torsion, tear, microscopic examination, and fire testing.

COMPRESSION TESTING

Compression loads are pushing loads that cause opposing stresses in materials, as illustrated in Figure 5–1. Pushing loads tend to flatten a metal, causing it to become shorter in length and larger in diameter. For ductile steels the compression strength is considered to be approximately equal to the tensile strength, but for nonductile materials the tensile strengths are much lower than the compression. Brittle materials, such as common gray cast iron and concrete, are normally used in compression, but not in tension alone because of the unpredictable variables in the materials. In this regard, concrete is frequently used in beams where tensile stresses are present, but tensile stresses are only one of the three common stresses (tensile, compression, and shear), as illustrated in Figure 5–2.

Testing Procedures

In compression testing the specimen is first measured for length and diameter and the area calculated. Several types of machines are available for compression testing, such as the universal testing machine (Fig. 5–3) and the special type of

FIGURE 5–1 Mild carbon steel specimens before and after compression loading. Extreme ductility provides for plastic flow at room temperature.

FIGURE 5–2 The roadway pushes downward on these steel beams, which places compression stresses on their top sides and tensile stresses on the bottom flanges. The concrete beam is in tensile stress on its bottom side while compression stress exists in the top surface. Vertical shear exists in the concrete beam where the column's edges contact the beam. Rolls allow for movement of the bridge under varying loads.

compression tester shown in Figure 5–4. After initial data is posted, the specimen is placed in the center of the compression chamber to avoid eccentric loading (Fig. 5–5). Face shields are then secured unless an explosion barrier surrounds the compression chamber. The shields stop flying particles which are emitted under high stress loading. For ductile materials no danger usually exists; however, some ductile metals quickly transform to brittleness as work hardness begins. A Lexan (thermoplastic) shield of sufficient thickness allows visibility and ample protection for many types of compression tests. With respect to compression failure and compression strength, care must be exercised at the point of failure to record

FIGURE 5–3 The model DS 300 Riehle screw-powered universal testing machine with furnace. This 300,000-pound capacity tester can test specimens at any temperature up to a variable temperature range. (Courtesy of Wilson Instrument Division ACCO, Bridgeport, Connecticut)

FIGURE 5–4 A compression tester with a 250,000-pound crushing capability. Special anvils are used for heavy loads, and the chamber is enclosed within an explosion barrier. Technicians are recording data during a test.

the proper load. Failure occurs either by shearing action or crumbling in brittle materials or by continuous flattening in very ductile metals (Fig. 5–6). During the test on a round section of yellow brass (1 inch long and ¾ inch in diameter), for example, the specimen will immediately bulge at its midpoint due to opposing

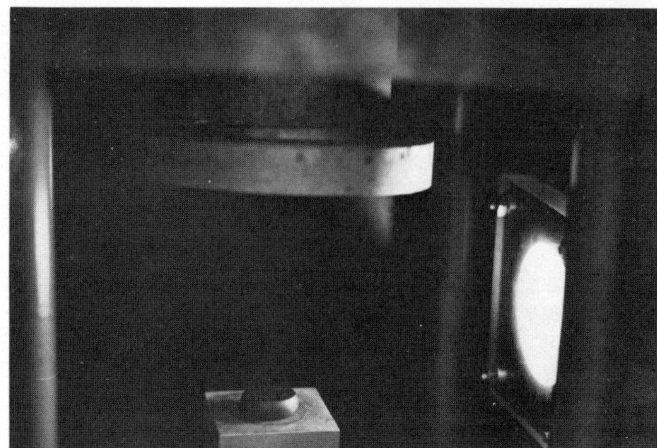

FIGURE 5–5 A steel specimen has been compression tested. Note that each specimen must be carefully centered to prevent eccentric loading. A 3/4-inch thick Lexan shield covers the observation window.

FIGURE 5–6 High malleability in this 1-inch diameter steel specimen allows plastic flow to compression failure.

stresses there (Fig. 5–7). The barrel-shaped specimen will continue to elastically increase in diameter to the yield. If the external load is released at its elastic limit, the original measurements of the specimen will be duplicated as it releases its stored energy. This elastic zone is where the part functions as in alternating compression loads. Continued loading brings on yielding. Work hardening and fracture or flattening follow, marking the compression strength in pounds per square inch.

Compression Loads Are Buckling Loads

A word of caution is given with respect to buckling loads in compression. As length increases with a constant diameter, the buckling danger and failure also increase. Therefore, compression loads must remain below the buckling load.

FIGURE 5–7   This yellow brass specimen elastically bulges at its midpoint. Continued loading causes work hardening and permanent deformation.

The buckling load is related to the geometry of a column as well as to the compression load and to how the column is attached at its ends. Attachment may be by bolts or by welding if metal columns are used and by reinforcing steel rods from the flooring if made of concrete. One of the stiffest column designs for a given set of values is the H shape. Such a shape resists twisting and buckling tendencies caused by the compression load. Tension stresses do not relate to the cross-sectional geometry in the same respect as compression because of the absence of twisting and buckling. A solid round bar and a tube, for example, with equal cross-sectional areas will exhibit the same tensile strength; that is, they will equally resist being pulled apart. The column with an equal cross-sectional area, on the other hand, must have its area spread out as far from its centroid (center of gravity) as is practical under compression loading in order to resist turning and buckling, as well as crushing.

SHEAR TESTING

Tension, compression, and shear are the three main engineering forces. Tension and compression forces are collinear forces which act at right angles to the part's cross-sectional area and are known as *axial* forces. On the other hand, parallel but noncollinear forces to the cross section of the part are shearing, or cutting, forces (Fig. 5–8). Such forces act as sliding forces and attempt to cut a material just as a pair of scissors cuts paper. Shearing forces must slide past each other to avoid tearing; and the closer the opposing but sliding forces get, the greater the tendency to cause shear and failure. There is then a critical distance which must be maintained between the two opposing forces so that shear will not transform to compression. Therefore, engineering designs must allow ample material to resist the shearing actions of weight or pressure, even though shearing strengths are the weakest of the three main strengths. As an example, common structural steel (0.35% C) has a cold-rolled tensile strength of approximately 72,000 psi, a compression strength of 72,000 psi, and a shearing strength of

FIGURE 5–8 The pin attached to the flywheel is in shear stress at the face of the wheel because the rod and wheel cause cutting stresses.

42,000 psi. Shear strength in this example is four-sevenths of the tensile, and this relationship can be used to approximate shear strengths of several other ductile steels. With regard to cold rolling, the metal has been shaped at a temperature below the recrystallization temperature of the metal. Hot working is performed above this point. In steels hot-working temperatures are higher than approximately 1400 °F (760 °C), or the red heat range.

Testing Procedures

During the shear test opposing loads cause cutting stresses across the part's cross section. Unless the area and strength of the metal are sufficient, the shearing stress will build up to a value which is slightly greater than the shear strength of the metal, and failure will occur without much warning. As in tensile testing, the load pointer will increase, dragging the follower pointer, as the load in pounds increases. If the load is released below the elastic limit in shear, no harm occurs to the part and it will spring back to its original position and dimensions. But if plastic deformation begins just beyond the elastic limit (a drop in the load pointer), shear failure will quickly follow unless the load is removed. Area of the cross section, microstructure, and kind of metal are key factors which control shear strengths of metals. In reinforced concrete the area and strength of the concrete are key factors in addition to the shear strength of the steel rods.

Because shear consists of two main types, vertical and horizontal, each must be considered in a design. As illustrated in Figure 5–9, vertical shear is at maximum intensity at the edge of the two columns supporting a loaded beam and is at zero intensity in the center of the span. When a load deflects the beam, maximum tensile stress occurs at the surface of the bottom flange of the beam, while maximum compression stress occurs at the top surface of the beam's flange. In this respect and with regard to geometrical shape vs stiffness, the I (wide gauge) beam has the essential engineering qualities to effectively sustain fluctuating loads within its elastic limit. Consequently, when maximum tension stress exists at the bottom of the beam under load and maximum compression stress

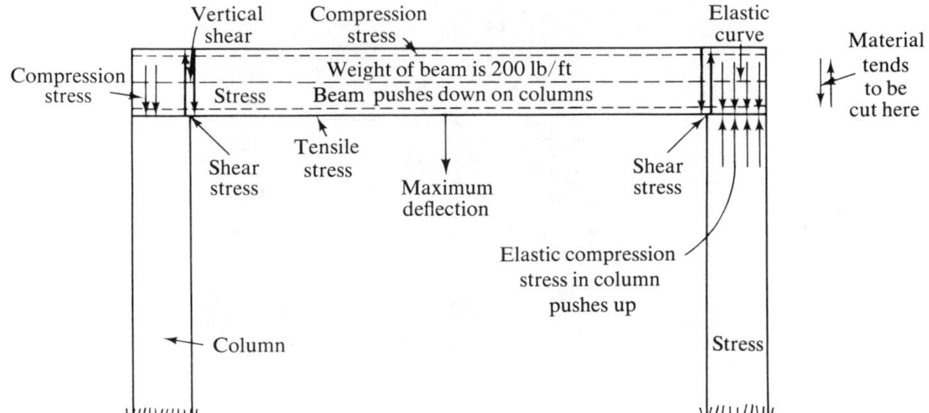

FIGURE 5–9   Multiple stresses, tension, compression, and shear, exist in a loaded beam, and compression stresses exist in the columns, all stresses being elastic.

exists at the top, then tensile stresses must diminish toward zero in the centroid plane of the symmetrically shaped I beam. From this central position, stresses change to compression and increase in intensity to a maximum compression stress level at the top surface. Such a situation points out the tendency of the metal along the beam's top to compress, or grow shorter in length, while the metal along the bottom surface tends to stretch and grow longer. Obviously, as stresses move toward the beam's center, opposite sliding tendencies exist, and these shearing stresses are horizontal (Fig. 5–10).

Single and Double Shear

Maximum vertical shear occurs at the edges of each column because of the weight of the beam pushing down and the resistance of the column pushing up. For sake of calculation simplicity, assume that the horizontal stress in this example is equal to the vertical. A cross-sectional view of the shearing stress points out a single shear value because there is only one load pushing down and one load or force or stress pushing up (Fig. 5–10). If the beam rests on its *x-x* axis, the flange, in contact against the flat surface of the column, then single shear exists. If the beam rests on a pad where two shearing actions could occur, then double shear exists in the bolt in the pad. Some bridge beams and pads are designed for multiple shearing resistance to increase the total shear strength of the bolt that holds the beam in place on the column and allows it to move when loaded (Fig. 5–10). Single and double shear are the two common forms of shear and may exist in any attitude with relation to gravity. In Figure 5–10 shear loads at the left reaction (column) are positive, and those at the right reaction are negative because the weight of the beam pushing downward causes the two columns to push upward.

Another example of shear exists in the riveted structure of aircraft wing skin. Sheets of aluminum alloy are riveted to the inner structural framework of the wing. When the movement of the wing causes it to bend downward, the top layer

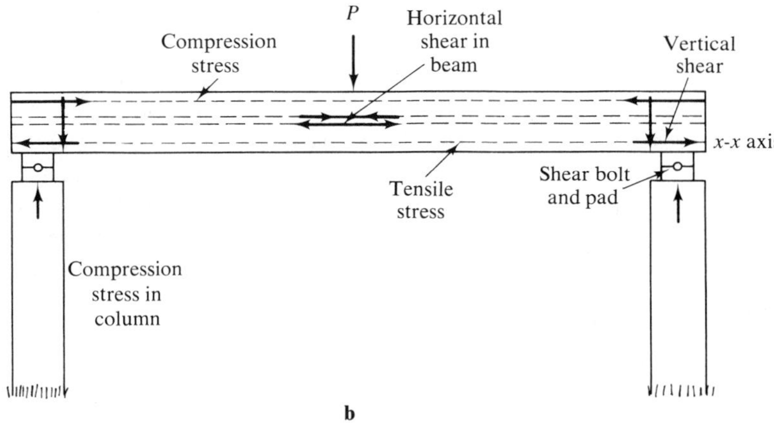

FIGURE 5–10   *a,* The loaded beam pushes down on the shear bolt while column stresses push up against the bolt, causing a cutting tendency by sliding forces in the bolt. *b,* The load *P* causes both vertical and horizontal shear stresses in the beam.

of metal is exposed to tensile stresses while the bottom side is exposed to compression stresses. As the wing moves upward the top side skin moves into compression while the bottom side is subjected to tension. But the surface skin in slow-flying aircraft is sometimes composed of several overlapping sheets which are held together by rivets. As illustrated in Figure 5–11, bending stresses in an aircraft member causes tensile, compression, and shear stresses in the joint. The rivet is caught in shear as the stresses in the surface sheets change from tension to compression, both stresses attempting to cut the rivet across one cross section. The load required to cut one rivet is equal to the cross-sectional area of the rivet ($\pi D^2/4$) multiplied by the shearing strength of the metal. Stronger joints use a plate and flush-type rivet so that double shear will be required (Fig. 5–12).

FIGURE 5–11 Bending forces in this aircraft structural member induce tension, compression, and shear stresses concurrently within the member.

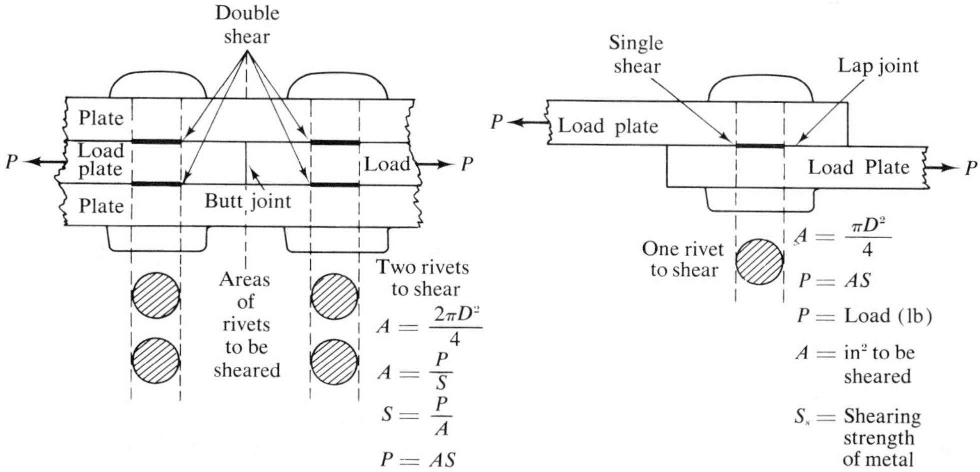

FIGURE 5–12 Shear forces in a butt and lap joint result from tensile forces in the main load-carrying plates.

Punch and Shear Relationships

When a force is applied parallel to the cross section of a material, shear actions develop. If the shape of the force is such that a punching stress develops, a hole results in the part under load when the stress is greater than the strength of the material. For example, Figure 5–13 illustrates a single punch moving through a structural steel plate. The load required to shear the plate is equal to the circumferential area of the structural plate ($\pi D t$) multiplied by the alloy's shearing stress of approximately 42,000 psi.

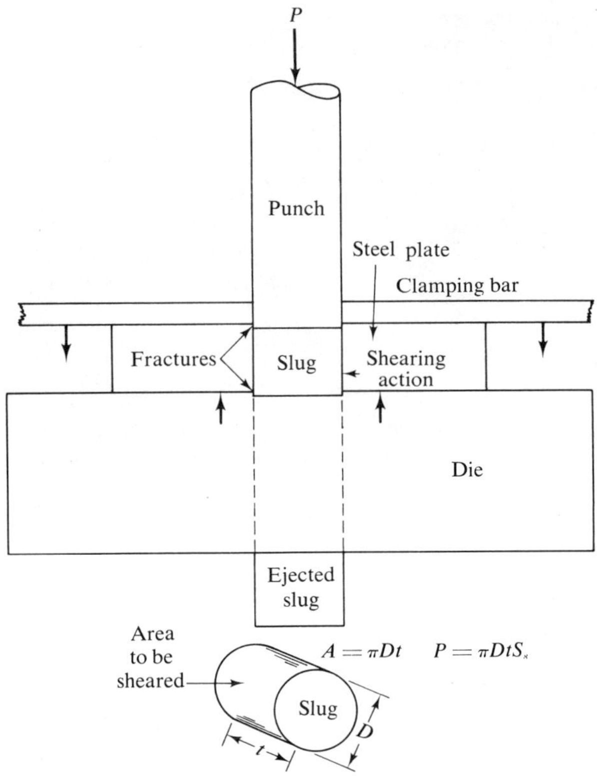

FIGURE 5–13   Punch and shear action occur as the punch descends onto the plate which rests on a die.

IMPACT TESTING

Impact testing (Fig. 5–14) is performed to help determine a relative measure of toughness or impact resistance of a metal. The finding are relative to each other when several metals are tested, but these findings do not give a true measure of the same metal and microstructure in an operating condition because of the effect of the notch and a significant reduction in actual toughness. The findings from the test possibly signify a minimum toughness factor for the specific metal because impact-loading conditions in service are less severe due to the absence of the notch which increases stress. Dynamic loading of a metal or other material is quite different from the several static loading situations; the dynamic load incorporates the sudden application of the load. Static loading as in tensile, compression, and shear, normally allows loads to steadily increase to maximum, and time is an important factor. With time as a function of loading, the metal will have an opportunity to adjust its microstructure to better resist the increasing load. Dynamic loading acts as a fast-moving projectile and allows no time for normal microstructural adjustment.

Multiple stresses imposed around the notch of a specimen are of several kinds and magnitudes. The idea is to remove the specimen from the swing of the

FIGURE 5–14 The SATEC Model SI-1 Impact Tester, manufactured by SATEC Systems, Inc. of Grove City, Pa., measures the energy required to break standard specimens in either Izod or Charpy tests. The tester works on the principle of the pendulum. Thus, the angular distance travelled by the pendulum after it strikes the specimen is a function of the energy required to break the specimen. (Courtesy of SATEC Systems, Incorporated)

hammer under any circumstance and record the loss of kinetic energy as the hammer swings through the fixture which holds the specimen (Fig. 5–15). The energy consumed in removing the specimen from the path .of the hammer is recorded on the impact tester's dial as foot-pounds (Fig. 5–16). The greater the numerical value of foot-pounds, the tougher the metal. Of course, as in all environmental testing, temperature is an important function. Unless otherwise stated, all types of testing are conducted at room temperature.

Fracture and Microstructure

A given load falls from a given height (impact load), contacts the specimen, and failure occurs. An examination of the failure zone shows either a ductile fracture, a brittle fracture, or some type of combination of ductility and brittleness. The fracture leaves the evidence of chemistry and microstructure when all other variables are held constant. When temperature becomes a variable, the fracture is analyzed to determine the effects of temperature while other variables are held constant. A harder metal that exhibits a fairly ductile fracture requires more energy to fracture while brittle fractures require less energy. Brittle fractures present a kind of cleavage appearance, and less brittle fractures include evidence of plastic flow due to some degree of ductility. Very ductile metals

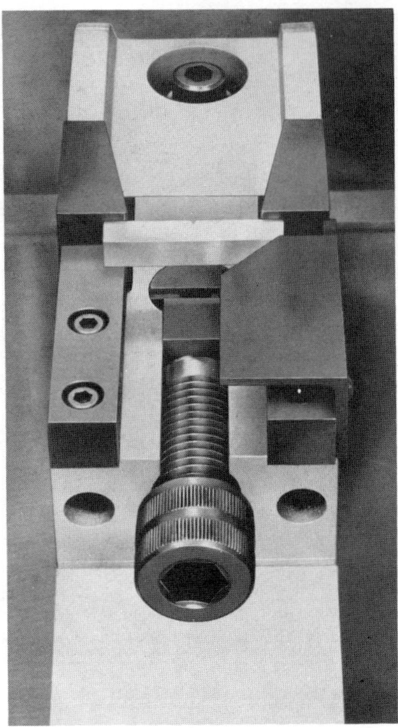

FIGURE 5–15   This photograph shows a Charpy test specimen in place on the anvil of a SATEC SI-1 Impact Tester. One of the two shrouds designed to prevent broken specimens from interfering with the free swing of the pendulum has been removed to provide a better view of the notched bar specimen. (Courtesy of SATEC Systems, Incorporated)

leave their evidence in the form of folded and stretched metal. Cleavage breaks offer little toughness, therefore, some kind of compromise between the cleavage and severely deformed metal is required when toughness is desired. With regard to temperature, the transition temperature relative to the shape of fracture is that temperature which produces approximately 50% cleavage and 50% plastic deformation. The lower the transition temperature, the safer the metal for use at temperatures below room temperature. Face-centered cubic lattice structured metals are not seriously affected by a transition temperature. When the part's operational temperature is above the transition temperature, safer operating conditions prevail. An explanation of the transition temperature is given in a later chapter.

Types of Impact Testers

Two main types of specimens for impact testing are available, the Izod and Charpy. Both types are illustrated in Figure 5–17. Each type contains a notch, but the Charpy has three different notched specimens. The Izod specimen is placed in the vise as a cantilever beam so that the notch is toward the hammer (Fig. 5–18). Either the V-notch, U-notch, or keyhole Charpy specimen is placed in the fixture as a simple beam so that the hammer strikes the specimen from the

FIGURE 5–16 The hammer is caught near top of its swing after breaking the specimen.

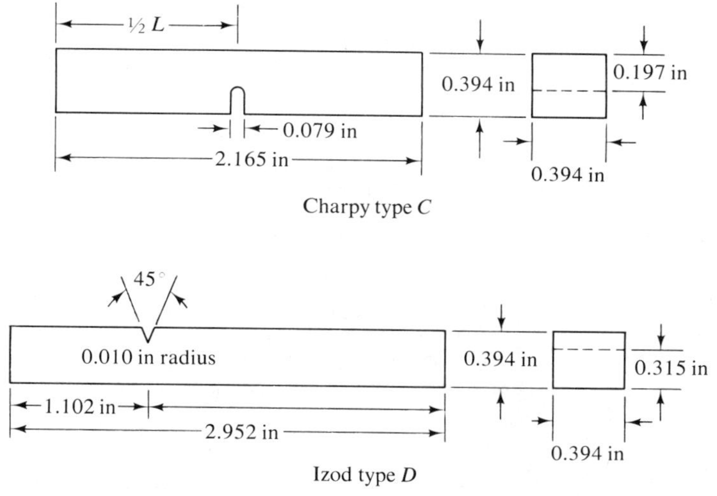

FIGURE 5–17 Charpy and Izod impact test specimens.

rear (Fig. 5–19). The standard impact tester applies the impact load of 120 ft-lb by means of the pendulum. The energy needed to fracture the specimen is determined by multiplying the weight of the pendulum by the height difference of the pendulum before and after striking the specimen. A free fall offers no resistance to the hammer (except friction) and no absorbed energy is recorded on the dial. When a specimen is struck, the impact value is recorded on the dial in foot-pounds. Results of these notched specimen tests provide comparison values of toughness among different metals under varying environments. The higher the foot-pound value, the tougher the material.

TORSION TESTING

Torque is a twisting moment applied to a material whereby the material is in torsion. A *moment* is the product of the force and the perpendicular distance from the force to the axis of tending rotation. An electric motor applies torque

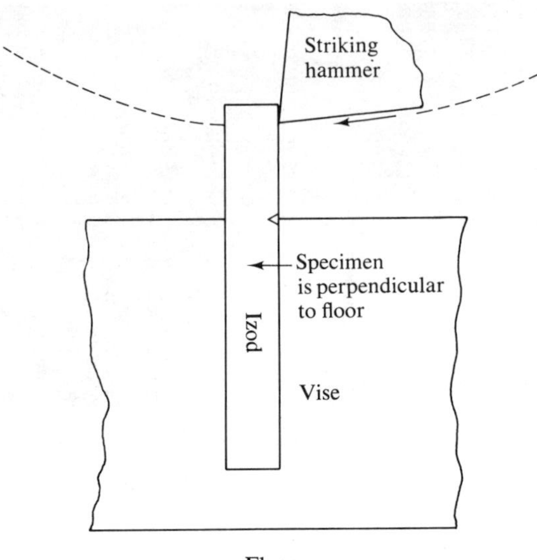

FIGURE 5–18   An Izod impact specimen being struck by a swinging hammer.

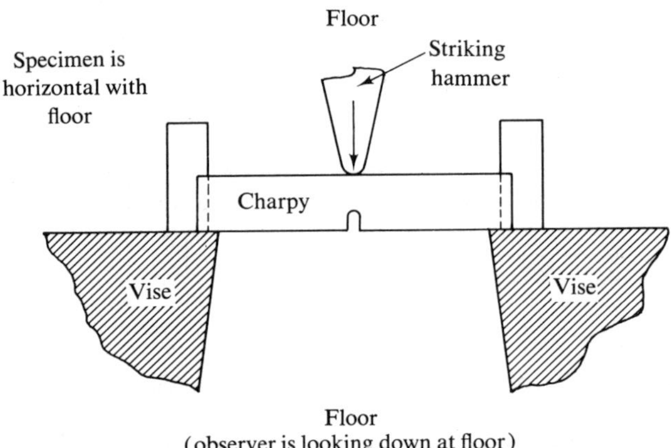

FIGURE 5–19   A Charpy impact specimen being struck by a swinging hammer.

to a shaft as the shaft turns a flywheel. When a wrench is placed on the bolt of a machine (Fig. 5–20) along with a turning force, metal in the bolt begins to twist elastically and subsequently turns when the torque is greater than the resistance of the bolt. The technician in Figure 5–20 is tightening a bolt to the required torque as indicated on the wrench. Due to the nature of torsion, an external force in torque at one end of a shaft causes shear stresses along the length of the shaft in a helix type of pattern at the surface. According to Hooke's law, stress is proportional to strain within the elastic limit of the metal and, therefore, the modulus of elasticity in shear can be determined. For steel the E_s

FIGURE 5–20   A locking bolt on this vertical milling machine is tightened to the required torque with the use of the torque wrench.

value is approximately 12 million psi. But in the case of torsional forces which tend to turn a particular shape, stresses are greatest at the surface of the shape. For a round shaft, as an example, internal stresses are highest on the shaft's surface and proportionally decrease to zero at the center of the shaft. A purpose of torsion testing is therefore to determine the modulus of elasticity in shear, the shearing yield strength, the ultimate shearing strength of the metal, safe shearing stresses, and angles of twist in shafting.

Stress Is Maximum at the Shaft's Surface

Because the center of a solid shaft offer no resistance to twisting forces, the metal may as well not be there. If the central portion of the shaft were removed and placed about the axis of the shaft to increase the diameter and produce a tube with a cross-sectional area equal to the original shaft, more torque could be applied to the tube than to the original shaft. Then, when more strength is required in a shaft, more metal is placed farther from the central axis and this axis is only space when the tube is used. Torque and shearing stresses are related to the polar moment of inertia and are associated as follows: torque is equal to the shearing stress multiplied by the polar moment of inertia, and this value is then divided by the radius of the shaft. The polar moment of inertia in a solid round shaft is equal to $\pi r^4/2$ and for a hollow shaft is equal to $\pi(r^4 - r_1^4)/2$. Torque formulas are helpful for calculating the angle of twist in a shaft to know how much torque can be applied to a shaft without exceeding the metal's elastic limit. The polar moment of inertia is the moment of inertia of the cross-sectional area of the member with regard to the axis perpendicular to the plane of the area. To further define this property, the moment of inertia is the product of an area of a member and the square of the distance from the axis to the member's centroid.

The torsion testing machine (Fig. 5–21) includes a chuck arrangement for gripping and twisting the prepared specimen with its gauge length marks, a measur-

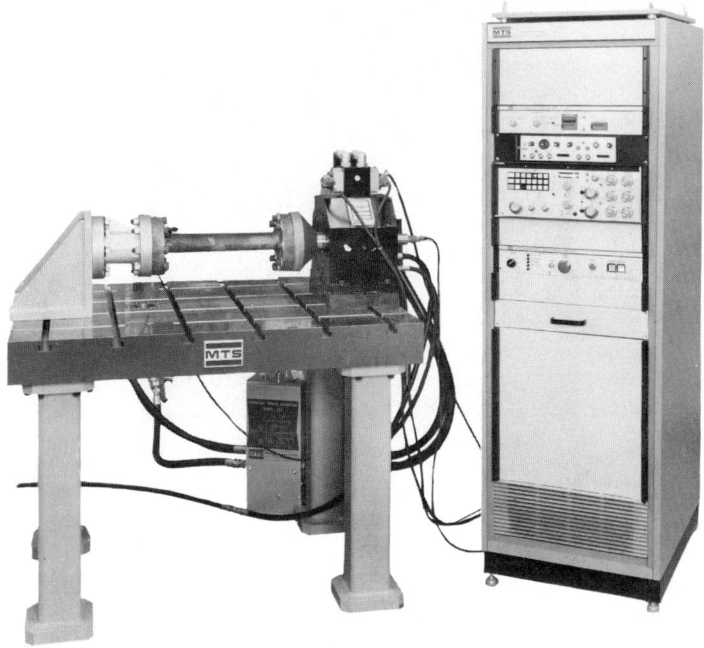

FIGURE 5–21   A torsional fatigue test on a generator shaft. The test system will provide up to 10° angular displacement at 10 Hz. (Courtesy of MTS Systems Corporation)

ing device for indicating torque, and another device for measuring angular displacements of the gauge marks, such as for calculating the angle of twist in a shaft. Shearing strain is calculated by dividing shearing stress by the modulus of elasticity in shear. The modulus of elasticity in shear is equal to the shearing stress divided by the shearing strain. Shearing stress is found by multiplying the strain by the modulus of elasticity. There are many occasions when torsional testing is used. It is very useful for testing various parts which are subjected to twisting forces, such as the spindles of drilling tools, automobile axles, and machinery shafting. A typical example of power transmission involves the large hollow shafts of ships where torques are concerned with thousands of horsepower.

BEND TESTING

Bend testing is mainly performed for the purpose of determining a degree of ductility in a metal, whether the metal is in the "as produced" condition or welded. Because ductility is the ability of a metal to be plastically deformed without rupture, several conclusions can be drawn from the various bend tests. During the test one of the two surfaces is placed in tension while the opposite side is in compression (Fig. 5–22). It is the tension side of the specimen that reflects ductility the most because of the chance of tears and cracks occurring during stretching beyond the metal's yield strength. When the internal stress exceeds the metal's tensile strength due to inability to deform further, metallic separation occurs.

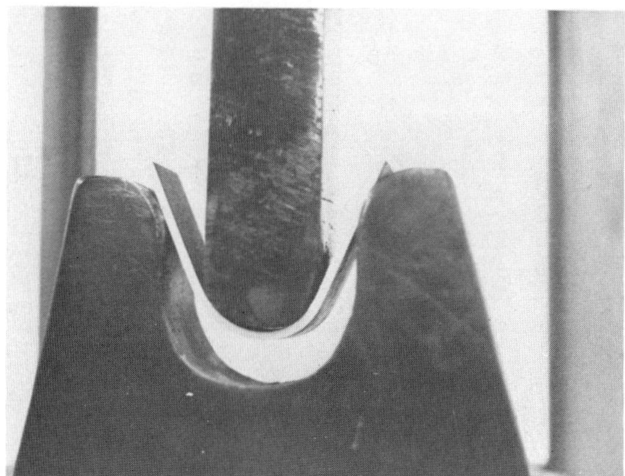

FIGURE 5–22 A prepared specimen undergoing the bend test, a measurement of ductility.

Testing Procedure

Individual specifications are available for conducting a particular bend test, especially those in the ASTM series. However, for general consideration, the specimen must be prepared in such a way that it represents a typical condition of the material. If the specimen does not represent an average or typical metal condition, it is useless to make the tests. Because bending forces create maximum stresses at the two opposite surfaces of the metal, only typical surface conditions must be present. Deep scratches, oxidized metal from flame cutting, pits, and cracks are not typical and will therefore produce lower values for ductility. The specimen is normally one-half as thick as its width and the radius and degree of bend are specified. Too small a radius creates excessive surface stress on the tension side, and too large a radius is insignificant. The typical bend for ductility appraisal is often accomplished in a machine similar to Figure 5–23 where the ram forces the specimen into the die while allowing a rolling action to occur on the outside surfaces of the metal. Springback must be considered because the specimen is forced a small distance beyond the desired angle to allow release of the contained elastic energy. Any crack on the surface side of the metal is considered failure; however, the size of crack is considered when drawing conclusions. Cracks along the two edges are normally not considered unless the edge surfaces were rough at the start. In other words, oxide-contaminated edges and rough edges from shearing will influence surface cracks at 90° to the metal's edges.

Welded Specimens

When the specimen is to be tested by the longitudinal face or transverse face and root bend tests, surface-welded metal is removed by machining. Therefore, the test provides evidence regarding the strength and ductility of the metallurgical condition of the weld without surface geometry alterations. As an example, a typical specimen approximately 6 in long and 1½ in wide may be ⅜ in thick when the radius is to be ¾ in for a 180° bend. Welded specimens are bend tested either with or without the bead, depending on the requirements.

FIGURE 5–23 A bend-testing machine with a specimen ready for the test.

Other ductility tests include the cup test in which the flat plate is clamped in a fixture while a ball-shaped penetrator deflects the specimen to fracture. Of course, the greater the ductility and plastic flow capability of the specimen, the greater the permanent deflection without rupture. For example, spheroidized steels and annealed copper will deflect a large distance before annealing is required to prevent tearing the metal. This test is used to predict the cupping ability of the metal for manufacture into seamless cans and large shell casings. Spheroidizing is a heat-treating operation that imparts maximum ductility in a steel.

Beam Deflection Testing

Another bend test refers to safe operational deflection testing, as is the case in a loaded bridge beam (Fig. 5–24). Several factors enter into conclusions which originate from the static magnitude of the load, the maximum moment under maximum load, the modulus of elasticity of the beam material, and the geometry of the beam's cross section with its dimensions. As has been pointed out, any bending load causes tension stresses at one surface and compression stresses at the opposite surface of the part. Because a structural beam must function under safe conditions, it becomes a matter of determining where along the elastic strength continuum of the material that the design is to be placed. A good practice is to set the operating stress at one-third or one-fourth of the tensile strength, and this will place the operating stress well within the elastic strength of the material. With regard to fluctuating loads, about one-half or one-third of the tensile strength is assigned for the safe operating load unless otherwise specified.

Loading Remains Within the Material's Elastic Limit

When elastic deformation in engineering design is compared to plastic deformation, such as in ductility testing, the observer recognizes where safety exists and

FIGURE 5–24 These steel beams must safely hold the unknown loads which are intermittently thrust upon them. Designers, engineers, and technicians must know materials capability and loading conditions for the structure to be safe.

where failure begins. A steel bridge withstands the heavy loads of trucks moving across its structural spans because the designer has planned the worst probable loading conditions into the design for any one instant of time and has geometrically designed the size and shape of the bridge's steel members to safely hold the traffic for years to come. Scientific deductions from inductive experimentation are safely reflected into the total stress system of the bridge (Fig. 5–25). Day after day the system of beams and columns safely deflects under changing loading conditions because the engineer knew the magnitude of the failing loads. He was therefore capable of limiting loads on the beam system by a combination

FIGURE 5–25 The design of these roadways has been approved for traffic. Every member has been carefully calculated with regard to its share of the load. The complex system reflects man's capability to predict and to uphold the prediction.

of engineering factors, including material and condition of material, maximum probable tensile-compression–share-stress values, maximum moments and deflections, and the cross-sectional geometric patterns of stress flow. The engineer is also trusted with the factor of safety, which absorbs all those unknown loading conditions which may occur and tend to overload the design.

TEAR TESTING

Tear or peel testing is a limited type of test that a manufacturer may use to determine some particular quality of a material. The test pertains mainly to metallic components that have been joined by adhesive bonding, spot welding, or related joining operation. For example, two spot-welded strips of steel sheets are tear tested to determine the quality of the spot weld. Because spot welding incorporates pressure with fusion of the metal, the nugget has strength characteristics in excess of the metal immediately adjacent to the fusion zone. When the two strips of welded metal are pulled in opposite directions until separation occurs, observation shows acceptance of the joint as the metal tears away from the stronger welded nugget. Defective spot welds break across the weld. It is therefore a matter of setting standards for each kind of test and then performing the test. Conclusions are drawn from what is observed, and processing modifications are made accordingly.

FATIGUE TESTING

Fatigue in a material includes flexible loading and time; possible fatigue failure also includes the tension factor, which may induce progressive failure of the material to the remaining ultimate strength. Statically loaded structures are not normally associated with significant levels of fatigue. But if the cyclic alternations of loads increase from intermittent static loading to a more pulsating situation in which elastic loading is more rapidly reversed in some manner that includes tension forces, then fatigue is a factor. (A fatigue tester is shown in Fig. 5–26.) Changing tension loads and time cause fatigue to develop in a material, and the resultant reactions frequently generate a material separation at a stress value far below the strength of the material. It is the stress level that does not bring on material failure that designers seek, while on the other hand, by knowing the material and the failing cyclic loading, the engineer, technician, and designer can hopefully predict the safe design that ignores progressive material failure by not giving it a chance to begin.

Purpose of Fatigue Testing

The purpose of fatigue testing is to study the long-term effects of cyclic loading where one of the loads involves tension. Such a testing process compresses time into months and possibly days. As an example, a helicopter can be cyclically loaded many thousands of times by proper techniques and testing equipment, including a special fixture with movable parts and loading capabilities. Strain gauges connected to critical areas of the structure and tachometers, along with accessory meters,

FIGURE 5–26   Sonntag Fatigue Testers, like the one shown here, are designed to test machine elements, structural components, and samples of materials with dynamic loads in tension, compression, bending, torsion, or combined stresses. One end of the specimen is held fast by a fixture atop the machine and the other is attached to the vibrating platen at the center of the table top. (Courtesy of SATEC Systems, Incorporated)

instruments, and cathode ray tubes, enable the engineering team to speed up the flight and landing processes so that the stresses of several years of combat operations are compressed into several days or less time. If a particular part effectively withstands a given number of cyclic loadings under operational conditions, it may be assumed that the part will endure for an infinite period of time. On the other hand, if the failing number of cycles is known for the part under controlled operational conditions, then the engineer can evaluate both the failing and safe fatiguing conditions and calculate a safe operating load below the failing number of cycles to sustain infinite operation under fatigue conditions.

Fatigue to Failure

Fatigue conditions include tension loading, varying the load, and time. When complete tension to compression loading occurs, the worst possible type of cycling conditions may exist, especially if the load is great enough. The metal attempts to separate as elastic tension stresses appear and leave. After a period of time, however, this elastic stress causes overstress at a particular region in the metal, and a submicroscopic size separation of the metal occurs. As alternating cycles continue, the crack progresses across the material to sudden failure without warning. A typical fatigue testing machine involves a rotating beam driven by a given torque. The load is attached to the protruding specimen so that the specimen bends with each revolution. The series of bendings places loads from specified tensile to compression values on the specimen. When and if failure occurs, the number of cycles is recorded as well as the load which provided the proper stresses. As an example, a prefered specimen was laboratory tested and lasted for 8,498,431 variable loadings at operational stresses before fracture occurred. This value was sufficient to discard the design. In this case either the operational stress must be lowered or the part's geometry must be changed in order that infinite cycling can exist. Geometry change includes adding more material at specified areas of stress and possibly smoother curves where directional changes occur.

Types of Fatigue Testers

Several types of fatigue testing machines are available—flexure, torque, vibration, and corrosion testers. The fundamental principle of fatigue testing is to duplicate operating conditions by both cyclic loading and time. Standardized testers include the mechanisms to induce desired loads on the specimen. Flexure testers provide cams or other devices that place predetermined loads in tension onto the specimen. When variable or complete cycling loads are needed, adjustments are made to move the specimen through loading cycles from tension to compression and back to tension for the time period. Or variable tension loads can be planned into the machine to incorporate a minimum static load. Actually, any shop-produced fixture that can simulate the fatigue environment is effective when careful calculations have been built into the device.

Fatigue and Corrosion

When corrosion joins with fatigue, a powerful and potential killer of metals exists. Either corrosion or fatigue is serious enough when the metal is operating within maximum safe limits. When they occur concurrently, they attack the metal in different aspects. Locally prepared environments have shown effective testing results in decreasing the dangers generated by corrosion and fatigue. As in other testing, the specimen and environment must represent the actual operating conditions, including temperature. Because corrosion continuously decreases the amount of metal available for sustaining the load, a gradual rise in stress occurs and, in turn, the probability of fatigue or progressive metal failure increases along with a shorter life cycle. Figure 5–27 shows the combination of corrosion and fatigue which brought sudden failure to the part.

MICROSCOPIC INSPECTION

Magnification equipment is often necessary to present a clear view of a material. For example, in order that the microstructure of a metal can be studied, the surface is properly prepared and observed at magnifications up to 2000×. Magnifications up to 600×, however, are adequate to show the intimate relationship of the grains and their interiors and afford the technician the opportunity to help verify why mechanical properties respond as they do. Pictures of these microstructural conditions, called photomicrographs, provide records to reflect the hardness, strength, and processing results of the metal from its beginning phase of processing to the finished part (Fig. 5–28).

FIGURE 5–27   This locking pin in a trailer hitch showed no external signs of deterioration. Suddenly it fractured, showing that most of the inside steel had been corroded away.

Prior to microscopic examination of a metallic specimen the surface of the specimen is polished to a mirror finish and etched with a particular chemical to cause destruction of one constituent faster than another. Such a procedure leaves one constituent or phase clearly observable in relationship to the surrounding material. Etching a pure metal will do little but reveal the grain structure because only one material exists in a pure metal. However, when two or more mixed constituents are present in a metal, etching causes one or more of the materials or solutions to lie in relief. The case with ferrite and iron carbide is shown in Figure 5–28. The pearlitic microstructure is composed of the light colored, soft ferrite lying among the hard, dark colored plates of iron carbide. When diluted nitric acid is placed on the polished surface of the steel specimen, a chemical reaction occurs that removes ferrite from the carbide plates and exposes the size, quantity, shape, distribution, and arrangement of the Fe_3C particles as they are mixed in the ferrite.

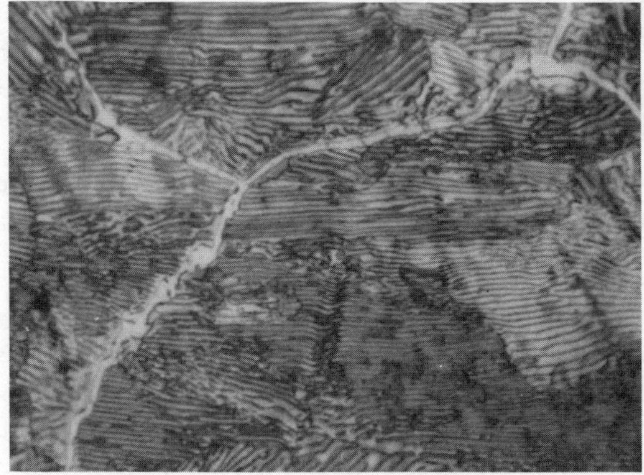

FIGURE 5–28   High carbon steel (1000×) that has been furnace cooled from just above the Ac_3. The lightly colored material at the grain boundaries is excess cementite. Pearlite consumes the grains. The etchant is Nital.

Similarity exists throughout the entire specimen, so a survey view provides data for calculation of approximate hardness, tensile strength, and in plain carbon steels, the approximate carbon content.

Polishing Procedure

The specimen must be typical of the metal from which it is removed. That is, no heat or cold work must disturb the specimen. Large specimens can be ground and polished as procured, but small specimens require mounting. Very small pieces of the specimen are mounted in a plastic slug in order to contain and hold the specimen for polishing. Two general types of plastics are available for mounting specimens, the thermoplastics such as lucite and the thermosets such as bakelite. For example, different colored bakelites (opaque) are available for color coding metals in different classes or different microstructural conditions. Figure 5–29 shows two typical mounting presses with several mounted specimens and associated equipment.

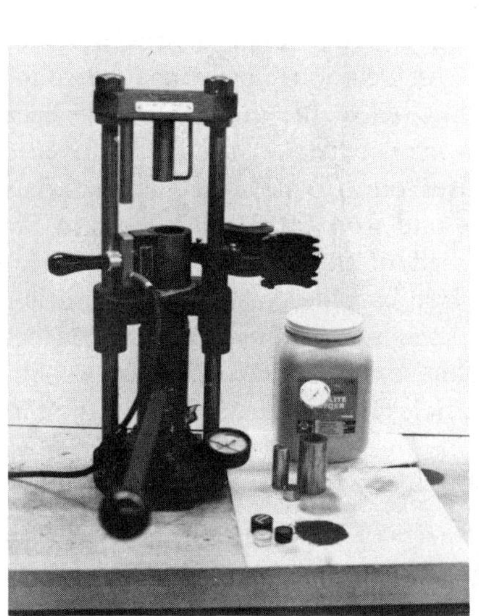

FIGURE 5–29 Specimen mounting presses used in the preparation of metallic samples for subsequent microscopic examination. (*b,* Courtesy of Buehler Ltd.)

The hacksaw is often the first tool used to obtain specimens. Hacksaw teeth marks on the specimen are removed by sanding belts, and larger prior scratches are sanded away by progressively finer sanding strips of abrasives. Rough polishing will include the use of abrasive grit paper number 120 and number 180; finer polishing necessitates the use of grit sizes 280, 320, and 600. Good practice requires that sanding be maintained in straight lines and at 45° to previous grit scratches. By turning the specimen 45° during each finer grit application previous scratches are removed, and subsequent scratches become finer and finer until number 500 or 600 grit produces a mirror finish. Abrassives finer than number 600 are ineffective in removing significant quantities of metal; therefore, levigated alumina in water is placed on the polishing wheels, to produce a high polish on the specimen's surface. Levigated alumina is an extremely fine-ground aluminum oxide and is harder than metals or carbides. Even though only a minute amout of metal is removed, the specimen's surface, whether mounted in plastic or not, becomes as reflective as a mirror and free of observable scratches and cold work.

Etching Procedure

A 3% nitric acid in alcohol etchant is prepared for most steel specimens. Time at exposure to the acid depends on the condition and chemistry of the metal. For pearlite, as an example, about five seconds is needed to bring about the removal of ferrite in order to see the carbide arrangement. Quick neutralization with water followed by air drying is essential. From the moment of drying nothing is to touch the etched surface because it is now ready for examination under the microscope; any contact with the etched surface destroys the view. With regard to etchants, Table 5–1 indicates the several reagents needed for some of the various metals. As

TABLE 5–1 ETCHING REAGENTS FOR VARIOUS METALS*

Metal and Alloy	Etchant	Composition		Procedure	Use
Aluminum	Sodium hydroxide Water	NaOH H_2O	10.0 g 90 ml	Swab for several seconds	General purpose
Aluminum	Hydrofluoric	HF H_2O	1.0 ml Bal	Swab for several seconds	General purpose
Aluminum	Keller	HF HCl HNO_3 H_2O	1.0 ml 1.5 ml 2.5 ml Bal	Immerse for 15 seconds	General purpose Copper, aluminum
Copper	Ammonium persulphate	$(NH_4) 2S_2O_8$	10.0 g 90 ml	Immersion	General purpose Copper, brass, bronze
Copper	Ferric chloride	$FeCl_3$ HCl H_2O	20 g 25 ml Bal	Immersion	General purpose Copper, brass, bronze
Copper	Chromic	CrO_3 H_2O	1% Bal	Immersion	General purpose
Copper	Electrolytic	NaOH $FeSO_4$ H_2SO_4 H_2O	5 g 30 g 100 ml 2000 ml	0.1 A at 9 V for 10 seconds	Contrasts beta and alpha phases
Lead	Combinations	Glacial acetic HNO_3 Glycerol	1 part 1 part 4 parts	Immersion in warm solution	General purpose
Magnesium	Glycol	HNO_3 H_2O Ethyl glycol	1 ml 24 ml 75 ml	Immersion	General purpose

TABLE 5–1 Continued

Metal and Alloy	Etchant	Composition		Procedure	Use
Magnesium	Citric	Citric acid	5 g	Swab	Grain boundaries
		H_2O	95 ml		
Nickel	Nitric-glacial acetic	HNO_3	50 ml	Immersion	Nickel-copper
		Glacial acetic	50 ml		
Nickel	Aqua regia	HNO_3	5 ml	Immersion	Ni-Cr-Fe
		HCl	25 ml		
		H_2O	30 ml		
Steel, carbon	Nital	HNO_3	3 ml	Immersion	General purpose
		Methyl alcohol	97 ml		
Steel, carbon	Picral	Picric acid	5 g	Immersion	General micro-structure
		Methyl alcohol	95 ml		
Steel, alloy	Nitric-hydrochloric	HNO_3	10 ml	Immersion	High alloy
		HGl	25 ml		
		Glycerol	25 ml		
Steel, austenitic	Nitric-hydrofluoric	HNO_3	5 ml	Immersion	Austenitic
		HF	1 ml		
		H_2O	Bal		
Steel, grain size	Hydrochloric	HCl	5 ml	Immersion	Austenitic grain size
		Picric	1 g		
		Methyl alcohol	100 ml		
Tin	Nitric	HNO_3	5 ml	Immersion	Tin Tin-cadmium Tin-iron
		Methyl alcohol	95 ml		
Titanium	Nitric-hydrofluoric	HF	2 ml	Immersion	Alpha
		H_2O	98 ml		
		followed by			
		HF	1 ml		
		HNO_3	2 ml		
		H_2O	97 ml		
Titanium		KOH	10 ml	Immersion	Beta
		H_2O_2	5 ml		
		H_2O	20 ml		
Zinc		CrO_2	200 g	Immersion	General purpose
		Na_2SO_4	15 g		
		H_2O	1000 ml		

* Expose polished surface to chemical until reflection is removed.
  If specimen darkens too much, repolish.
  Dry picric acid may be explosive.
  Handle all chemicals according to instructions.
  Water dilutes chemicals.

the metals change and as conditions in the metals change, it will be necessary to use different etchants. Dipping the specimen in a dish of etchant is one of the common methods of application. Should the specimen become too dark because of overetching, it is necessary to revert to the polishing wheel and then to re-etch. Too short an etch, on the other hand, may not provide the necessary contrast and the view may be too light. Another dip in the etchant is then required so that an adequate or balanced light and dark view is obtained. Practice is necessary to obtain the desired etch. No formula is available to provide the etch-time cycle with complete accuracy. Nonferrous metals are polished and etched by the same general procedures as steel.

Microscopes

The typical metallurgical microscope is equipped to provide up to 2000 magnifications. Normally, however, magnifications less than 1000× are sufficient to draw conclusions about a specimen. As shown in Figure 5–30, the prepared specimen is

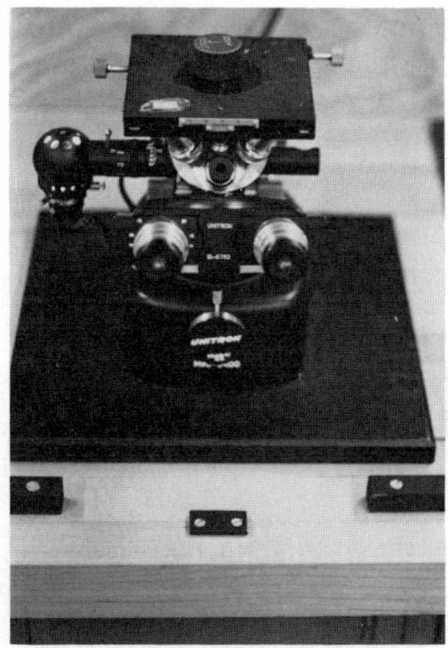

FIGURE 5–30   A metallurgical microscope. The specimen is resting on the stage and is ready for examination.

placed on the stage of the microscope directly above the particular objective. Because artificial light with a filtering and control system is part of the microscope, ample light is available for reflecting the specimen's surface condition through the objective and into the ocular, or eyepiece. The amount of magnification is equal to the initial magnification of the objective multiplied by the initial magnification of the eyepiece. Focus is accomplished by bringing the stage downward to a very close distance from the objective. Fine focus is then established with the use of the micrometer knob. By manipulating the light control and filter systems along the fine focus, a clear view of the microstructure is observable. Longitudinal and lateral movements of the stage provide a cross-sectional survey and summary of the microstructure. Colored filters are available when needed.

What to Look For in the Microscope   Some knowledge of the chemistry of the constituent enables the observer to plan a search technique. For example, when carbon steels are examined under magnification, light and dark areas are observed. Knowledge of the chemical agent (3% nitric acid) and its effects on the constituent enables the technician to come to certain conclusions. Light colored areas in the carbon steel specimen are ferrite, and dark colored areas are carbides. Also, dark lines are grain boundaries. As the carbon content increases, the carbide content increases. If pearlite is present, the steel had to be cooled slowly from austenite. The amount of pearlite, therefore, points to the amount of carbon content. As pearlite increases, ferrite decreases up to the eutectoid. Beyond the eutectoid of 0.8% carbon, a light colored network of cementite (free iron carbide) appears and increases as the carbon increases. If the steel had been cooled rapidly from austenite, a needlelike structure of martensite would be present.

More expensive microscopes are equipped with cameras so that photomicrographs can be produced. In fact, both the polaroid and 35 mm cameras are available on this type of microscope, and the motion picture camera is available for taking pictures of precipitating chemicals as they move from solution. Both xenon and tungsten lighting are available. The microstructure observed in the microscope is compared with available photomicrographs which are displayed on the wall (Fig. 5–31).

FIGURE 5–31   A technician is ready to place an etched specimen on the stage of the metallograph for visual scanning and subsequent photography.

The Electron Microscope   When pictures of microstructures require magnification greater than 2000× the electron microscope is used (Fig. 5–32). This electron microscope has a voltage potential of 100,000. In this respect, magnifications beyond 30,000× are often used for research purposes, and a vast array of new information is being provided by routine scanning at these very high magnifications. Should magnifications extend into the many billions, the basis of matter might be seen. Such a view of a metal would present an outstanding degree of symmetry, even though many errors and deviations would be noticed. The atoms would be pulsating within their radii and electrons would appear somewhat like smoke among the central nuclei. Some vacant spaces in the organized lattice would be seen where an atom occasionally had failed to take its natural position along the long rows of atoms. And even more remarkable, much of what would be observed would be space.

FIGURE 5–32 This electron microscope has a magnification of 2000×
and a resolution of ≅ 10 A°. (Courtesy of Baylor University Medical Center)

Grain Size

The microscope gives the technician an opportunity to study a metal's grain size
and the condition in the grain. For example, when a steel's grain size is desired,
examination at 100× compares the view with ASTM micrograin size standards
from number 00 to number 14. The etchant is a hydrochloric-picric acid combination in ethyl alcohol. The austenitic grain standards provide comparison charts to
observe the grain size of the applicable specimen. As an example, an ASTM
number 1 grain size contains a nominal number of one grain per square inch at
100×. A finer grain of number 8 contains a nominal number of 128 grains per
square inch at 100×. Also, the ASTM nonferrous grain size standards for copper
alloys are used for comparing these nonferrous grain structures at 75×.

Microstructures

Structures within the many grains of the metal are observable after proper polishing
and etching. For example, a eutectoid steel's grain size is calculated to be mostly
number 5 in size at 100×. Being fully annealed from the austenite it is reexamined
at approximately 500× to bring out the remainder of the *microstructure,* the
structure between the boundaries of a grain. At this higher magnification the
laminations of carbide and ferrite present a mother-of-pearl appearance, thus the
structural condition is called pearlite. It is about RB 93 in hardness (Fig. 5–28).
Should this same steel be quenched in water from austenite, the ferrite and carbide
would be trapped in solid solution. The needlelike structure called martensite, about
RC 65 in hardness (Fig. 5–33), would result.

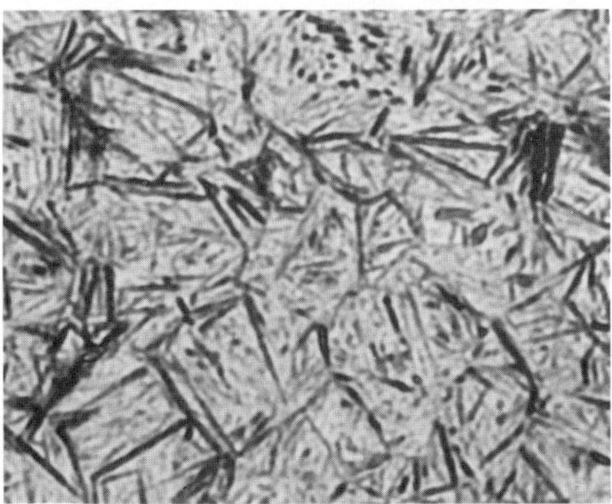

FIGURE 5–33   High carbon steel (1000×) that has been water quenched from just above the Ac₃. The light colored material is martensite, and the dark needles are bainite. The etchant is Nital.

FIGURE 5–34   A 1¾-inch thick magnesium weld showing 100% penetration and a nearly straight line fusion zone accomplished by the electron beam. (Courtesy of Electron Beam Welding, Incorporated)

The microscope is used to study the cross section microstructure of a weld. Figure 5–34 shows the straight-line penetration of an electron beam welded section of magnesium. Figure 5–35 shows the technician operating a 3 kW electron beam welder. Because this type of welding is confined to a vaccum chamber, the size of the parts to be welded are controlled by the size of the welding chamber.

IDENTIFICATION TESTING

In many cases, metal stocks are accidentally mixed before color coding is accomplished, or the wrong color codes are used. Such a situation becomes frustrat-

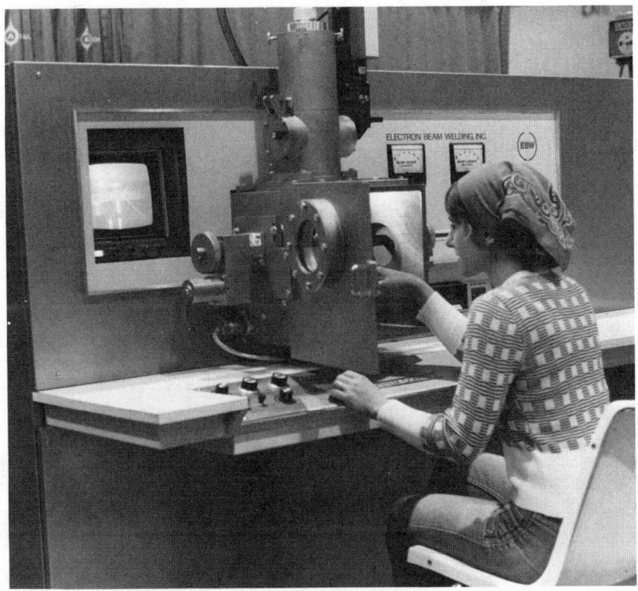

FIGURE 5–35   A technician prepares to inspect a part with the electron
beam welder. The chamber is a 12-inch cube and requires 45 seconds for
pumpdown to 5×10^{-4} torr. Its power is 3 kW, 60 kV, 50 mA (Courtesy of
Electron Beam Welding, Incorporated)

ing because many steels, for example, look alike after rolling. However, there are
several ways to straighten out this situation. One of the quickest methods and the
most accurate is the chemical analysis. Plants having this capability can quickly
separate the metals so that color coding can be accomplished, usually on one end
of the bar or sheet, and this end is obviously used last. An eddy current tester can
also be used to separate the metals with a very high degree of accuracy, even with
regard to strength or hardness. But, very often, neither of these facilities is available, so the metallurgist or technician is placed in a position of relying on personal
skill. In this respect, there are several quick and easy tests for separating metals,
but none of these is completely accurate. The degree of accuracy is often sufficient
to get the job done, however. This level of choosing a chemical condition in a metal
must also be acceptable for potential strength purposes should the guess be slightly
in error.

Spark Testing

The spark test is a quick and easy method for separating some kinds of ferrous
stock and also some nonferrous metals. Samples from the metal in question are
placed against the periphery of a rotating grinding wheel for the purpose of producing a spark shower. Best results are obtained when a steady pressure is placed
against the surface of the wheel so a spark stream about 12 inches long is emitted.
Once the "feel" of the pressure is obtained, other specimens are relatively compared by using the nonvariable pressure and area of metal contact with the wheel.
Even heat-treated parts give off a different spark stream when compared to the

same non-heat-treated specimen. Observation at about four inches from the wheel's surface provides a good shower from which judgment can be made. As the chemistry of the steel changes, the spark shower also changes in color, volume, and geometry of the spark emissions. Such a testing procedure is more valid when known specimens are compared to the unknown.

Description of Spark Bursts   Spark streams include all kinds of colored configurations. It is the quantity, size, shape, and color of the emissions that provide approximate conclusions. For example, low carbon steels emit few spark bursts along straight shafts, while high carbon steels emit numerous bursts along short, straight shafts. Some of the spark configurations include forks, shafts, appendages, dashes, sprigs, and bud-type arrows (Fig. 5–36). Along with these configurations,

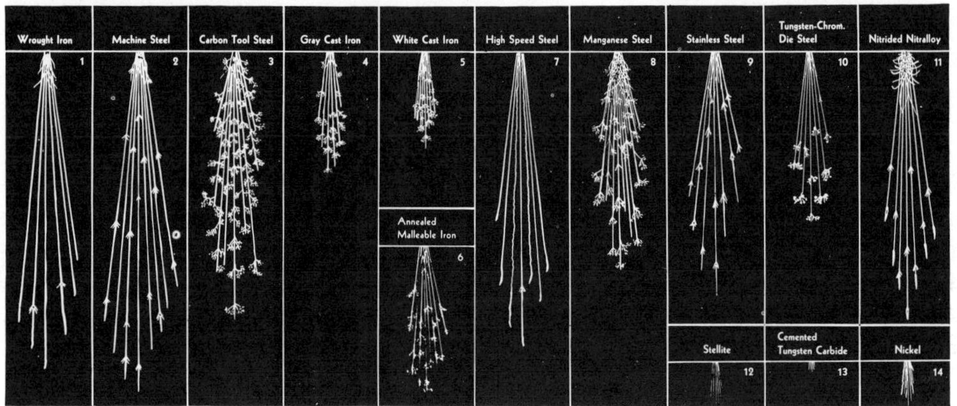

Metal	Volume of Stream	Relative Length of Stream, Inches†	Color of Stream Close to Wheel	Color of Streaks Near End of Stream	Quantity of Spurts	Nature of Spurts
1. Wrought iron	Large	65	Straw	White	Very few	Forked
2. Machine steel (AISI 1020)	Large	70	White	White	Few	Forked
3. Carbon tool steel	Moderately large	55	White	White	Very many	Fine, repeating
4. Gray cast iron	Small	25	Red	Straw	Many	Fine, repeating
5. White cast iron	Very small	20	Red	Straw	Few	Fine, repeating
6. Annealed mall. iron	Moderate	30	Red	Straw	Many	Fine, repeating
7. High speed steel (18-4-1)	Small	60	Red	Straw	Extremely few	Forked
8. Austenitic manganese steel	Moderately large	45	White	White	Many	Fine, repeating
9. Stainless steel (Type 410)	Moderate	50	Straw	White	Moderate	Forked
10. Tungsten-chromium die steel	Small	35	Red	Straw*	Many	Fine, repeating*
11. Nitrided Nitralloy	Large (curved)	55	White	White	Moderate	Forked
12. Stellite	Very small	10	Orange	Orange	None	
13. Cemented tungsten carbide	Extremely small	2	Light Orange	Light Orange	None	
14. Nickel	Very small**	10	Orange	Orange	None	
15. Copper, brass, aluminum	None				None	

†Figures obtained with 12" wheel on bench stand and are relative only. Actual length in each instance will vary with grinding wheel, pressure, etc. *Blue-white spurts. **Some wavy streaks.

Worcester, Massachusetts · NORTON COMPANY · Santa Clara, California
PREPARED FROM STUDIES BY THE RESEARCH LABORATORIES OF NORTON COMPANY
FORM 176-19PAME-3-63 HP
PRINTED IN UNITED STATES OF AMERICA

FIGURE 5–36 Characteristics of sparks generated by the grinding wheel. (Courtesy of Norton Company)

the quantity and color of the shower is analyzed to help draw conclusions. High speed steel, for example, gives off long red shafts with a burst now and then. As the chemistry of this metal changes with a loss of tungsten, the spark shafts brighten somewhat from dark red. Manganese promotes both spark burst intensity and a whitish color. Nickel also increases the whitish color while causing some bright bursts and sprigs at the end of the straight shafts. Cast irons spark less than steels because of the higher carbon content. Gray iron has less intensity of bursts than the harder white iron due mainly to the graphite in the gray iron. Chromium and molybdenum introduce arrows at the end of the straight shafts and the color brightens toward orange. Chromium alone promotes whitish colors and tends to cause shorter shafts. Vanadium introduces breaks in the shaftlike arrows at the buds. Iron nitride promotes white intensity and bursts at the wheel with a few bursts at the end of straight shafts. Wrought iron is mostly shafts with some light intensity at the wheel, along with a brownish, straw-colored shaft which whitens at the appendage of the shaft. Stainless steel promotes the straw-colored shaft which whitens at the forked bursts. When titanium is pressed intensely against the wheel, white shafts appear. Chromium-nickel-molybdenum nonferrous alloys promote small orange shafts. Tungsten carbide promotes small red shafts, and nickel causes the shaft to become orange in the absence of carbon. Most other nonferrous metals are nonsparking; however, some alloys of copper-nickel promote orange shafts. It must be kept in mind that this test is only an approximation and in no way is a substitute for a reliable test.

Magnetic Testing

The magnet is also used to help in the evaluation of metal identity because all ferrous metals are magnetic except the austenitic types. Further, nickel and cobalt have small degrees of magnetic attractions which can help identify the metal. The austenitic steels will spark, but have no magnetic attraction because they are face-centered cubic in structure as compared to all other body-centered, ferrous metals.

Fire Testing

Magnesium, titanium, and uranium burn under certain conditions. The magnesiums are especially combustible; for example, if several chips of various metals are placed on fire bricks in the presence of a flame, the magnesium will quickly catch fire and rise in temperature to about 3200 °F (1760 °C), which is hot enough to melt a steel container. Other metals are also combustible in finely powdered form such as dust in the air. (Sulfur dioxide or sand will extinguish magnesium fires.)

Chemical Analysis

As has been pointed out, chemical analysis is quick and accurate if the necessary equipment is available, but sometimes a simpler chemical spot test is sufficient. A drop of a certain acid on the surface of a specified metal produces a given reaction; on other surfaces a different reaction occurs. Knowing the reaction helps a technician to identify the metal. For example, a drop of hydrochloric acid on zinc quickly produces hydrogen, while a drop of the acid on stainless steel causes no significant reaction. High carbon steel is quickly etched with nitric acid, and caustic soda quickly destroys aluminum and its alloys.

Color and Weight Tests

All of the steels are basically the same silvery gray color. One-third zinc (satin gray) and two-thirds copper alloy (brown) produce a yellow nonferrous metal known as brass. Increasing the zinc content changes the yellow toward white or steel gray. Decreasing the zinc changes the color toward brownish red. Titanium is heavier than aluminum while their colors are similar, but titanium gives off white shafts from the grinding wheel and aluminum does not. Magnesium is similar to aluminum in color but is much lighter in weight and will fracture if bent cold. The low melting alloys are zinc-gray in color and are heavier than the titaniums, aluminums, and magnesiums. Gray iron reflects the graphite gray color, while fractured white iron is more like steel in color.

Miscellaneous Tests

Small specimens of mixed metals can be heated to bright red temperatures and water quenched to determine hardenability. Temperatures up to 1700 °F (927 °C) should be used to allow for the alloy steels if any are present. A compairson of the hardness values before and after quenching infers an approximate chemistry. With the use of additional tests the carbon steels can be separated from the alloys, and the carbon steels themselves can be separated into low, medium, and high carbon types. Stainless austenitic steels can be identified by using a combination of the grinding wheel, magnet, and microscope. Extra etching may sometimes be necessary to confirm the evaluation, and grain size can be quickly approximated with the fracture test by fracturing a notched specimen with the hammer.

Questions

1. Differentiate among compression, shear, and impact testing.
2. Describe the difference between single and double shear.
3. When is the torsion test used?
4. Describe several factors identified during the bend test.
5. Why are welded sections destructively tested?
6. Describe how progressive failure occurs.
7. What are the essential factors of fatigue loading?
8. Why is microscopic inspection often essential during the quality control process?
9. List and describe several short types of tests which may be used in metal identification.
10. Describe the main elements in a stress-strain diagram. What is the purpose of the diagram?
11. Describe how a metal fails in shear.

12.   What is ductility? How is ductility calculated in a metal?

13.   How can the magnet help to separate mixed sections of metals?

14.   How can magnesium be ignited? How can it be extinguished?

15.   What is the main precaution to be taken during compression testing?

16.   Why can shear loads be single or multiple?

17.   How can the spark test be used to separate mixed stocks of steel?

18.   Describe the difference in shear failure between a rivet caught between tensile and compression forces and a plate failing under a punching load.

19.   Differentiate between the Izod and Charpy impact tests.

20.   Differentiate between ductile and cleavage-type fractures.

Nondestructive Testing

Inspection criteria for quality control of manufactured parts include destructive and nondestructive testing processes. Obviously all parts cannot be destructively tested; therefore, the nondestructive testing processes are used where stress is an important factor in the safe functioning of the part. Too often, a part or an assembly is placed into an operating design and quick failure of the part occurs. On the other hand, some nondestructively tested parts have lasted for years and then, as if on a countdown, sudden failure occurs. Testing and inspection methods and techniques cannot guarantee infinite safe use of the part or assembly, but the level of confidence resulting from appropriate nondestructive testing processes is so high that safe design loading is predictable. Where human life and high costs are involved, nondestructive tests are mandatory to help assure safe operating conditions. If 90% of the irregularities and discontinuities are found during inspection 95% of the time, a high level of safety exists. In this respect, it must be assumed that all of the large defects are known and that only the very small ones are not detectable, but facts show that some discontinuities remain in the part because there is no such thing as a perfect metal. However, this level of confidence may not be acceptable in certain situations until the criteria include the smallest size discontinuity to be found. The failure to find the larger metallic separations must be eliminated. The search results must then state that under design loads the part is safe.

Nondestructive testing is a process that measures properties of a material without changing the characteristics of the material. In this respect, it is essential to know the mechanical properties in addition to the significant types, sizes, shapes, and orientations of discontinuities. A discontinuity is an interruption of some kind within the metal; in other words, a discontinuance of the metal at the edge of a crack, hole, or inclusion. Nondestructive testing is a science in many of its operations because the test methods follow the laws of physics. In other aspects, many of the techniques and evaluations are deductive because nondestructive testing allows individual ingenuity a chance to function. With regard to measurement, it includes the identification of strength of the metal, considering the presence of any discontinuities and irregularities and their significance to applicable stress. Nondestructive testing is further concerned with subsequent evaluation of the metal in terms of safe design loading capability and service reliability. Nondestructive testing processes are applicable to all solid materials.

PURPOSE OF NONDESTRUCTIVE TESTING

A main purpose of nondestructive testing is to find defects in the metal, either at the surface or inside, and evaluate their presence with regard to safe operating performance. Also, a purpose of the several tests is to state that the metal can be relied upon to do the intended job. But time is always a function of performance; therefore, performance may be inferred as infinite. The criteria must also state that tests will be made at appropriate intervals along the line of time to further assure the continued safe performance of the part. In other words, service testing is frequently essential. In this respect, discontinuities often grow in size over a period of time, and as the separation grows the level of safe operating conditions is reduced because of fatigue and possible critical fracture. Besides indicating discontinuities, nondestructive testing establishes thickness of metals and locates surface and subsurface irregularities. Nondestructive testing is also used to identify metals and their several mechanical properties, such as hardness and strength. Correlation of these values with stress patterns fits into prediction capabilities and subsequently into reliability. High levels of validity not only support predictability, but serve as management tools in cost accountability, integrity of the manufacturer, safety, and profit.

Testing Is Necessary During and After Manufacture

Nondestructive testing of stressed metal parts is often essential prior to manufacture, during manufacture, and after manufacture. On the other hand, nondestructive testing is also applied to parts and assemblies already in operating environments. Nondestructive testing is used basically to find solidification defects in a metal before the metal is shaped, to find processing flaws during the manufacturing process, to assure validity of the finished part after manufacture and from time to time while the part is in service. For example, the steel for a forging which is to be used in an aircraft landing gear assembly is tested before and after forging to determine that no forging bursts or significant inclusions exist in the metal. If bursts, inclusions, or significant surface irregularities exist, no further processing, such as machining, must be performed because the inherent internal burst or inclusion will follow the part throughout its lifetime. Inherent defects can quickly lead to progressive failure. If the metal remains sound after forging, then further processing is accomplished. During additional processing, such as bending operations, tests may again be made to assure the continued existence of valid metal. In this regard, tears develop when forming loads exceed the metal's tensile strength. When the part is completed, another test is made, this time to verify that the materal lacks significant discontinuities. Incidentally, it seems foolish to manufacture a part, assemble it, and then find it unfit for structural loading.

As time goes on during the service life of an installed part, nondestructive testing processes may again be applied to detect the presence of new discontinuities or the growth of previous ones which have been extended due to fluctuating stresses of service use. These fatigue-induced cracks are often more difficult to find than are birth and processing cracks and other flaws. The early detection of small discontinuities in an aircraft wing spar, for example, is essential in order to prevent the projected catastrophic destruction of the wing and other sections of the assembly. It is a matter of simple logic to verify that the growing crack reduces the quantity of available metal needed to resist the growing stress and sustain the load. When the remaining metal is insufficient to hold the load because of crack growth, fracture of the metal is instant and its carrying stress is quickly released onto the adjacent structure, possibly causing an overload and further failure.

REQUIREMENTS OF NONDESTRUCTIVE TESTING

Nondestructive testing uses all known forms of energy in its sophisticated search for flaws and irregularities in and on the surface of solid materials. Energy is beamed into and through thin sections of material as well as through thick sections of parts such as pressure vessels, warships, and heavy castings. Four requirements of nondestructive testing are necessary to find and analyze discontinuities in these materials—equipment, qualified personnel, a search procedure, and evaluation criteria. Existing equipment has a high level of capability, but it is a problem to find qualified personnel to use the equipment to its maximum capabilities. There is also a lack of understanding of the several types of criteria to be used in search methods, but the biggest problem is to provide effective evaluation of the findings. The assignment of partially qualified personnel to an inspection job impedes a valid analysis of equipment findings; the meter or display presents an accurate indication, but it must be effectively evaluated to be useful. Evaluation includes a description of the indication in its three dimensions and its relationship and orientation to the flow of stress, and a decision to accept or reject the part in its present condition. Personnel charged with interpreting nondestructive testing results must have intensive education, training, and experience. A background of physics, chemistry, mathematics, metallurgy, materials, manufacturing processes, and design principles is mandatory.

When human life or great loss of time and money depends on safe operating conditions of materials, nondestructive testing is required. Certainly, when heavy loads are placed on a part or a structure, as illustrated in all of the examples of stressed parts, there must be assurance of material reliability and safe operating conditions. Failure of the manufacturer to perform the essential nondestructive tests may be disastrous. However, many of the manufactured products today are covered by specification and code requirements for nondestructive testing processes, and the nondestructive tests are mandatory. The builder of a pressure vessel to be used by the petroleum or nuclear industries, for example, is bound by code requirements for nondestructive testing. The fabricator has no choice. Inspections are laid out in specific terms, including the inspection procedures and the specific qualifications of the welder. A highly recommended welder cannot touch this design without certified credentials, and then his previous work is closely examined destructively and nondestructively by several methods, such as tensile and bend tests and by X-ray and ultrasonic inspections. Unfortunately, a highly recommended welder may be astonished to find out that his time in the field is not always a good enough qualification. The saying that "welding is welding" is the mark of the layman who has shown his lack of knowledge.

NONDESTRUCTIVE TESTING AND NONSTRESSED PARTS

Manufacturing processes are ultimately involved in producing both stressed and nonstressed parts. The nonstressed part manufacturer is not concerned with load-carrying capability of the part or assembly because no significant loads are imposed on these parts. Consequently, nondestructive testing is foreign to many of the design processes of nonstressed parts. Such items as common gray iron castings, washing machine cases, commercial hardware (bolts and nails), metal stools and desks, stepladders, cheap tools, machinery covers, cooking utensils, and related items are not exposed to heavy loading conditions and therefore need only to maintain their produced shapes.

The transition in design to heavier loadings may occur somewhere, however, and if it does, then danger becomes a potential and possibly an actuality. Lack of knowledge by manufacturing personnel of the load-stress relationship can lead to production of more highly stressed parts without adequate quality assurance and product reliability. For example, the substitution of a commercial hardware bolt in an engine assembly can result in instant failure of the bolt because the substituted bolt is only half as strong as the original bolt and may even contain a seam or inclusion.

It is hoped that all metal manufacturers will soon become cognizant of the nondestructive testing processes in order that safer parts will be distributed. For example, a small metal stool can easily hold the weight of the heaviest human being, but the same stool will instantly collapse when a typical industrial load is placed on it. It is the lack of knowledge, among some responsible manufacturers, of load-stress-strain-fracture characteristics that imposes inexcusable failures of materials onto the consumer. And it is the consumer who pays, sometimes with his life. So, when is nondestructive testing needed? No formulas are available to draw a line, but knowledge of materials, manufacturing processes, load-stress relationships, and fundamentals of nondestructive testing give the technician the essential education to deductively and inductively apply the principles of nondestructive testing when it is needed. Ignorance of safe design is unacceptable because life and property are deeply involved.

STRESSED PARTS AND THE NEED FOR SAFETY

Once loads of significant magnitude are exerted onto materials, reaction in the part is expected instantly. Naturally, reactions must be effective while being elastic. As an example, the aircraft is a flying container of stressed parts (Fig. 6–1). Lift forces from the wing flow to attaching bolts to transfer stress to the fuselage. Failure of the bolts to instantly react elastically produces plastic flow in the bolt or bolts, and possible failure follows. If fracture occurs, stress is released into nearby bolts and

FIGURE 6–1   The B-1 intercontinental strategic heavy bomber is powered by four General Electric F-101 turbofan engines giving it supersonic speed with a maximum takeoff load of 400,000 pounds. It includes many of the materials which have been discussed. (Courtesy of Rockwell International)

other structures. As for the load, thousands of pounds are held by the bolt or series of bolts and these are in a direct series of stress patterns; consequently, the remaining bolts may become overloaded and catastrophic failure results as the wing disintegrates. Another example includes the failure of an aircraft rivet or a group of wing rivets, which places additional stress on the remaining rivets and induces overloading conditions and possible wing failure. A cracked landing gear can be catastrophic during a landing because too little reaction occurs, only collapse. A laminated shell of a large pressure vessel (Fig. 6–2) can result in an explosion of

FIGURE 6–2   The joint in this pressure vessel was welded by the submerged arc process. During the welding operation the shell began to divide into two layers, the result of a large inclusion. Efforts to repair the joint are observed at the far right.

severe magnitude. A cracked automobile crankshaft can ruin the whole engine as its parts shatter. The metal in the breech ring of a cannon must be sound in order to contain the tremendous firing pressures (Fig. 6–3). A cracked weld in a pipeline is capable of releasing thousands of gallons of oil into the environment, and a cracked steering column can suddenly cause loss of control of a crowded bus. Further, railroad car wheels (Fig. 6–4) generate cracks. Unless these cracks are found before the critical fracture size is generated, a catastrophe may occur as the wheel distintegrates. Railroad rails (Fig. 6–5) require periodic inspection because they, too, generate cracks from the rolling impact of the wheels and the expansion and contraction stress of temperature changes.

FIGURE 6–3   This breech ring weighs 850 pounds and is an example of unidirectional solidification of the metal. (Courtesy of Watertown Arsenal)

FIGURE 6–4   After thousands of miles of rolling on steel rails under very heavy loads, train wheels often generate cracks which can lead to catastrophe if not found in time.

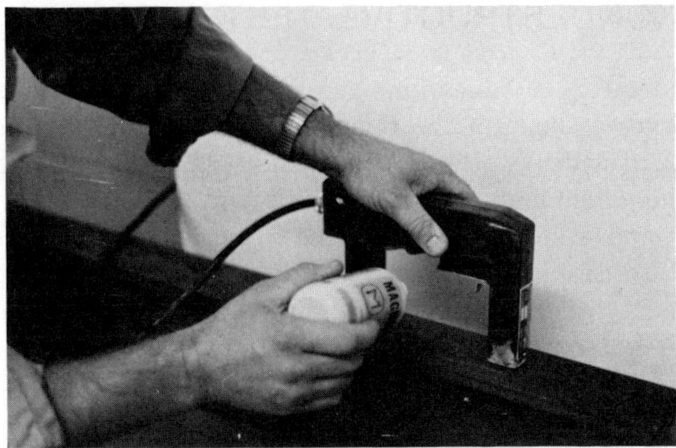

FIGURE 6–5 A surface crack in a railroad rail is confirmed by magnetic particle inspection.

NONDESTRUCTIVE TESTING AND LIGHTLY STRESSED PARTS

A third group of parts is found among those which sometimes fit in between the stressed and nonstressed. The noncritical, or lowly stressed, group may sometimes be exposed to heavy loading at low frequency levels, such as building columns designed from round pipe. Depending on the particular design and loading conditions, nondestructive testing may or may not be conducted due to cost factors, no code requirements, and the gamble that the metal is "good enough" under the circumstances. In this respect, most of the products we observed around us were possibly produced in absence of nondestructive testing, and these objects and structures function safely. Critical inspections for common pipe castings are often skipped, but many castings are exposed to at least one nondestructive testing method. Some of these cast pipes are porous and contain blow holes, yet no type of leak tests are made. In this regard, a pipe is a type of pressure vessel; when fluids flow through the pipe, leaks may ultimately occur.

Another large area where nondestructive testing may or may not be used is in welding operations on commercial products such as framework of mobile homes, automobile trailer hitches, and high-rise apartment building porch railings. Faulty welds in all of these examples have contributed to personal injury and some deaths. To inspect or not to inspect is often the prerogative of the manufacturer, fabricator, or welder. To merely visually inspect a weld that is to be stressed and stamp the seal of certification without some form of nondestructive test could be a dangerous act. Human eyes cannot see within the welded metal. The presence of a large void caused by porosity or the presence of a large piece of slag within the cast metal could lead to sudden fracture at some future time. The melting of electrodes within a metallic joint does not guarantee their diffusion and proper mixing with the base metal. Multiple passes by an unskilled welder can capture inclusions. However, quality control with its growing number of inspections and product reliability is becoming more important to manufacturers since the good name of the company is essential.

KINDS OF NONDESTRUCTIVE TESTS

To help assure adequacy of education, training, inspections, and valid certifications relative to materials evaluation the American Society for Nondestructive Testing plays a key and a critical role, as reflected in its publication *SNT-TC-1A*. As pointed out in this document, several methods and inspection techniques are used to test, inspect, and validate a part being inspected. These include penetrant, magnetic particle, ultrasonic, eddy current, radiographic, leak, and neutron radiographic procedures. Qualifications of the inspecting technician are explicitly described. Because a large percentage of manufactured products in the stressed category are governed by specifications, codes, and standards, it is becoming more difficult to place parts on the market without the stamp of certification. In other words, it is illegal to sell unsafe items which are controlled by codes. Many of the documents of codes originate from the federal government and are in consonance with OSHA. However, there is still a great need for code control in more areas of fabrication, such as in some categories of welding. The engineer and technician are responsible for carrying out the nondestructive testing of materials in compliance with the governing inspection document.

LIQUID PENETRANT METHOD

Parts to be inspected by the liquid penetrant method must be thoroughly cleaned by steam, solvent, or detergent processes. Wire brushing and blasting cause metal folding and crack concealment, therefore, these cleaning methods are not recommended. Further, the parent metal must be exposed to the process; therefore, plated and painted parts must have the coatings removed. Machined parts must be free from burrs, and these may be chemically removed followed by water neutralization. Due to the potential danger of surface cracks, all efforts must be used to allow the penetrant the opportunity to find these discontinuities.

Application of Penetrant

The principles of liquid penetrant inspection are based on the ability of a penetrating liquid to enter a surface crack. The penetrant is a special type of oil with a high penetration capability. The oil contains either a suspended fluorescent powder or colored dye. When a part having a surface crack is exposed to the oil, penetration into the crack occurs (Fig. 6–6); tight cracks naturally require a longer time for effective penetration. Flow into the discontinuity is by capillary action whereby both adhesive and cohesive forces exist. The penetrant may be applied by dipping, swabbing, or spraying the part at room temperature. The part may be metal, ceramic, plastic or other nonporous material. Soaking is important to allow for adequate penetration.

Water Washing and Drying

After a reasonable time lapse for soaking purposes, the part is exposed to washing so that all surface oil is removed. Spray washing is commonly done at pressures below 40 psi and the spray angle is less than half of a vertical attack. Immersion

FIGURE 6–6   This crankshaft has just been removed from a penetrant bath in order to inspect the surfaces for cracks.

washing is commonly performed; obviously, it is essential to remove all surface traces of penetrant in order to prevent false indications during inspection. But, washing must not be so thorough that the penetrant is removed from the crack. Following the washing procedure, the part is dried at a temperature below 180 °F (82 °C) and is allowed to set, or dwell, for a short period of time.

Developing and Inspection

Dwelling allows bleeding of the penetrant from the crack toward the surface and is accelerated with the use of a special powder developer. Surface attraction of the penetrant occurs and marks the outline of the crack or hole. Because the oil contains a dye or fluorescent powdered particles, visibility is magnified and a nearly perfect image of the discontinuity exists. Dye penetrants come in several colors and mark the shape of the crack. However, when a black light is reflected onto the part's surface, the crack fluoresces because of the reaction between the floating particles and the light rays from the special lamp (Fig. 6–7). Black light is in the spectrum of ultraviolet and exists in the vicinity of 3600 angstrom units.

Stationary Inspection   Equipment for the stationary type of penetrant inspection includes a long bench-type of arrangement, such as that shown in Figure 6–8. A space is needed to assemble the parts, followed by a tank which contains the penetrant. Soaking of the part requires ample time for penetrant entry into any irregularity open to the surface. Next in line is a drain arrangement to allow the excess penetrant a chance to flow from the surface. After an adequate soaking period, the part is washed to remove all remaining surface penetrant. Because penetrants differ in their contents, different washing methods are available. Next is an inspection station used to determine that all excess penetrant has been removed. Immediately, the part is placed in the drying oven and is then removed so developing can be performed. This process uses a dry developer for the purpose of extracting some of the crack-contained penetrant. Wet developing also requires drying. Because the developer pulls the penetrant to the surface from a crack, visual observation is enhanced due to the bleeding action at the crack's surface.

FIGURE 6–7   A cracked part shows how the dye penetrant process magnifies the size of the surface crack for visual inspection. When the fluorescent penetrant process is used, this crack will appear light colored.

FIGURE 6–8   A fluorescent penetrant inspection station which is used to find surface cracks in solid materials.

Testing in the field is accomplished with portable equipment such as a spray can kit with all the essential supplies, including the black light. Inspection is performed by both white and black light. Dye penetrants clearly outline the cracks, and black light magnifies the crack's contour and supports visual inspection. Often, it is necessary to inspect very large parts which cannot fit into penetrant tanks. These are therefore inspected with portable equipment (Fig. 6–9).

Evaluation

The penetrant process is highly reliable with a high level of probability that surface cracks will be found when qualified technicians are used. After finding the flaws, evaluation follows. Unless the technician understands the principles of metallurgy and manufacturing processes, he may reject a part which really may be acceptable,

FIGURE 6–9 Portable black light testing equipment provides for inspections away from permanently installed equipment.

or he may accept a part that contains a nucleus for critical fracture. Economics is also part of inspection and so are the probabilities of complete failure due to stress concentration and internal overload if a cracked part is accepted. Evaluation asks the following questions. Shall acceptance be agreed upon with stipulations that timed followup inspections be made to trace any progress of crack expansion? Shall the part be rejected and many dollars be quickly turned into scrap? Or, shall the part be accepted without reservation?

Before any of these decisions is made, if a crack is present, an analysis must be outlined to show the relationship of the crack's geometry to the orientation of the pattern of stress, the probable direction of crack growth if fatigue is a factor, the reduction in the safety factor, or why the crack is insignificant. Answers to these factors must include loading conditions. Both static and dynamic loading conditions must be reviewed in addition to all operating conditions in the particular operating environment such as temperature and atmospheric chemistry. The task of accepting or rejecting the part is a great responsibility because there is no such thing as a perfect commercial material. The task then becomes one of allowing certain types and sizes of discontinuities to pass inspection. However, one test may be insufficient to determine acceptance. Consequently, one or more other test methods may follow the penetrant test as either a redundant test or as a separate test to determine the depth and character of the crack.

MAGNETIC PARTICLE INSPECTION

Another method of nondestructive testing uses magnetic lines of force to indicate surface and near-surface cracks and other flaws in ferromagnetic metals. The limitation is immediately observed as the lack of testing capability for the nonferrous metals and austenitic steels. However, due to the tremendous number of parts to be stressed which are made from the ferrous group of metals, the search for discontinuities in these metals by magnetic inspection is extensively used. With regard to surface conditions of the parts, the same general cleaning processes are used as are applicable to the penetrant processes.

The principle of flaw detection is based on the ability of the metallic part to become magnetized. Direct current is usually used to maintain a constant flux. In some situations alternating current is used, but the generated north-south poles around discontinuities continually alternate, and this pulsating flux field is evident by the vibration of the metallic particles at the crack. If a metallic separation occurs within the steady direct current flux pattern, north-south poles will immediately be statically established (Fig. 6–10). As soon as the part is magnetized, a ferrous

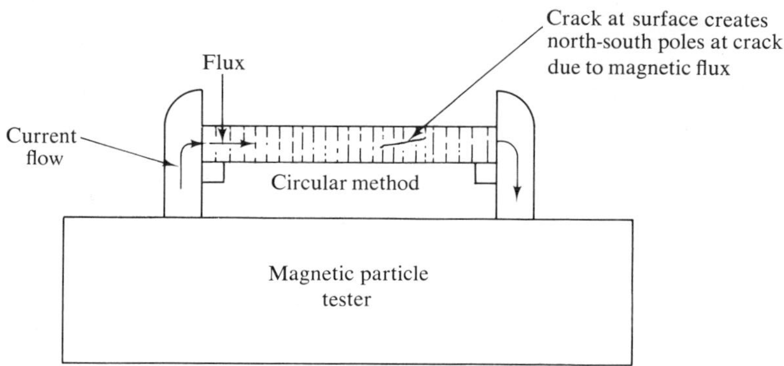

FIGURE 6–10   North-south poles are established at surface cracks when magnetic flux intercepts the cracks.

base powder is poured onto the part, usually by means of a liquid in which the powder is suspended. In the presence of the small magnets which are induced at any irregularities, the powder accumulates and clings to the part in the broken flux fields, thus outlining the irregularity. Sometimes a colored dye is used, or assistance is provided by black light whereby the magnetized powdered particles fluoresce in the presence of the light ray.

Unfortunately, discontinuities and other irregularities may lie in any position relative to stress flow, axis of the part, or processing directions. Often, however, knowing the direction of processing can lead to clues for intensive searching techniques. Two main searching techniques, the circular and longitudinal, are used due to a possible random orientation of the flaw in the part. Because magnetic lines of force are positioned at 90° to the flow of electricity, and because these flux patterns are most effective when intercepting a flaw at 90° to their patterns, both techniques of testing must be used. As a result, most flaws are subject to identification because either or both of the testing techniques will locate them. Strongest indications emerge when the flaw is at a right angle to the flux flow. Consequently, both techniques ultimately intercept the surface and near-surface flaws. Acceptance or rejection of the part is then a matter of interpretation of the indication.

Circular Testing

Circular testing is performed by allowing current to flow through the part, causing a flux field to form a circular pattern aroung the part (Fig. 6–11). As illustrated in Figure 6–11, two surface cracks are outlined as the flux cuts across the cracks and creates north and south poles. The inspection liquid is then poured

FIGURE 6–11 Circular method of magnetic particle inspection as illustrated in Figure 6–10. The crack, shown near the arrow, was found by longitudinal testing.

or sprayed onto the part, and the irregularities cause deviations in the flux patterns. Strong magnetic attractions occur along the length and width of cracks. When the powdered indication is wiped away and new liquid is poured onto the crack, a strong and detailed outline of the crack is again established. Depth of cracks is not directly portrayed, but the amount of powdered material at the crack helps to reflect the magnetic intensity residing in the crack. Very narrow cracks, however, are hard to detect and may appear only as a hair, but they are indeed metallic separations. A near-surface crack presents a weak indication as compared to the surface crack, and the powdered material is more easily wiped away. A repeated exposure to the liquid again presents a weak accumulation of the powder, reflecting the possible need for further inspection by another method such as ultrasonic.

Flaws Parallel to the Axis of the Part   The circular or head shot process is needed when flaws exist in near parallel positions to the longitudinal axis of the part. Magnetic lines of force cut across all discontinuities and reflect their presence from very weak indications when the discontinuities are at 90° to the part's axis or parallel to the flux flow, to very strong indications at those discontinuities which exist at 90° to the flux flow and are parallel to the axis of the part. The part may be held between the jaws of a viselike fixture which acts as electrodes, or a separate pair of electrodes may be used as prods to allow current to flow across a weld, for example. In this situation current moves from one prod through the weld or other shape to the other prod. Prods may be of the fixed type whereby a fixed area is inspected (Fig. 6–5), or separate prods may be used. A circular field is established around the prods while current is flowing. Prods are placed at various orientations to the suspicious area to allow inspection and detections of flaws existing at random locations.

Current Must Be Controlled   Hollow shapes such as tubes are tested and inspected by placing an electrical conductor inside the part, the conductor being clamped between electrodes. In this case flux cuts across at right angles to the longitudinal axis of the part. Because the intensity of the flux field is governed by the amount of current flow, careful planning is essential in order to provide enough current, but not too much. When excessive current flows through a metal, the metal becomes heated and may even become overheated. Of course, heat treated steels may require scrapping. Proper surface contact sometimes is provided by a copper mesh placed between the end of the part and the contacting electrode. Such an arrangement prevents arcing and allows current entry all across the area of the part. Too low a current may not provide the needed flux field. With respect to voltage, most equipment has a built-in voltage control mechanism which regulates the voltage input as the amperage is changed, depending on the volume of metal involved as well as shape of the part and its metallurgical condition.

Longitudinal Testing

The other main technique for flaw detection is the longitudinal procedure. In this technique a coil of copper wire is placed around the part to be tested. Current flow through the coil causes flux to be established along the longitudinal axis of the part, the flux being at right angles to the flow of current in the coil (Fig. 6–12).

FIGURE 6–12   Longitudinal inspection of an automotive crankshaft. This method searches for surface cracks at 90° to the circular method.

Again, the amount of current needed depends on the flux requirements of the part. In the longitudinal procedure cracks and other flaws which lie at right angles to the part's axis are strongly indicated, while weaker indications are presented as the flaw becomes oriented toward the axis of the part. Because the wire of the coil is wound in a helix fashion, the central portion of the coil becomes nearly saturated with flux. Coils may be of different shapes, or a coil of conductive wire may be merely wrapped around a ferromagnetic part to induce a strong flux field in the part. Again, as the flaw becomes oriented toward the flux pattern, weaker indications arise along the contour of the crack. Sometimes, a swinging field may be established to provide strong flux fields at all surface flaws. However, in either testing technique, alternating flux fields rest close to the part's surface while direct current fields induce deeper flux patterns.

Evaluation

As in penetrant testing and inspection, evaluation must follow. And as has been pointed out, machines and equipment are usually capable of finding the flaws and the engineer and technician are frequently depended on to evaluate them. Analyses of flaw indications result in acceptance or rejection of the part. Again, in order for valid decisions to be made technicians must be knowledgeable in the metallurgical factors of both metal production and processing. In addition, the load-stress relationship must be understood along with strength requirements of the design. Furthermore, the fatigue relationship to flaw growth must be considered in the evaluation; hopefully it is a valid prediction of remaining serviceable time. Because both penetrant and magnetic particle testing are used in the search for surface cracks, evaluation criteria are basically similar. Because a crack reduces the amount of area available for load carrying capability, this important aspect must be acted on with decision.

Equipment

Several types of testing equipment are available for magnetic particle testing. Most equipment is stationary (Fig. 6–13), but a substantial quantity is portable

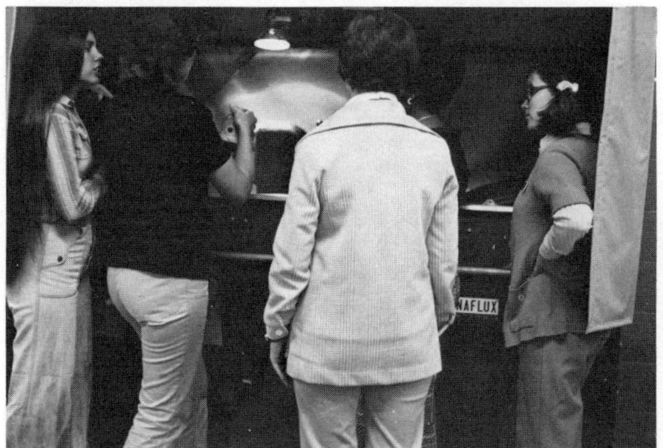

FIGURE 6–13 A stationary type of magnetic particle testing equipment which is suitable for inspection of hundreds of differently shaped parts. The technician is placing a part in the tester for a combination inspection of magnetic particle and fluorescent penetrant.

because of needs in the field. For example, numerous bridge joints, column and beam welds in buildings, powerplant and nuclear vessels, pipelines, and similar applications need inspection after welding. Unless the joint is adequately tested, there is no assurance that the joint is safe. Different types of prods are used as well as special coils and loops of cable. When many parts are to be tested, an automated facility is established. With regard to stationary equipment, several manufacturers produce a machine that combines both the circular and longitudinal capabilities along with essential attachments and accessories.

Currents up to 5000 A are common in standard testing machines and larger currents are available. Also, the black light accessory is available as well as colored particles to increase the flaw detection capability. Alternating current usually feeds the machine and part of this is subsequently rectified to direct current whereby both alternating and direct currents are available. After parts are inspected, they must be demagnetized by alternating flux fields which reorient the metallic domains in random fashion. Successive impulses of decreasing alternating currents gradually provide a nonmagnetized part. At this point in the inspection process caution must be used to prevent corrosion of the cleaned steel. Often, a light coating of oil is applied to the surface of the part to retard corrosion.

Magnetic Rubber Testing

A modification of the conventional magnetic particle testing method is the magnetic rubber technique used in nonaustenitic, ferrous crack detection inspection. The principles of the test depend on the ability of magnetic lines of force to slowly draw a rubber syrup mixture containing a finely ground iron powder into a surface crack. If a cracked part is suspected on the vertical side of a large member, for example, a clay dam is built around the region in order to retain the liquid rubber. When the magnetizing prods are placed in contact with the metal at each end of the dam and then energized for a short period of time, poles at any surface crack draw the rubber deep into the roots of the crack (Fig. 6–14). When the

FIGURE 6–14   A steel part is subjected to a magnetized force in order to draw liquid rubber into any surface discontinuity within the central splineway.

solidified rubber is subsequently removed, a crack is identified as a type of hair or protrusion hanging onto the mass of solid rubber. For example, a bolt hole is poured full with the liquid rubber, given a magnetizing treatment, and the rubber allowed to air harden. The solid mass of rubber is then removed from the hole. Any oddly shaped pieces of rubber hanging onto the expected mass may be indicative of a crack within a thread.

EDDY CURRENT TESTING

A more recent method of nondestructive testing is the use of eddy currents. The principle of this test method is to induce eddy currents in an electrical conductor or part with a high frequency test coil for the purpose of examining surface con-

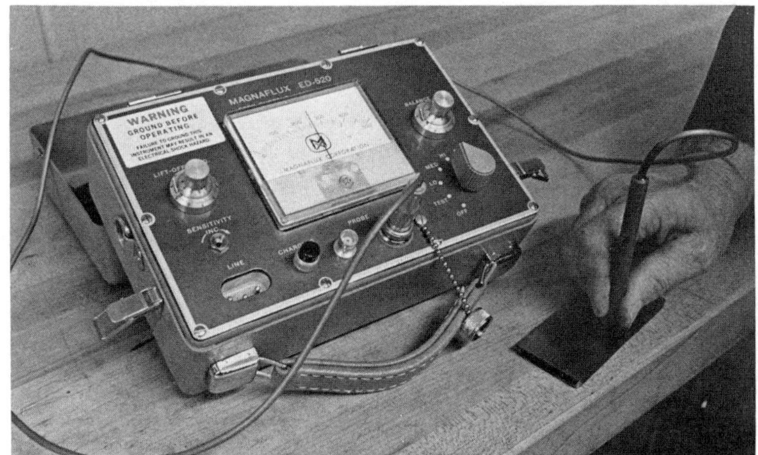

FIGURE 6–15   An eddy current tester is used to separate different kinds of
metals.

ditions of the part (Fig. 6–15). The density of the induced eddy current is proportional to the coil's magnetic field. Eddy current is analogous to the secondary
current induced from the primary windings of a trasformer. As long as the primary
voltage is present the secondary potential is present and it becomes active when
in the proximity of a conductor such as metal. These induced currents are like
other electrical currents in that free electrons are influenced by electromagnetic
fields. Irregularities, such as discontinuities, in a part to be tested cause an impedance to the flow of eddy current, and this change of current is observed with
an instrument readout. Because alternating current is used the frequency may
be changed to suit the particular testing situation. Even though frequencies are
in the vicinity of several thousand alternations per second, the lower frequencies
enable eddy currents to penetrate deeper into the metal while the higher frequencies
tend to exist near the surface.

Eddy Currents Exist at the Surface of a Metal

Basically, the eddy current test method is used for surface and near-surface
flaws. Eddy currents move with greater ease close to or at the surface of electrical
conductors, therefore, subsurface irregularities in excess of ⅜ inch in depth are
not normally detectable. Electrical conductivity and frequency, as well as coil
coupling and permeability, influence the effectiveness of the test. A flux field is
established at each frequency change of direction, and it is this alternating flux
pattern that cuts across a metal part and establishes an eddy current flow in the
part. In turn, the eddy current's magnetic flux filters back into the initial flux field
of the coil and compromises this initial flux pattern. It is this compromise that
signals surface irregularities, and it is surface and near-surface conditions that
establish the environment for the flexible movements of eddy currents. These
currents are similar to fluids in that they bend and compress into numerous flow
patterns and continuously fluctuate as the surface conditions change. Being unable
to jump a discontinuity, the eddy current detours around the metallic separation

and moves toward its original pattern. Such changed conditions in current flow initiate changes in the electromagnetic field of the current, and these changes are recorded in flaw detection instruments in terms of what is sought.

Special Instruments Record Signals

Oscilloscopes show interferences of the eddy current by cathode ray tube responses to coil probe movements across the surface of the part or into a hole. Semipermanent trackings of the search coil's movements are displayed on the screen and are erasable on command. Oscilloscopes are used in the search for discontinuities within tubes at long distances from the controlling instrument, such as tubing in boilers and related designs. Other detection equipment includes dial-type instruments for measuring coil responses as parts are tested. Eddy current techniques adapt very favorably to automation processes in manufacturing. The probing test coils may be made into an endless number of configurations.

Due to the nature of eddy currents and their complete dependence on the primary flux coming from a high frequency coil, such factors as closeness of the coil to the part and geometry of the part must be considered when designing the coil. Coils may be made in numerous shapes and sizes (Fig. 6–16); some are used inside a part such as a tube or bolt hole, some are used around a part such as a rod or bolt, and some are used for various surface applications. This method of nondestructive testing has many sophisticated aspects along with a high degree of reliability in sorting different metals by chemical and mechanical properties, differentiating among heat treated metals, measuring thickness of metals within its range capacity, detecting discontinuities, and several other related capabilities. In the field of salvage, the test discriminates among those metals which have been exposed to fire effects. In order to perform in these varying tasks test coils are made to suit the part to be tested with the flow of the eddy current in the direction of the winding of the coil. For example, a small coil is placed within a

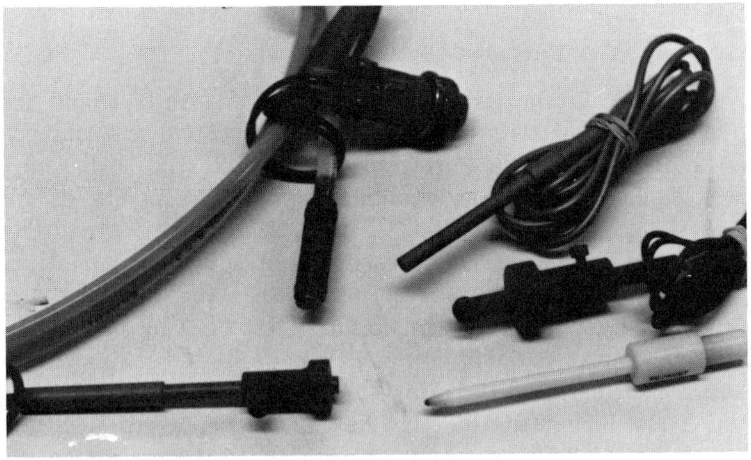

FIGURE 6–16   These are examples of eddy current probes which are used in the many types of eddy current inspections.

deeply threaded hole to allow eddy currents to flow around the surface of the hole. A crack in the wall will be detected by the crack detector instrument.

Impedance Testing

Even though several types of eddy current tests are commonly used, such as impedance, phase analysis, and modulation, many simpler applications use the impedance testing technique. In this type of testing a static test is made. For example, voltage, frequency, impedance, and amperage provide the essentials for the test. Such testing rests on the total change occurring in the impedance of the test coil as it engages the part and induces current. Impedance is then an electrical property which relates the voltage to the current, and the characteristics of the current vary with conditions in the conductor or part. These material conditions include the geometry and size of the part being tested, the physical soundness of the part, the permeability or microstructural condition of the part, and the part's conductivity.

The Need for Standards

The use of standards in penetrant and magnetic particle testing would require thousands of expensive shapes. However, in eddy current testing the use of standards is often a necessity. The nature of the test implies a comparison with known conditions. For example, heat-treated parts which have been mixed can be separated by hardness with a high degree of accuracy. Deviation of feed-back data from the standard indicates different conditions in the test specimen. Sound standards are given a zero reading on the crack detector's instrument dial. Feedback from a nearby conductor gives a deviation. Evaluation of the deviation follows.

ULTRASONIC TESTING

Any solid material which contains a discontinuity or material density differential can be tested with high frequency sound waves to detect and identify the location of the discontinuity or density differential. The location of defects deep within a material can be accomplished by fewer methods than those used in the search for surface defects. The ultrasonic method of flaw detection in a material is very effective in locating the several types of discontinuities and other irregularities. A high frequency sound beam is transmitted into the material in an organized search pattern so that all of the material will eventually be tested (Fig. 6–17). Mechanical type sound waves move through a solid material and bounce back to the receiver from the back side of the part or from a flaw when the pulse-echo technique is used. An alternate to this reflective technique is accomplished by means of "through transmission" whereby a separate receiver receives the transmitted signal or its modification caused by the interception of a flaw. When the sound beam intercepts an irregularity such as a crack during pulse-echo testing, the path of part of the energized beam is instantly reversed back to the receiver and displayed on the cathode ray screen of the oscilloscope along with the front and back side beam indications. A simple cathode ray display is illustrated in Figure

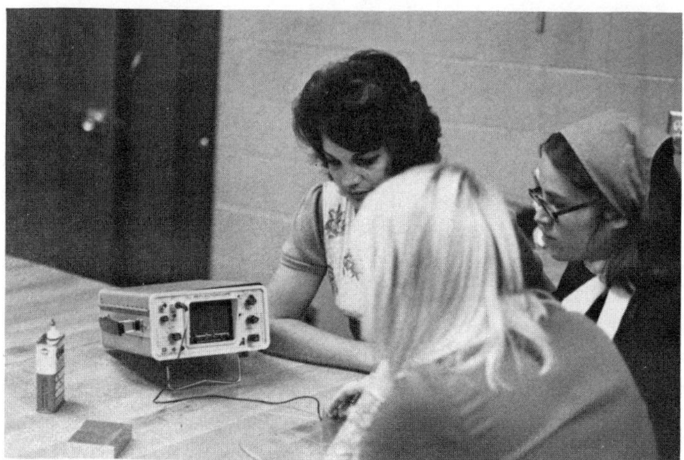

FIGURE 6–17   A typical ultrasonic testing instrument showing the use of the transducer. The small central indication is a crack in the part being tested.

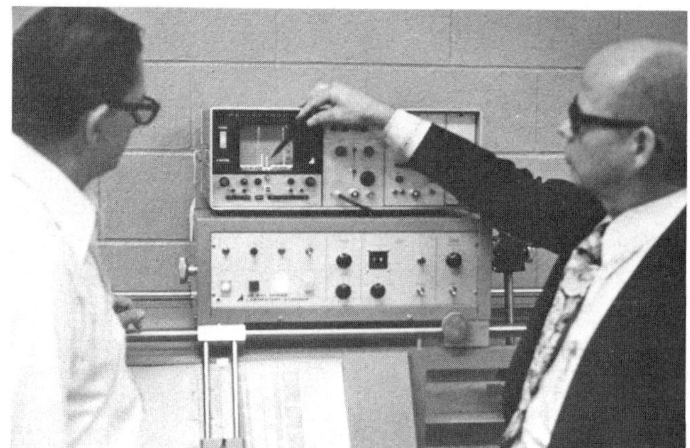

FIGURE 6–18   The presence of discontinuities in a metallic part is illustrated on the screen of this ultrasonic testing unit.

6–18. The four blips, from left to right, represent the top surface of the part, a flaw of some type, and the back side of the part. This type of scanning procedure is quick and efficient.

Variable Frequencies Are Required

Electrical voltage ranging from 100 to 2000 volts is tranformed into high frequency vibrations in the range from 20,000 to 25 megacycles (Mc). The choice of frequency depends on the material, depth of scan, definition of indication desired, and other factors. Sound waves move through solid matter from the interface to a charged transducer. These mechanical vibrations are quickly received in the transducer and are subsequently portrayed on the cathode ray tube as a visual display. Because the crack or other irregularity may be oriented in any plane relative to the path of the sound beam, an organized technique of search is essential so that the flaw will not be overlooked. With reference to frequency, for example, deep flaws are detected by lower frequencies and shallow-type flaws by higher frequencies. Small flaws require higher frequencies for clear identification, therefore, a compromise is made in both frequency and search technique. Sound waves pulsed into the surface of metal at 90° to the surface search longitudinally for subsurface defects, but may not identify defects which are parallel to the beam. An angular beam or shear wave may then be used to detect defects lying in a plane vertical to the surface because the angular beam reflects from a longer area of the defect. Consequently, several search techniques are often used before the inspection of a part is complete. Shear waves detect flaws not normally found by the longitudinal search. Both techniques are required because a part has three dimensions, and a discontinuity lying in a particular plane may be recognized only as a weak signal. This situation is somewhat analogous to the arrow moving head-on to the observer. When the length of the flaw is parallel to the beam of the sound waves, the beam recognizes only the diameter of the flaw and its length is not indicated. On the other hand, a surface sound wave, such as a lamb or Raleigh, can be pulsed into the part. Lamb waves are high frequency sound waves which travel near the surface in thin materials, while Raleigh waves move along the surface of the material. When either wave intercepts a flaw in the metal, a signal is portrayed on the screen of the oscilloscope. The great advantage of using several search techniques is the location of flaws at great distances, such as 10 to 20 ft from the transducer or at adjacent sections of metal.

The Transducer

As shown in Figure 6–19, the transducer is the device which sends and receives the electromechanical pulse of energy. In order for electrical energy to be trans-

FIGURE 6–19   This *C* scan ultrasonic unit not only shows the presence of a discontinuity on the screen, but automatically prints the display on the chart below. The part being inspected is submerged in the water tank and the transducer moves across its surface.

formed into mechanical energy, it must pass through a special material such as quartz or a special type of ceramic. These materials have piezoelectrical properties; that is, they have the ability to convert electrical impulses into mechanical vibrations and, in turn, transform the vibrations into electrical energy. In addition to quartz crystals, barium titanate and lithium sulfate make fine stable transducer material. Geometry of the transducer allows it to be moved across the surface of the specimen so it is always in proper contact. Contoured transducers as well as angular types are available. The transducer is connected to the surface of the metal with a couplant such as oil, grease, or water. As the energized transducer is subsequently pushed across the surface of the part, a series of blips are projected on the screen of the oscilloscope. Because the sound beam pulses continuously as the transducer moves, at least two blips are presented on the screen of the oscilloscope, the first signal reflecting the face or front side of the part and the last signal reflecting the back side of the part. Signals or indications observed between these two blips require investigation as being possible discontinuities. Calibrated cathode ray screens provide data for interpreting the location and size of the defect. When time becomes a measured distance along the base line of the viewing screen, location of discontinuities is established.

Multipurpose Characteristics of Ultrasonic Tests

Ultrasonic testing not only finds surface and subsurface flaws in a material, but it also is used for purposes such as locating steel rods in concrete floors and highways. Rods, beams, pipe, and other materials offering density differentials are quickly located. Even holes in wood are detectable when the proper searching techniques are used. A very handy technique searches for dimensions. Thicknesses of parts as well as hardness values of a metal and other mechanical properties such as strengths are very accurately measured by ultrasonics.

When the principles of metallurgy and manufacturing processes are understood, the technician is able to apply sophisticated techniques of ultrasonic searching to a part and form a valid conclusion with regard to serviceability. Even though no commercial metal is perfect, very small irregularities are often acceptable, but these must be evaluated with reference to kinds of loads and directions of stresses to which the part is exposed. In many instances these small indications are evaluated as being insignificant, but sometimes a particular small flaw may be labeled as one to be watched in subsequent inspections. Such an evaluation is a warning of potential danger ahead; therefore, these types of indications are placed in a controlled category. In time, the controlled indication is often cause for rejection of the part due to projected ultimate failure.

When large parts are to be examined or for other special reasons an automated C scan searching technique is often used. The part, for example an aircraft wing spar or helicopter rotary wing, is submerged in water and subsequently tested (Fig. 6–20). As the transducer systematically moves over the surface of the part, the recording device draws the search pattern on paper as the oscilloscope portrays the movement as a display. Cracks and other irregularities are shown on the oscilloscope's screen as well as permanently recorded on the two-dimensional chart.

FIGURE 6–20 A composite bonded assembly is being inspected by the squirter technique and through transmission with the *C* scan ultrasonic unit. (Courtesy of McDonnell Douglas)

Standard Test Blocks

Standard test blocks are needed for purposes of comparison and evaluation of detected flaws. When the size and geometry of known flaws are compared to those indicated in the test specimen a valid means for decision making exists. Because flaws may exist at varying depths in a metal and may lie in any plane with respect to the search beam, several sizes of test blocks (Fig. 6–21) are needed. These blocks can be locally made but must be finished with precision. For example, the holes (5/64-in diameter) have flat bottoms to simulate the flaws at a given depth from the part's surface.

RADIOGRAPHIC TESTING

The oldest systematic process in nondestructive testing is the radiographic, and this reliable method originated from the use of gamma and X radiation. When

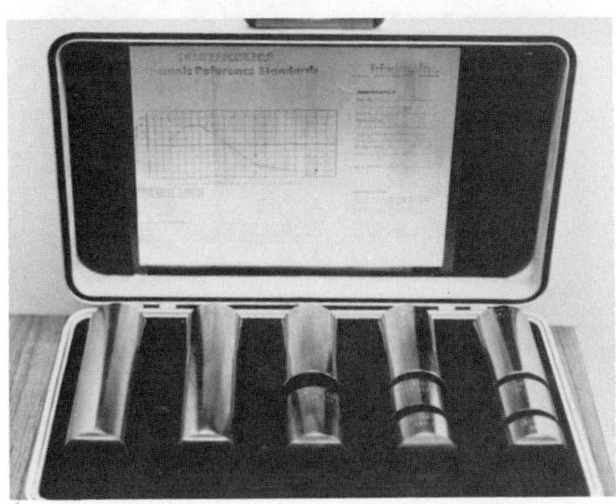

FIGURE 6–21 Ultrasonic reference standards with certification chart.

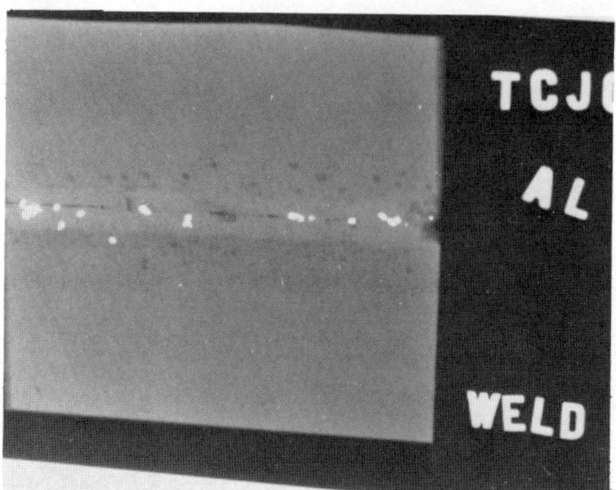

FIGURE 6–22   Radiograph of a weld showing inclusions as light spots. The dark line in the weld represents incomplete fusion.

powerful rays of penetrating energy bombard a material and are recorded on film, a density picture of the inside of the material occurs (Fig. 6–22). A homogeneous metal, for example, will cause a homogeneous recording on the silver bromide film, while a crack in the metal is recorded as a density differential and appears as a darker colored outline in the surroundings of sounder metal (Fig. 6–23). Flaws inside a solid material cannot be detected by several of the previously mentioned testing methods, but penetrating radiation portray the flaw on film.

Characteristics of Radiation

Both X and gamma rays have similar characteristics. They obey the laws of light in regard to their straight-line paths. (Man's senses cannot detect radiation, however, and the inspection process includes hazards to health.) These energy waves have no weight, mass, or electrical charge. They ionize matter, travel at the speed

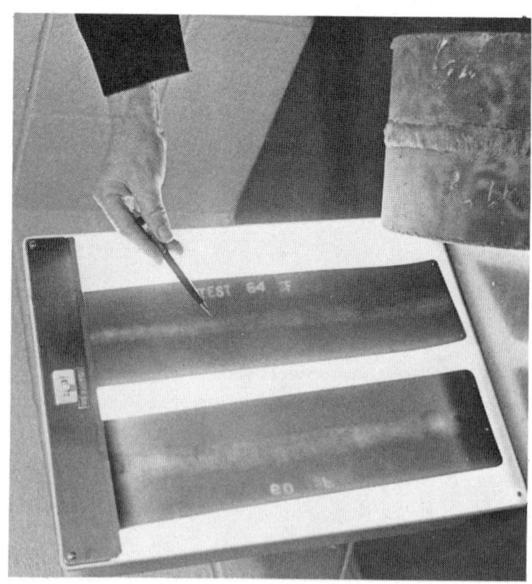

FIGURE 6–23   Radiograph of a welded steel plate.

of light, and are electromagnetic. When X and gamma radiation contacts a special film, the film is exposed. Since matter absorbs energy, the ray's depth of penetration is greatly controlled by the mass and density of the material. Radiation is indirectly proportional to its wavelength, and its penetration depends on its intensity, which is measured in roentgens per hour. The roentgen (r) is the amount of X or gamma radiation that produces an electrical charge of one electrostatic unit in 0.00129 g of air. As the high energy beam moves through a metal containing a hole or crack, for example, the density difference between the metal and crack will be recorded on film, the discontinuity being revealed much like the break in a bone exposed by X ray.

X radiation is emitted from an electronic tube (Fig. 6–24). When a highly heated stream of electrons from a cathode is bombarded against the angular face

FIGURE 6–24   A dismantled X-ray tube showing the cathode on the left where electrons are emitted and are subsequently bounced off the anode on the right to become X rays. Electrodes are enclosed in a thick glass case in the presence of a vacuum.

of an anode, the deflected particles of energy are called X rays. The intensity of the X rays is increased by increasing the voltage at the anode. When the lead-shielded X-ray tube's window is focused onto the target material, a highly effective inspection system is available. On the other hand, the gamma ray is emitted by a decaying radioactive material (Fig. 6–25) and constantly sends parts of itself into the surrounding environment. In time the dying piece of matter loses half of its nuclei and this half-life measurement becomes an important property of the radioactive material.

Comparison of Gamma and X Radiation   Some materials lose half of their original nuclei in only a few seconds, some such as Iridium 192 lose a half-life in about 75 days, and others require thousands or millions of years to disintegrate to the half-life condition. Obviously, as repeated half-life cycles occur there is always a smaller quantity of matter remaining, and the remaining amount gradually becomes ineffective. The strength of the original source material is measured in curies, the curie (Ci) being the amount of disintegration occurring at the rate of 3.7×10^{10} disintegrations per second. With regard to control of this disintegration or radiation, X rays can be turned off by stopping the cathode

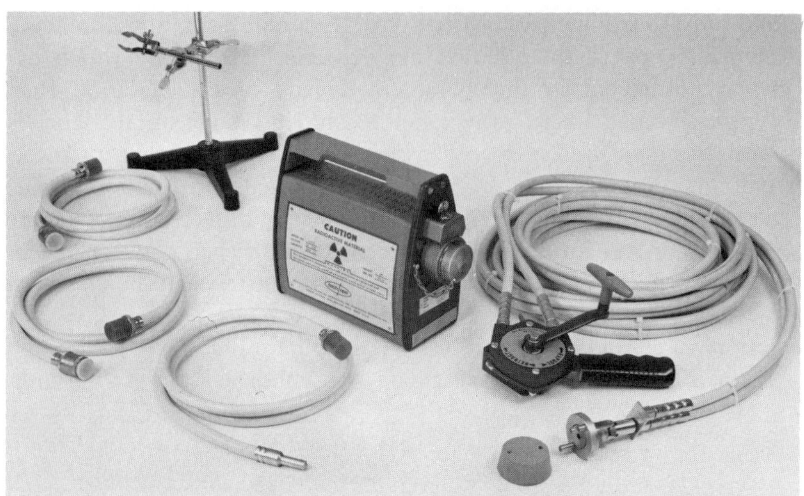

FIGURE 6–25   A typical Iridium 192 Gamma Ray Projector system consisting of main shield (container), control cables with handcrank, three sections of source guide tube and source positioning stand. (Courtesy of Tech/Ops Radiation Products Division)

emission of electrons, but the gamma ray cannot be turned off. However, the window of the thick lead container can be closed at will. Opening and closing the window is equal to turning the energy on and off because when the radiation is contained the lead walls absorb this radiation and prevent penetration. Thick plates of steel are commonly inspected with Iridium 192 (Fig. 6–26).

FIGURE 6–26   A 100-curie Iridium 192 source being used with lead "collimator" for radiography of a 3-inch wall steel vessel to be used in Canadian Heavy Water Plant. (Courtesy of Tech/Ops Radiation Products Division)

The alpha particles of radiation are not harmful, but the X and gamma rays are dangerous, deep penetrating rays and must be carefully controlled. Control means to contain the radiation within a given space such as in a box, jug, or room. Because the X-ray machine or isotope is contained within the setup room (Fig. 6–27), controls are placed outside the room. When the machine is in

FIGURE 6–27 Control of radiation processes is maintained outside the shielded room.

operation and to indicate the presence of radiation in the room, a red light appears near the door. The shielding material within the room is usually sheets of lead, but poured concrete in much greater thickness is a substitute. For most of the soft ray inspections, a 12-ft square room is lined with ⅛ in. to ¼ in. thick sheets of lead that always overlap at their edges. Nails or bolts with lead-capped heads are used to attach the lead to the walls. A swinging or sliding entry door is also lead shielded, the overlap again being provided at all edges of the door's framework. Because streams of rays can move under a door, the bottom of the door must move within a shielded groove. An alternate room design has a zig-zag open door entry; the change of direction stops the radiation. All designs, however, must consider the bouncing effects of radiation.

Radiation Must Be Controlled   Because radiation can be dangerous, necessary controls and safeguards are established. Numerous control requirements originate from the federal government and subsequently through state and local governments. Because of the possible leakage of shielding materials and installation errors, personnel must be ever cognizant of the radiation danger potential. Several monitoring instruments are available (Fig. 6–28) to indicate and to control or reduce this leakage. For example, the Geiger counter and similar instruments act as an alarm in the presence of radiation, and the dosimeter accumulates the amount of radiation as does the personal badge. Periodic reading of the badge indicates the radiation received by both the badge and wearer. The daily or periodic accumulation of radiation eventually leads to a maximum safe level, while further accumulation endangers the person's health. With respect to radiation leakage, all personnel working in the exposure area are subject to a safe permissable dose. This dose is not harmful to the person in his lifetime, however,

FIGURE 6–28 A radiological survey meter is used to check radiation leakage beyond the shielded walls of the X-ray room.

consideration must be exercised with regard to the long-term total accumulation effects.

Making a Radiograph

When a radiograph is needed, the technician chooses the appropriate kind of film and places it on the opposite side of the part from the source of radiation. To help assure an acceptable radiograph, a penetrameter is placed on the part to be inspected (Fig. 6–29) to provide a measure of correct procedures used during the period of radiation exposure. The penetrameter, a specially prepared sample of metal similar to the part, contains precision drilled holes, the diameters of which are related to mass of the part. When the holes are observed on the radiograph, the procedures may be assumed to be correct if density or darkening of the radiograph is correct.

FIGURE 6–29 A steel gear is ready for the X-ray inspection. The part rests next to the film and a lead sheet is on the bottom. The penetrameter on the part's surface will reflect the quality of the X ray.

Because X and gamma rays diverge with increased distance from their sources, the area covered is greater and intensity is diminished as distance is increased. Therefore, the inverse square law is used to calculate exposure factors and safety practices. The intensity of the radiation beam varies inversely with the square of the distance from the source and a spreading of the beam occurs. Also, the part's geometry causes changes in the resulting shadow picture. After establishing the focal distance and alignment of the focal spot and film, the technician sets the machine for correct milliamperage, voltage, and time. A push of the control button sends the bursts of energy through the part to the film for a given time period. At the end of the time period the technician retrieves and develops the film in a dark room. Interpretation and evaluation demand knowledge of many factors which are discussed throughout this text.

All kinds of parts are radiographed, and they range from a small printed circuit board to forgings and castings weighing several tons. As the size of the part increases, the energy and intensity of the source usually increase. X-ray machines are available in the range from ten thousand to several million volts (Fig. 6–30).

FIGURE 6–30   An in-motion radiographic test of a composite bonded assembly. (Courtesy of McDonnell Douglas)

Common sources of radioactive materials are available in designated curie sizes, 25 curies often sufficing, and include Iridium 192, Thulium 170, and Cobalt 60. Again, it is strongly pointed out that only knowledgeable personnel should handle and work in areas where radiation is used.

Questions

1. What is nondestructive testing?
2. How is the search for a flaw organized?
3. Describe the dye penetrant method of testing.

4. Why is the fluorescent penetrant method of testing used?

5. When is the magnetic particle method of testing performed?

6. What types of metals are capable of being inspected by the magnetic particle method?

7. How is eddy current used to find a flaw on the surface of a metal?

8. Why are deep flaws not found by the eddy current method of testing?

9. List some advantages of eddy current testing with respect to penetrant inspection. Also, list some advantages of penetrant inspection with respect to eddy current testing.

10. Describe the principle of ultrasonic testing.

11. Differentiate among the longitudinal, shear, and surface ultrasonic wave tests.

12. Describe the principle of the pulse-echo ultrasonic test.

13. What is "through transmission" with regard to ultrasonic testing?

14. Name some advantages of the C scan ultrasonic test.

15. What is the purpose of radiation testing?

16. Differentiate between X and gamma radiation inspection techniques.

17. List some precautions to be taken prior to testing with radiation.

18. List some advantages of radiation inspection techniques with respect to other testing techniques.

19. What is the purpose of magnetic rubber testing?

20. Defend the need for nondestructive testing in industry.

Environmental Conditions and Metals

An environment is the region of space surrounding a functioning part or assembly. The environment includes an atmosphere or lack of atmosphere, such as a vacuum. Also included in the environment are temperature conditions, gravitational forces, and loading forces which are exerted onto the operating materials. Normal environments include those earthly conditions where seasonal temperatures and normal pressures exist in the proximity of the material. Abnormal environments exist where unusually cold or hot temperatures are present or when corrosive conditions exist. Most man-made things function in normal environments, but many things function in either subzero or elevated temperatures. Corrosive conditions are sometimes present in the hot environment, also. Consequently, before a part can be designed, its operating environment must be analyzed so that proper materials can be used to help assure correct and sustained performance over long periods of time. Some metals become stronger as temperatures are reduced, however, most become weaker. Then too, the modulus of elasticity, a constant factor at or near room temperature, changes as temperatures change. Basically, as temperatures are reduced to below zero, the aluminums and other face-centered metals increase in their stiffness or modulus of elasticity. This is also true for other mechanical properties. A 10% increase in the modulus of elasticity has been noted at −300 °F (−173 °C) for the aluminums. On the other hand, as temperatures are increased above room temperature the modulus of elasticity is reduced for both aluminums and other metals. For example, the modulus of elasticity value at room temperature is approximately ten million psi, while at 500 °F (260 °C) it has decreased 20% for the aluminums.

COLD AND CRYOGENIC ENVIRONMENTS

Strengths and other capabilities of materials often change drastically as the temperature environment decreases to subzero conditions. Because most applications of engineering designs function in temperatures closely related to those of the four seasons and even more often at temperatures between freezing and 100 °F (38 °C), it is customary to think of strengths of materials as being set as the result of heat treatment and of materials as equally strong under other environmental conditions. This is not the case, however, because as temperature is decreased from the average room temperature conditions to below 0 °F (−18 °C), drastic changes take place in some materials when they are stressed at the same loading as would occur under normal temperature conditions.

Cold temperatures are considered as being those temperatures normally associated with freezing conditions, such as 32 °F (0 °C) and colder. Subzero temperature conditions must include at least −100 °F (−73 °C) and possibly −150 °F (−101 °C) or −200 °F (−129 °C). The very cold temperatures from around −200 °F (−129 °C) to asbolute are known as *cryogenic*. Both earth and space environments include cryogenic temperatures. When engineering materials such as metals are exposed to the very cold temperatures, a separate set of mechanical properties must be reviewed before design commences. For example, body-centered cubic steels become brittle when impact loaded in subzero temperatures, while face-centered cubic metals usually exhibit high degrees of toughness. It can then be stated that as the temperature decreases to subzero conditions, most metals show a decrease in their strength properties normally occurring at room temperatures. Therefore, only a few metals are available for cryogenic designs.

Toughness of Face-Centered Cubic Metals

Industry is mostly concerned with ordinary temperatures which extend from 129 °F (49 °C) to −70 °F (−57 °C) because many metals are usable in this environment. The face-centered cubic metals exhibit excellent toughness in this temperature range. Some of these metals include aluminum, copper, austenitic steel, and nickel. The toughness of face-centered cubic metals originates with the atoms of the metals. It is the alignment of the corner and face atoms of their unit cells which accounts for their plastic nature at temperatures well into the cryogenic range.

Many face-centered alloys have equal or improved mechanical properties at subzero temperatures. Some body-centered steels withstand these low temperatures due to alloying with chromium, nickel, and molybdenum. Also, many of these steels have been manufactured into parts heat treated to tensile strengths in the range from 150,000–180,000 psi, and they serve equally as well at subzero and normal temperatures. Even though molybdenum is a body-centered cubic type of metal, its effect in stabilizing toughness in steels to around −100 °F (−73 °C), 25 ft-lb Charpy keyhole impact, makes it ideal for use in cold temperatures. High flying aircraft structures must remain safe where temperatures of −60 °F (−51 °C) exist, and molybdenum is used in their makeup. Aircraft landing gear systems and primary structural members are made of the nickel-chromium-molybdenum steels, but of the body-centered cubic types. The same metal cannot be used at −200 °F (−129 °C) where any impact occurs because the toughness factor is reduced to less than 15 ft-lb. This value is the realm of brittleness, and failure occurs in shock loading.

The properties of body-centered and face-centered cubic metals are reflected in their atomic design. The face-centered metals contain more atoms in their unit cells than the body-centered metals. As will be explained in subsequent parts of this chapter, the larger number of atoms in the face-centered cells allows more deformation of their design; consequently, more ductility and more toughness are exhibited by these metals.

EFFECTS OF LOW TEMPERATURE ON STEEL

At temperatures as low as -100 °F (-73 °C) most of the steels lose their strength and toughness properties. As an example, an AISI 1010 having a tensile strength of 54,000 psi at room temperature is extremely brittle at 0 °F (-18 °C), yet it will bend before breaking at room temperature. This unacceptable weakness of most of the body-centered steels at reduced temperatures leaves only the face-centered types for such adverse operating conditions. But when the AISI 304 face-centered cubic austenitic steel (Table 7–1) is exposed to

TABLE 7–1 TENSILE PROPERTIES OF SOME AUSTENITIC STAINLESS STEELS

AISI	Testing Temp. (°F)	Yield Strength (psi)	Tensile Strength (psi)	Elongation (%)	RA (%)
304	75	33,000	85,000	60	70
304	−320	57,100	205,500	43	45
304	−425	63,700	244,500	48	43
304L	75	28,000	85,000	60	60
304L	−320	35,000	194,500	42	50
304L	−425	33,900	220,000	41	57
310	75	45,000	95,000	60	65
310	−320	84,900	157,500	54	54
310	−425	115,500	177,500	56	61
347	75	35,000	90,000	50	60
347	−320	41,200	186,000	40	32
347	−425	45,500	210,500	41	50

Reprinted from "Selection of Stainless Steels" © American Society for Metals, 1968.

-425 °F (-254 °C) its tensile strength increases from 85,000 psi at room temperature to approximately 244,500 psi with a Charpy keyhole ft-lb value of 80 (Table 7–2) and an elongation in two inches of 48%. Data in Table 7–1 show that as the temperatures are reduced the yield and tensile values increase, even though excellent toughness values remain. These steels have universal usage in cryogenic temperatures and owe this outstanding capability to their atomic lattice network. Table 7–2 points out the unusual toughness values of four austenitic steels at several cryogenic temperatures. With regard to metal thickness, the toughness factors are not significantly influenced. There is, however, a toughness influence with respect to orientation of the metal's directional properties to the load in some cases, but again, all values are exceptionally good.

Austenitic Steels

The drastic changes in strengths of the austenitic steels as temperatures are reduced are again pointed out by an examination of data in Table 7–3 com-

TABLE 7–2 Low Temperature Impact Strength of Several Annealed Austenitic Stainless Steels*

AISI	Testing Temp. (°F)	Specimen** Orientation	Type of Notch	Energy Absorbed (ft-lb)
304	−320	1	Key	80
304	−320	2	Key	80
304	−320	2	Key	70
304	−425	1	Key	80
304	−425	1	V	91.5
304	−425	2	V	85
304L	−320	1	Key	73
304L	−320	2	Key	43
304L	−320	1	V	67
304L	−425	1	V	66
310	−320	1	V	90
310	−320	2	V	87
310	−425	1	V	86.5
310	−425	2	V	85
347	−320	1	Key	60
347	−320	2	Key	47
347	−425	1	V	59
347	−425	2	V	53
347	−300	1	V	77
347	−300	2	V	58

* Toughness is not appreciably influenced by thickness of plate
** 1, Longitudinal; 2, Transverse
Reprinted from "Selection of Stainless Steels" © American Society for Metals, 1968.

TABLE 7–3 Typical Room Temperature Tensile Properties of Annealed Standard Austenitic Stainless Steels

AISI	Yield Strength (psi)	Tensile Strength (psi)	Elongation (%)
201	55,000	115,000	55
202	55,000	105,000	55
301	40,000	110,000	60
302	40,000	90,000	50
302B	40,000	95,000	55
303	35,000	90,000	50
303Se	35,000	90,000	50
304	42,000	84,000	55
304L	39,000	81,000	55
305	38,000	85,000	50
308	35,000	85,000	50
309	45,000	90,000	45
309S	45,000	90,000	45
310	45,000	95,000	45
310S	45,000	95,000	45
314	50,000	100,000	40
316	42,000	84,000	50
316L	42,000	81,000	50
317	40,000	90,000	45
321	35,000	90,000	45
347	40,000	95,000	45
348	40,000	95,000	45

Reprinted from "Selection of Stainless Steels" © American Society for Metals, 1968.

pared with data in Tables 7–1 and 7–2. A common characteristic of these steels is their universal softness at room temperature along with large ductility values, but with quick work-hardening capability. This factor is illustrated in the large difference between the yield and tensile strengths. Because these steels are not heat treatable for increased strength, work hardening is the means used at room temperature to provide higher strengths. Because these steels have low and high temperature resistant properties, they are used where varying temperatures occur, such as for the exhaust manifold shown in Figure 7–1 or the pressure vessel in Figure 9–32, which is used in the transfer of liquid nitrogen.

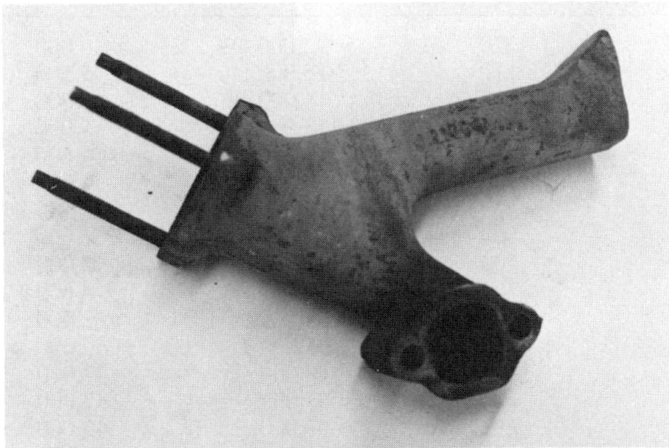

FIGURE 7–1   This exhaust manifold is made of austenitic steel and functions under adverse conditions of temperature and corrosive gases.

EFFECTS OF LOW TEMPERATURE ON TITANIUM

Another tough metal, the titanium alloy Ti-6Al-4V is also used to house liquid hydrogen in pressure vessels at temperatures as low as −320 °F (−196 °C). Even at this temperature the metal is exceptionally tough in impact resistance. The alloy Ti-5Al-2.5Sn shows a yield strength of 169,000 psi and a tensile of 181,000 psi at −320 °F (−196 °C). To show its versatility, this same metal is used for jet engine compressor blades and for heat-treated main landing gear trunions. Common heat treatments produce strengths in this alloy of 150,000 psi, but with half the weight of steel. Other uses of this titanium alloy include rocket motor covers, missile components, and high speed aircraft parts where temperatures run as high as 1000 °F (538 °C) on leading edges of air foils.

EFFECTS OF LOW TEMPERATURE ON ALUMINUM

The aerospace industry requires strong but lightweight metals as well as some that will withstand adverse temperature conditions. Aluminum and many of its alloys function exceptionally well in the presence of cryogenic temperatures, especially in space applications, but are very limited in strength characteristics when temperatures exceed 300 °F (149 °C). In a few alloys, strength drops off rapidly at 500 °F (260 °C). As in some of the titanium alloys, austenitic steels, some of the copper alloys, and many nickel alloys, strengths of the aluminums usually increase as temperatures are reduced, accompanied with acceptable ductility and toughness values. Generally, for the aluminum alloys maximum tensile and yield strengths occur at -320 °F (-196 °C) or lower, and as temperatures increase the tensile and yield strengths decrease (Table 7–4).

TABLE 7–4 TYPICAL TENSILE PROPERTIES AT VARIOUS TEMPERATURES FOR ALUMINUM ALLOYS

Alloy and Temper		Temperature (°F)	Tensile (psi)	Yield (psi)	Elongation (% in 2 in)
1100	0	−320	25,000	6,000	50
		−112	15,000	5,500	43
		− 18	14,000	5,000	40
		75	13,000	5,000	40
		212	10,000	4,600	45
		300	8,000	4,200	55
		400	6,000	3,500	65
		500	4,000	2,600	75
		600	2,900	2,000	80
		700	2,100	1,600	85
2024	T4	−320	84,000	61,000	19
		−112	71,000	49,000	19
		− 18	69,000	47,000	19
		75	68,000	47,000	19
		212	63,000	45,000	19
		300	45,000	36,000	17
		400	26,000	19,000	27
		500	11,000	9,000	55
		600	7,500	6,000	75
		700	5,000	4,000	100
7075	T6	−320	102,000	92,000	9
		−112	90,000	79,000	11
		− 18	86,000	75,000	11
		75	83,000	73,000	11
		212	70,000	65,000	14
		300	31,000	27,000	30
		400	16,000	13,000	55
		500	11,000	9,000	65
		600	8,000	6,500	70
		700	6,000	4,600	70

Courtesy of the Aluminum Association

Mechanical Properties Are Affected

In conjunction with increases in strength for the aluminums as temperatures are reduced from room temperature, ductility may increase somewhat in the 1100 alloy, an unusual mechanical property situation. Small variations in ductility, however, occur at these low temperatures, some alloys experiencing only a small loss. As an example, the alloy 2024 at -320 °F (-196 °C) has an ultimate strength of 84,000 psi and a yield of 61,000 psi with an elongation in two inches of 19%. At room temperature the ultimate strength is 68,000 psi and the yield is 47,000 psi with an elongation of 19%. The aluminum alloy 1100 in the annealed condition while at -320 °F (-196 °C) has a tensile strength of 25,000 psi and a yield strength of 6000 psi with an elongation in two inches of 50%. At room temperature, its tensile is only 13,000 psi, while its yield is not greatly affected, being 5000 psi with an elongation of 40%. As the temperatures increase in all the aluminums, strengths decrease rapidly.

Table 7–4 presents data for three aluminum alloys which have been tested at various temperatures ranging from cryogenic to 700 °F (371 °C). The alloy 7075-T6 is the strongest of those shown, having a tensile strength of 102,000 psi at −320 °F (−196 °C). However, this alloy has poor elongation values at low temperatures. The alloy 2024-T4 has a tensile strength of 84,000 psi at −320 °F (−196 °C) with a fairly good elongation factor of 19%. As temperatures are increased in metals, ductility increases and strengths decrease. It can be seen that beyond about 300 °F (149 °C) most aluminums cannot be used in heavy loading situations. However, a new group of aluminums known as the hot alloys has recently appeared, and these are stable under lower loading conditions up to 500 °F (260 °C).

IMPACT AT LOW TEMPERATURE

Caution must be used in the design process with respect to notch values because the presence of the notch in a part is the presence of a stress raiser. In actual engineering designs these stress raisers are eliminated as much as is practical, but data from the impact tests are relative to all specimens tested. Consequently, the relative toughness values among the different materials are available. If the notch were not present, the test specimens would be mangled by deformation or fracture occurring anywhere, and no significant and relative data would be available. Engineers, technicians, designers, and architects must rely on positive data for their designs. These data come from actual operating conditions in simulated environments, but the impact test must be studied carefully because many factors govern its results. Obviously, if the notch were absent, the part would exhibit a tougher characteristic. But somewhere along the temperature reduction process, some ductile metals begin their change to brittle fractures when subjected to impact loading and these transition temperatures must be carefully noted.

THE TRANSITION TEMPERATURE

The transition from ductility to brittleness is accelerated by the notch, but other factors also indicate the movement of the metal to brittleness and sudden fracture, for example, when temperatures are reduced from normal room temperatures. In tension a ductile metal fails across its cross section by a type of shear which produces the cup-cone plastic separation. On the other hand, when rigidity is present in the metal and no plasticity exists while the metal is under tensile loading, a cleavage or straight-line break occurs across the cross section or across the flow of stress. In this respect face-centered cubic metals have the exceptional ability to allow their unit cells to move long distances under deforming tensile loading, all the way to plastic fracture, whereby insufficient metal is left in the cross section to sustain the load. But when temperatures are reduced in the body-centered type of metals a transition phase develops in the metal's ability to flow plastically because its ductility and impact resistant values are affected. The transition temperature begins when a noticeable drop in ductility occurs. When stress concentration, as in a stress raiser, exists at the moment of rapid loading along with a decrease in temperature, brittle failure may occur. When temperatures lower than -100 °F (-73 °C) are present and when the non-face-centered cubic structured metals are exerted to sudden loading, cleavage of the granular structures occurs. The face-centered cubic metals, however, exhibit a variable ductility factor but still remain dependable in resisting impact.

When Shear Failure Changes to Cleavage

A close examination of fractured impact specimens indicates a range of plastic flow to complete rigidity. As the amount of plastic flow is reduced in the fracture as the temperature is reduced, the grains will reflect less deformation. Continued temperature reduction promotes rigidity of the atomic lattice to a point where noticeable shear failure changes to cleavage; the temperature at which this point is reached is the transition temperature. Elastic fracture, not plastic, occurs in cleavage failures. When approximately half of the fracture shows cleavage a definite change has occurred in fracture characteristics. Again, the temperature at which this happens can be considered the transition temperature for that metal. In this regard, metals having low transition temperatures are best suited for designs in cold environments; the lower the transition temperature the more reliable the metal for use in cold temperature applications. The movements toward the transition temperatures are very small in face-centered metals, even though most of these metals fail by plastic flow and do not show brittle failures. Another indication of the presence of the transition temperature is an approximate 50% reduction in impact values. The transition temperature may be considered as being halfway between tough and brittle conditions.

MECHANICAL AND METALLURGICAL ASPECTS AT LOW TEMPERATURE

Low temperature behavior of metals is influenced by the mechanical and metallurgical aspects of the metal under stress. Mechanical factors originate from fixed conditions and are influenced by either rapid acceleration of concentrated stress, impact loading, or static loading. The V-notch impact test shows complete concentration while the keyhole causes a somewhat reduced stress pattern. However, when either test is applied over a large range of metals at constant temperatures, a relative toughness spectrum can be established. Severe stress concentrations must be avoided in operational designs. Under static loading conditions impact is not involved; therefore, the main engineering stresses of tension, compression, and shear prevail. Even so, a metal that is subject to cleavage failure shows no outward signs of pending failure as the point of rupture occurs.

The metallurgical aspect involves many variables. When a metal is anticipated for use at subzero temperatures the following factors must be considered: chemistry of the metal, microstructure of the constituents, geometry of the shape, temperature operating range, kind of loads and methods of application, and safety factors. Chemistry of the material relates to manufacturing processes beginning with the chemistry of the metal in the ladle and subsequent gas analysis. With regard to microstructure, the pure metals and solid solutions of the face-centered cubic types promote toughness in the cryogenic temperature range. The microstructure favors the solution types in addition to the precipitation types of alloys such as heat-treated aluminum alloys. When hardness is required in the steel, then a martensitic structure is essential. In all cases where impact is involved, except at elevated temperatures, the fine-grained structures are tougher. As for shape of the part, designers must simply eliminate stress raisers at critical areas in the part or assembly. Generous fillets at corners and additional material at stress concentration points are required. An analysis of loading conditions after all physical tests of the operating environment have been made leads to a design which incorporates proper safety factors.

ELEVATED TEMPERATURE ENVIRONMENTS

Metals which operate in an environment where temperatures are substantially above room temperature exhibit a different set of mechanical and physical properties than those in colder temperatures. Most metals function in the normal temperature ranges associated with the four seasons, but a substantial quantity perform at either intermittent or constant temperatures above 400 °F (204 °C). As temperature increases while a metal part is in service, some of its strengths may be reduced whereby failure of the part may result. Strength reduction may occur as a result of rapid oxidation and chemistry change of the metal or through elastic loss and plastic failure. However, some metals may fail as a brittle fracture while hot because of chemical changes and granular disintegration. For purposes of this discussion, a demarcation temperature of 1200 °F (648 °C) will be used to establish the beginning of the elevated temperature range. This temperature is used because temperatures beyond this point begin to have disastrous effects on many metals, some being liquid at this temperature.

Few Metals Function Above 1200 °F (648 °C)

Manufacturing processes produce enormous quantities of tools, parts, vehicles, machines, and numerous items for consumer use. Most of these items are rarely exposed to temperatures beyond the limits of 400 °F (204 °C). Between room temperature and 400 °F (204 °C) little change occurs in most metals, excluding the low melting types, so the numerous mechanical properties pertain to conditions between 70 °F (21 °C) and 400 °F (204 °C). Several of the lower melting metals, such as aluminum and magnesium, begin to respond to temperatures of 300 °F (149 °C) and higher, while the ferrous group and many nonferrous metals are not significantly affected. As an example, many hardened tools such as gauges and cutting tools are tempered at approximately 400 °F (204 °C) and, therefore, do not change their properties when reheated to this temperature. Springs are tempered at 700 °F (371 °C) and do not deteriorate when reheated to this temperature. Automotive exhaust springs operate in a hot environment with no loss of strength, but 700 °F (371 °C) will ruin most cutting tools other than the high-speed, nonferrous, and intermetallic types. This temperature will soften aluminum, magnesium, copper, and brass and will cause some of the low melting metals such as solder to become a liquid. Some of the structural steel parts of aircraft will not be affected by a temperature of 700 °F (371 °C), but 1000 °F (538 °C) will remove much of their strength. High speed steel tools and heat resistant parts are not affected by temperatures of 1050 °F (565 °C) but 1200 °F (648 °C) begins to soften them. And so, at 1200 °F, most steels; all of the aluminums and magnesiums; copper, brass, and zinc alloys; and all low melting metals lose their strength to the near minimum values or become mushy and liquid.

Heat Resistant Metals

Metal parts which function at temperatures above 1200 °F (648 °C) concern the manufacturer because the availability of these high temperature resistant metals is limited. Temperatures beyond 1200 °F (648 °C) invite oxidation and scale to some of the metals as well as creep. (*Creep* is the long-time dimensional change in a metal under load.) Another problem is recrystallization. When environmental temperatures increase beyond 1600 °F (871 °C) only a few metals can function properly under load. A survey of the metals points out some of the common types which will withstand temperatures around 1600 °F (871 °C). These include nickel, rhodium, platinum, niobium, titanium, chromium, tungsten, tantalum, molybdenum and cobalt. All of these metals have high melting points and are therefore more suitable for resisting high temperatures. There are also the rare metals (Chapter 11) which have equally as high melting points, but these are more suitable for other uses, even though their great potential for resisting high temperatures exists.

Combinations of elements often produce outstanding alloys. An an example, an alloy containing a small amount of carbon, 10% nickel, 20% chromium, and the remainder iron, molybdenum, tungsten, and titanium is well qualified for operational temperatures around 1650 °F (899 °C) where both strength and oxidation resistance are essential. Another high temperature resistant alloy includes in its analysis about 22% chromium, 9% molybdenum, and 45% nickel. This alloy resists both creep and oxidation at 2150 °F (1177 °C). Such an alloy finds extensive use in exhaust systems and turbine wheels. Then, there are the modifications of the 18–8 type stainless steels which show excellent creep resistant performance up to 2000 °F (1093 °C). This group of austenitic steels includes approximately 18% chromium, 8% nickel, a small amount of titanium, a low carbon content, and the remainder being the basic elements in steel. A comparison of impact strength and hardness for several stainless steels after heating for prolonged periods of time is given in Table 7–5. For example, the type 321 stainless steel has a ft-lb value of 107 at room temperature, at 900 °F (482 °C) its toughness value is 101, and after functioniing for 1000 hours at 1200 °F (648 °C) still maintains the Charpy ft-lb value of 69. Such an endurance is equalled by only a few metals. And even more remarkable, this alloy loses only a small amount of toughness after 10,000 hours of exposure to 1200 °F (648 °C).

Refractory Metals   As operating temperatures of metals climb higher than 2500 °F (1371 °C), the choice of metal diminishes to only a few, including molybdenum, rhodium, tungsten, platinum, tantalum, niobium, various combina-

TABLE 7-5 EFFECT OF PROLONGED HOLDING AT 900–1200 °F (482–648 °C) ON ROOM TEMPERATURE TOUGHNESS AND HARDNESS ON SEVERAL STAINLESS STEELS

| | Room Temperature (°F) Charpy Keyhole Impact Strength (ft-lb) | | | | | | | Room Temperature Brinell Hardness | | | | | | |
| | | 1,000-hr exposure | | | 10,000-hr exposure | | | | 1,000-hr exposure | | | 10,000-hr exposure | | |
AISI	Room	900	1050	1200	900	1050	1200	Room	900	1050	1200	900	1050	1200
304	91	87	75	60	79	62	47	141	145	142	143	143	132	143
304L	89	93	76	72	85	71	63	137	140	134	134	143	143	143
309	95	120	85	43	120	51	44	109	114	109	130	140	153	159
310	75	—	48	29	62	29	2	124	119	119	130	152	174	269
316	80	86	72	44	87	49	21	143	151	148	170	145	163	177
321	107	101	90	69	88	72	62	168	143	149	166	156	151	148
347	56	60	55	49	63	51	32	169	156	167	169	156	169	123
405	35	—	36	26	—	39	34	165	—	143	137	—	143	143
410	33	—	41	27	39	3	21	143	—	114	154	124	143	128
430	46	—	32	34	1	3	4	184	—	186	182	277	178	156
446	1	—	1	1	1	1	1	201	—	211	199	369	255	239

Reprinted from "Selection of Stainless Steels" © American Soc ety for Metals, 1968.

tions of their alloys, and some rare metals. One of the highest temperature operational metals ever produced is an alloy of tungsten and a small amount of molybdenum. Effects of loading stresses and corrosive environmental gases, coupled with temperatures in excess of 4000 °F, establish a usable life for the metal measured in minutes or hours when exposed to such an environment. Such is the case in rocket engines and exhaust systems. Refractory metals serve their purpose in this example by helping to provide the means for the vehicle to obtain its trajectory speed and flight stability. Time at temperature is then the most limiting factor with regard to service life of the metal. There is also the case of a vehicle reentering the earth's atmosphere after being exposed to cryogenic temperatures in space. On impact with the earth's atmosphere numerous conditions change. The time between entry and normal earth conditions tests some of man's greatest attainments in refractories and other materials which are used in the heat shields of spacecraft. Both impact loading and thermal shock expose the shield and vehicle to potential destruction.

The Need to Resist Creep

An alloy should be fully annealed at about 2000 °F (1093 °C) to impart strength to metals at such high temperatures along with the ability to resist oxidation. This temperature assures the solution of most of the carbides and provides very large grains. Both of these factors help assure stabilization under adverse operational loads. A large grain structure reduces creep and the solid solution increases strength in some alloys by using the combined qualities of all the constituents. *Creep* is the elongation of a material under tensile load below the yield while at an elevated temperature over a long period of time. Other types of creep include compression, torsion, and bending.

As metals get hotter they become more plastic and their yield strengths are lowered. When the hot metal creeps or stretches plastically under its safe operating tensile load, it becomes permanently longer, just as if the load were beyond the yield strength. If the dimensional increase interferes with moving parts, such as the creep of a turbine blade as it turns within its housing (an increase in length), instant destruction of the assembly may occur. In this respect, an alloy containing 13% chromium, 4% molybdenum, 7% aluminum, some titanium as a stabilizer, and the remainder nickel, is used for making jet engine turbine blades which operate at approximately 1725 °F (941 °C). Increased resistance to creep, oxidation, and fatigue failure is provided by coating these super alloys with tungsten carbide. The tensile strengths of these blades are great enough to provide high r/m with multistress conditions and dimensional stability at 1725 °F (941 °C). Very few materials will equal this endurance (Fig. 7–2).

Creep   Creep, the gradual slip of material under load at high temperature, begins at weak points in the atomic lattice. Because both heat and force are constant factors trying to reorient the metal in a more compatible relationship with the stress pattern, some minor changes in lattice locations occur at the outermost perimeters of the organized unit cells; these changes are at the grain boundaries. Fortunately, the amount of slip is very small because fewer grains are affected due to the existing coarse grain pattern. Creep, however, is resisted by several factors.

FIGURE 7–2 These turbine blades are made of the best available metal which must resist high temperature, creep, corrosion, and high stresses.

First, because high temperatures usually promote solutions of constituents, any material that adds to the strength of the matrix will naturally reduce the creep tendency. The addition of a material which promotes the formation and stabilization of compounds which are well dispersed in the tough solution will further increase the strength of the metal and resist creep. On the other hand, while the metal is exposed to very high temperatures it is often attacked by hot oxygen at the same time that the temperature influences the rigidity of the metal. In order to reduce surface contamination of the metal at elevated temperatures, as well as to stiffen the solution, selected combinations of nickel and chromium are used. A small amount of carbon also tends to stiffen the solution and increases the strength. Cast metal is commonly used in the design of parts that need to resist creep, because extremely high operating temperatures relate to large grains and often to solution conditions of the microstructure.

Choice of Metal in the Design

Deciding what metal should be used for a particular design becomes a matter of logical deduction and scientific experimentation with the metals to "try to see what happens." Of course, the problem solution theory states that the problem must be identified first. Then the facts are to be gathered and analyzed. Possible solutions are presented, and it is at this time, under operational environmental conditions, that the best solution is offered. For example, the hot environments of furnaces generally require alloys of nickel, chromium, iron, rhodium, platinum, or molybdenum, and the higher temperature operating thermocouples of the furnaces require an alloy of platinum and rhodium. The chromel-alumel thermocouple is effective up to 2150 °F (1177 °C) for reasonable periods of time, and the platinum-rhodium type is acceptable up to 3400 °F (1871 °C) when properly encased in a tube. Electrically heated furnace elements, as well as hearth plates, crucibles, and heating elements for toasters, are made of the nickel-chromium alloys. Heating elements resist electrical current and become hot. Significant oxidation does not occur because the surface of the metal resists deep oxide formation. The platinum-rhodium element is used where temperatures do not exceed about 3200 °F (1760 °C) for a long period of time.

Jet engine turbine temperatures in excess of 1700 °F (927 °C) demand the best available metal because of the high temperature, oxidation, and multistress loading conditions, including tensile. Combinations of nickel, chromium, titanium, molybdenum, and aluminum satisfy the very tough environment. A section of a modern jet engine is shown in Figure 7–3.

FIGURE 7–3   This section of a jet engine represents an outstanding achievement of technology.

With regard to the use of high temperature and radiation resistant materials in nuclear power plants, the austenitic steels, special alloys, and graphite find a favorable place and have performed very well. There is evidence that radiation causes lattice adjustments in the metal which affect mechanical properties. However, it appears that this lattice change is greatly influenced by time under radiation, even though a self-healing adjustment later develops when the radiation is removed.

PLASTIC DEFORMATION IN ANY ENVIRONMENT

When an increasing load is placed on a metal, fracture eventually occurs when the load becomes greater than the strength of the metal. The fracture will be either a ductile, a partially ductile, or brittle break depending on several situations relating to the metal's ability to flow plastically (Fig. 7–4). Plastic flow in a metal results from the slip or movement of an atom or atoms along a slip plane in the lattice which separates the rows of atoms. The atoms again attain nearly normal conditions with respect to lattice orientation (Fig. 7–5) after slip occurs. When obstacles, such as foreign atoms, occur along the plane or planes of possible slip, a less ductile or brittle fracture results because plastic flow is small or does not occur at all. Plastic flow is then the sliding movement of one material past an adjacent material, just as cards slide past one another when the deck is held at the bottom

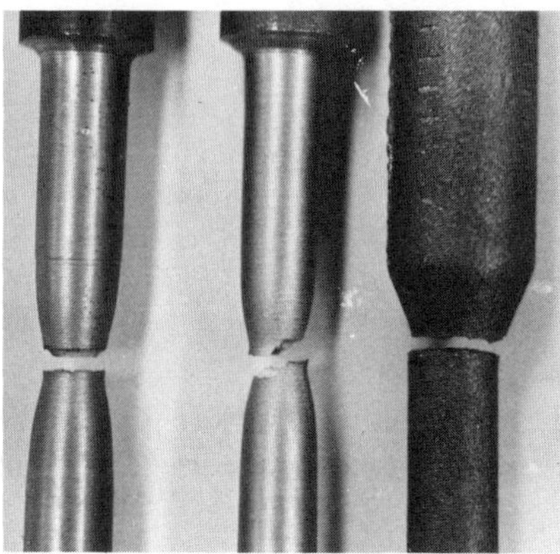

FIGURE 7–4   Three main types of metal fractures, ranging from ductile (cup and cone) to the jagged ductile to the straight, or brittle, type.

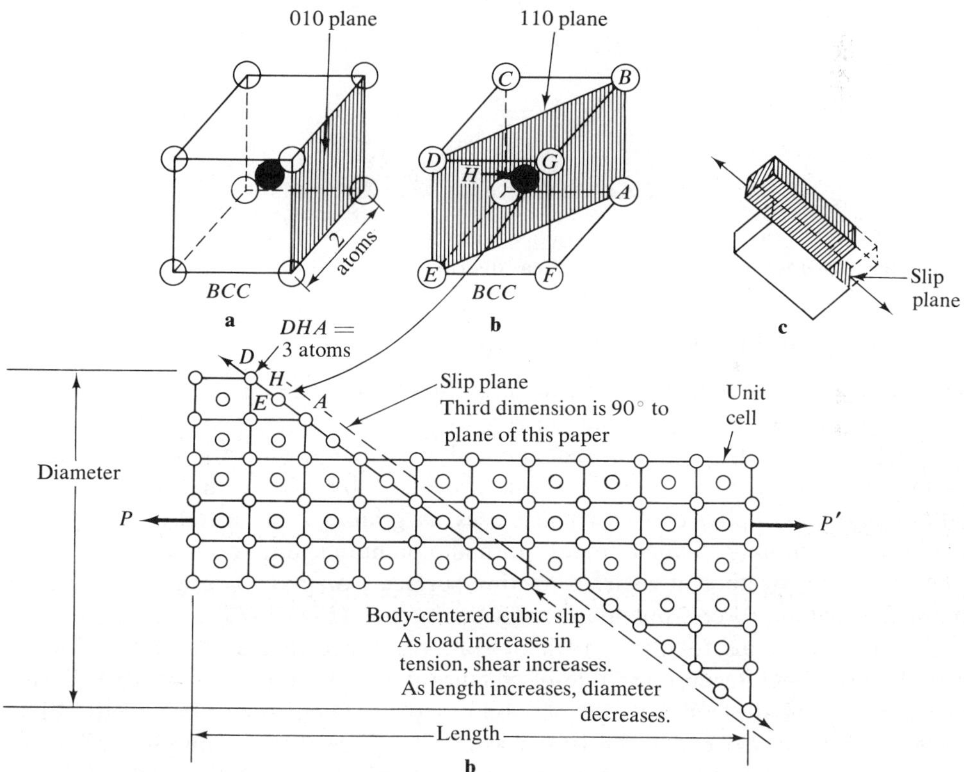

FIGURE 7–5   Slip planes in a unit cell: *a,* the body-centered cell; *b,* slip occurs along *DHA; c,* material slips along the plane.

and the top cards are pushed. The sliding cards move in relatively opposite directions according to two opposite forces pushing them. The pushing or shearing force must be greater than the resistance of the card to movement force before the card will move.

Deformation Occurs Along Slip Planes

A metal consists of many layerlike planes, or slip planes, along which portions of the metal slide smoothly and cause permanent deformation when two opposing but noncollinear forces are applied. The direction of sliding occurs along those slip planes which are best oriented for slip to occur and does not take place throughout the whole mass of metal at the same time. In this respect, an entire row of atoms does not normally move concurrently over the adjacent row of atoms, but an atom or small line of atoms moves consecutively from the origin in increments of movement to the end of the plane. In other words, the atom, which is the material, moves from location to location in steps along the plane (Fig. 7–6).

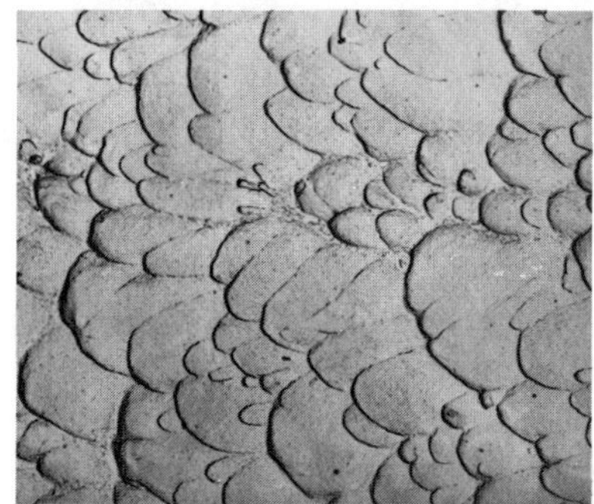

FIGURE 7–6 Electron microscope view (5000×) of a Ti-6AL-4V alloy showing fracture resulting from shear overload.

The Metallic Grain and Its Lattice

The slip planes and makeup of metals are different because their atoms are different; some atoms are larger than others and some are arranged differently. A material or metal is three-dimensional and is available in various sizes. These various shapes which are handled and processed by people and machines are nothing more than piles of very small pieces, or grains, of the metals. The metallic grains are randomly shaped under normal conditions and randomly arranged. They are held together at their contacting surfaces, or grain boundaries, by their powerful bonding forces, but the metal will fracture (Fig. 7–7).

Fracture of metals follows a path of least resistance that exists at the time of metallic overload. Because the line of fracture involves the atoms and their design in the layout of the cell, several possibilities of metallic separation occur. First, the separation may occur in the least organized region of the grain, at the boundary. Secondly, the crack may move along a better organized path, such as across the grain and through a path of preferably oriented slip planes in the cubical structure. At the time of metal solidification grains become three-dimensional cubics and quickly grow from the rapidly forming cells. Consequently, a grain of metal is an organized group of unit cells; the cells consist of vibrating atoms, pulsating within allocated spaces which are maintained by each atom. Within the atom is mostly space and this allows atomic activity and flexible control of its powerful force.

FIGURE 7–7   Electron microscope view (5000×) of AISI 4340 with a grain boundary fracture due to hydrogen embrittlement.

Interatomic Forces

Forces between atoms are atomic and cohesive and the summation of the positive and negative forces of the atoms demands stability within the flexible radii among adjacent atoms. Flexibility of the radius between two atoms is normal because of their constant vibrations. The mean radius between the atoms depends on surrounding environments which usually include variable conditions. And so, the atom can be moved from location to location in a lattice of atoms by means of pressure and slip along a weak plane where shear can occur. A row of atoms can also be extended many atom diameters by unit action of slip. Indeed, the atoms are small even though they range in size. As an example, a line an inch long represents approximately one-fourth of a billion hydrogen atoms resting side by side.

The Atom Responds to Energy Changes   Even though the atom is movable, its makeup is fairly stable because its positive nucleus keeps it apart from its neighbor, and its fast-spinning electrons help maintain an operating radius between neighboring atoms. Interatomic forces and radii remain stable in equilibrium conditions until disturbed by some foreign form of energy. When the forces of repulsion and attraction balance, the size of the atom is established and its energy is at a minimum. Consequently, any size change of the atom requires an increase in energy from some external source. Loss of electrons by ionization decreases the radius, while addition of electrons increases the radius. It must be kept in mind that the loss or gain of electrons influences the size of the atom and, in turn, the atom becomes ionized. The negative ion (atom) is therefore larger than its previous size, and the positive ion (atom) is smaller. Interatomic activity of the unit cells is dependent on surrounding environments and includes changing temperatures, varying pressures, and electrical conditions. For example, as the temperature increases, atomic radii increase and density decreases. Based on this phenomenon, the family of atoms resides as solids, liquids, or gases.

Lattice Imperfections

Because of the many variables occurring at the time of cell origin, the perfect lattice type of grain rarely forms. That is, as the atoms line up according to laws of solidification, a misfit of atoms occurs here and there. Corner atoms of cells normally share with adjacent iron cells and form long rows of atoms until blocked by collision with other forming lattice groups. During the solidification process many variables exist which originate from the rapidly changing temperature and the chemistry of the depleting liquid. Chance allows many dislocated atoms to form in the lattice, causing the lattice to be deformed in places. A dislocation, for example, is the absence of part of a row of atoms or the presence of part of an extra row of atoms (Fig. 7–8). Absence of some atoms from the lattice leaves

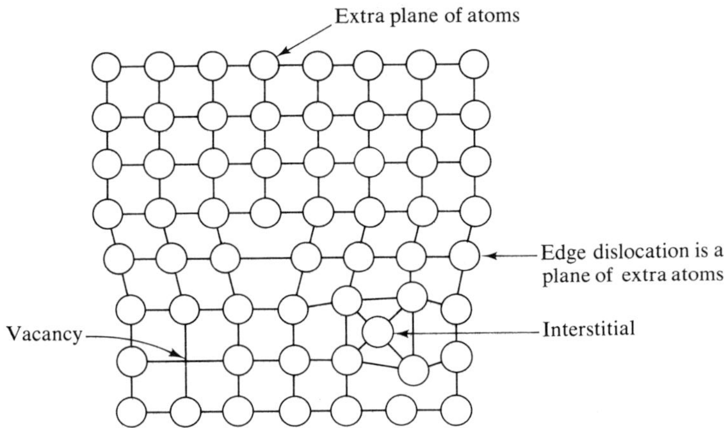

FIGURE 7–8   A dislocation in a lattice shows how the extra plane of atoms causes atomic shifting to the edges. Other grain imperfections include the interstitial and the vacancy, or void. Perfection in cell organization does not always exist.

incomplete bonds, interferes with normal radii between other atoms, and allows slip of the dislocated atoms along the lattice all the way to the surface when sufficient force is applied along the favorable slip plane. At the surface or edge of the metal a step, or protrusion, occurs leaving a permanent change in shape (Fig. 7–9).

Imperfections in the lattice networks of atoms, such as dislocations, exist as edge and screw types. The edge types occur at the end of the atomic plane (Fig. 7–10), and the screw types turn through the grain by switching to other planes (Fig. 7–11). Most dislocations involve a combination of the two types. Another imperfection is the presence of foreign atoms in the lattice, and the third defect is the vacancy, or void. Based on the ability of dislocations to allow slip, plastic flow of metals occurs when stresses push the dislocations through the slip planes of the lattice. Dislocations, or defects in the lattice, move to the end of the lattice plane in progressive steps while acting under the necessary force. Such shifting of the dislocations allows other atoms to move along the same plane and provides dimensional change during cold and hot working operations. In regard to working of a

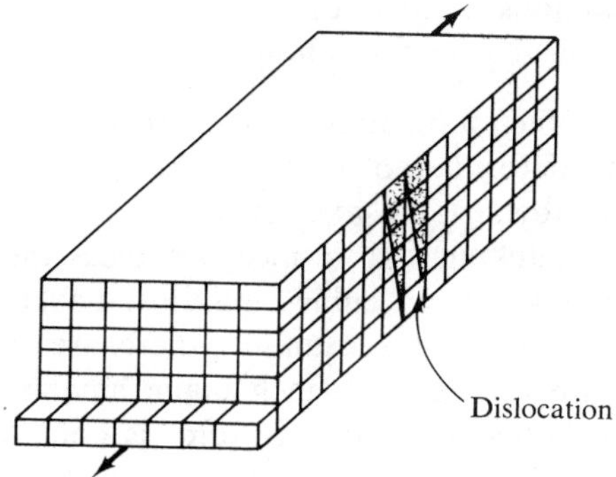

FIGURE 7–9   A step occurs along the face of the metal as the atoms are pushed to the surface of the face.

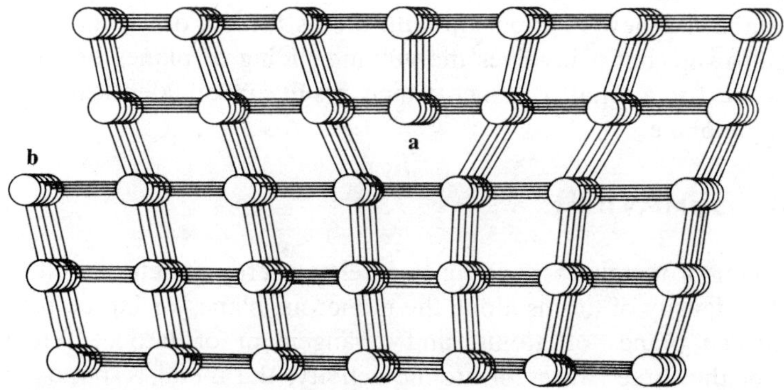

FIGURE 7–10   An edge-type dislocation showing unit cell deformation, original dislocation (a), and edge step (b).

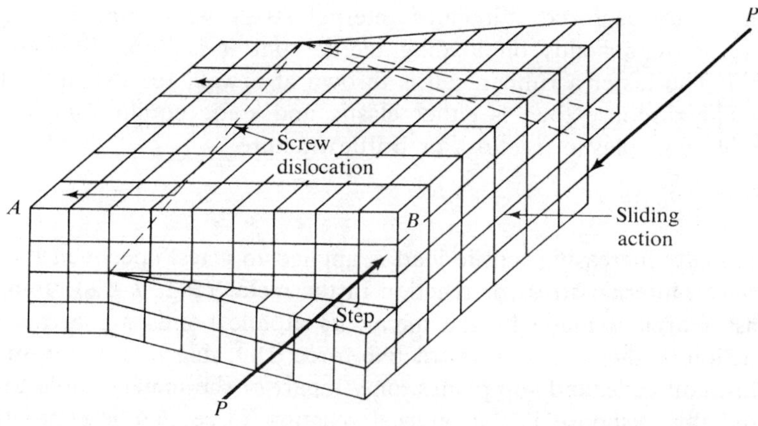

FIGURE 7–11   Screw type dislocation showing edge protrusion. Top cells are rotating along the atomic slip plane due to shear forces. A step parallel to AB results.

metal, deforming conditions also produce dislocations; therefore, metals include these imperfections in nearly any condition.

Slip and Twinning　　　When a metal is stressed beyond its yield, where permanent deformation begins, the deformation results from slip or twinning; the grain system rotates by deformation toward the axis of stress. In *slip* the direction is in the plane of densest atomic packing, and in most instances the planes vary among the face-centered cubic, body-centered cubic, and close-packed hexagonal systems. In the face-centered cubic system, for example, there are twelve slip system possibilities because there are four planes, each plane having three directions of slip. Accordingly, when the lattice structure of grains is complete in three dimensions, and when a sufficient number of lattices are exposed to the line of force, the external force must exceed the resistance of the metal before slip and permanent deformation can occur. When slip does occur it happens first in those grains and lattices requiring the least levels of stress. In this regard, the numerous dislocations provide sufficient orientation of slip planes for atoms to shear away from neighbor atoms and establish a new relationship with atoms farther down the atomic plane. *Twinning* is also slip, but it involves the volume sliding of planes of atoms parallel to another plane for a small distance which results in an identical image of the lattice across the plane.

METALLIC BEHAVIOR

Internal behavior of metals responding to foreign energy is determined from inter-atomic activity, density of atoms along the numerous planes or lattice rows, spacing distances between planes of atoms, and arrangement of lattice. When force is applied, one of the three causes of atomic activity, the metal responds either with elastic deformation, plastic deformation, plastic fracture, or brittle fracture. The response appears instant, but a series of possible selections is tried by the metal until either the selection is matched to the force or submission to destruction occurs because of the metal's inability to stabilize the force. A kind of stimulus-response mechanism occurs as force stimulates internal stress with immediate response. Speed of response depends on loading inertia, but it is approximately directly proportional. The faster a rubber ball is thrown at a wall, for example, the faster it returns. The response then is either elastic and static equilibrium, elastic and plastic equilibrium, plastic fracture, or brittle fracture.

Elastic Deformation

When a gradually increasing tensile load is applied to a test specimen the external load induces an internal stress or reaction in the metal (Fig. 7–12). Immediately, the response search is made by the metal; its atomic bonds are arranged in an opposing action to the potential destructive force. But, due to random orientation of the grains, unit cells, and slip planes, only a part of the metal is able to respond initially, and this response is the internal reaction force that is more favorably aligned at the time to resist the external force. This internal force is the sum of all the small resisting forces in various regions of the granular structure. As the external force increases, more elastic resistance occurs, and if the external force does not exceed the metal's ability to respond elastically, no significant

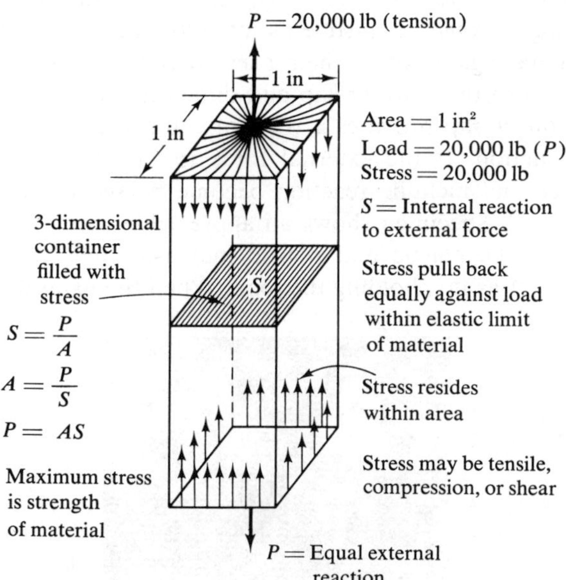

FIGURE 7–12   Stress is induced by external force and varies with the force.

plastic flow occurs in the metal. The metal rests in equilibrium when load and stress become balanced. As the external force is subsequently reduced to zero, the internally reacting stress is also reduced to zero. This elastic stretching of metals is typical of operating conditions in well-designed applications. The term given to the equal proportion of stress to strain or elastic reaction in metals is the modulus of elasticity. Accordingly, a metal may elongate or deflect a substantial amount under tensile load, such as a static condition in a bridge beam as it attains equilibrium conditions under load. Elastic bonds in the atomic lattice remain stressed as long as the force is exerted on the metal and return to zero stress as the load is removed.

Elastic to Plastic Deformation and Equilibrium

The second type of reaction available to the metal under influence of a gradually increasing tensile load is elastic and plastic deformation to some point at equilibrium. Again, as the load is applied, elastic bonds in the metal respond and continue to increase in response intensity in an effort to stop the destructive force from pulling the lattice apart. Elastic bonds can be compared to rubber bands—the longer the stretch, the greater is the force needed to increase the stretch. As the external force continues to increase and as the metal becomes longer in length (elastic stretch), it suddenly begins to yield, if ductile, to the external force by submitting to this force which is greater than it can immediately stabilize. The metal yields along selected planes of atoms in the numerous available lattices where dislocations exist, just as a playing card slowly slides along the surface of its neighbor to increase the stack's length. As each atom or dislocation system of atoms slides in shear past its neighbor row of atoms, it builds up a resistance to further slip in the form of increased stress, and this tends to balance the external force. Then, as the external force again increases and as the metal gets stronger by work hardening through adjustments of slip planes, additional slip planes throughout the metal again rotate their axes to more favorable alignments with the destructive force so that further slip and dimensional change can occur.

The phenomenon of slip provides a way for a metal to try and save itself from fracture and destruction. If the external force is halted, the slip and elongation stop, and the force and metallic reaction become balanced. At this point a measurement of the tensile specimen shows an appreciable increase in length, and this increase consists of the elastic and plastic increases, but the specimen has not fractured (Fig. 7–13) as it is holding the load. When the external force is removed,

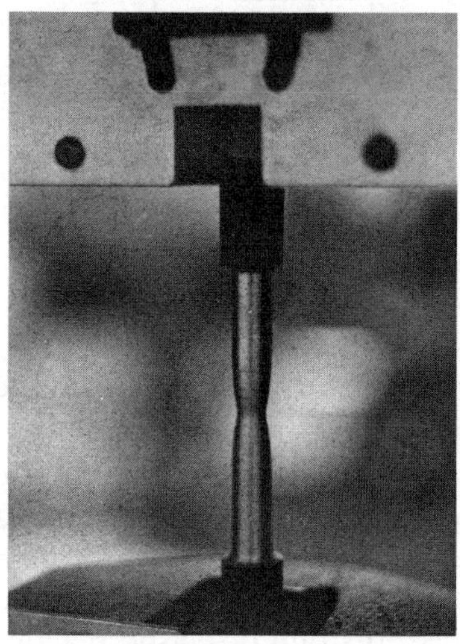

FIGURE 7–13 A tensile specimen has elongated under both elastic and inelastic stresses, but has not yet fractured.

springback occurs. A permanent length increase exists and the width shows a permanent decrease in dimension. Both length and width dimensional changes are therefore caused by plastic slip, not elastic, in various regions of the grains. Springback loss of dimension is elastic.

Plastic Deformation and Work Hardness

The third type of reaction available to a metal under an increasing tensile load follows the same procedure as the preceding example. As the elastic deformation stage is passed the metal begins to yield, and permanent deformation proceeds as rapid elongation commences. Reaction in the metal is fast; many unaffected grains begin to turn their boundaries in more favorable alignment with the destructive force. This rotation of grains occurs due to the ever increasing stress levels reaching the particular grain orientations with their additional quantities of slip planes. As each new plane moves into alignment an increase in force is required to cause further slip because of the reduction in available slip planes and the energy buildup. Slip then induces higher stresses which induce higher distortion and energy levels in the jagged planes that have already slipped.

Such a deformation of the organized lattice severely limits continued dislocation movements, and higher levels of rigidity appear. This rigidity is observed as extreme resistance to further permanent deformation by tensile force. However, if the force is strong enough to demand further metallic yielding, some increased deformation may occur as the metal fractures. If the same loading is carried out in compression procedures the same general phenomena occur. That is, if the force is compression and if the resistance is measured by depth of force penetration into the metal or the metal flattens, then a value of resistance to penetration or hardness is established. Consequently, plastic deformation in metals below their recrystallization temperatures causes increased levels of stress and hardness through atomic sliding along planes of slip.

Hardness Is Strength   An analysis of resistance to the deformation factor points out the rising stress levels needed to overcome rigidity. In this regard, stress is convertible directly to strength merely by dividing the external force by the area of metal opposing the force. Such calculations measure stress or strength in pounds per square inch of affected area. In summary, as resistance to deformation increases, rigidity of the metal increases in measurable units, and these units are hardness and strength values. An application of load onto a ductile metal then follows the elastic-plastic deformation route to equilibrium or fracture. When no further resistance is available in the metal to retard the rising force, the tensile or compression strength occurs. At this point in tension the metal may fracture or it may suddenly elongate in a pure plastic manner, that is, by rapid reduction in cross-sectional area and necking. When the stress in the remaining area is equal to the external force, ductile fracture occurs (Fig. 7–14). In compression loading and as the stress levels increase, the metal increases in cross-sectional area as it is reduced in length. When further resistance is impossible the metal suddenly fails in shear or extends rapidly in a pure plastic condition.

Brittle Fracture

The fourth type of reaction in a metal under load beyond the yield is brittle fracture. This type of fracture is inherent in brittle metals and in extremely strong metals or other materials. If a ductile metal like structural steel, for example, is remelted and alloyed with additional elements such as chromium, nickel, and more carbon, the resultant alloy will have a completely different mechanical property potential. Chromium carbide atoms establish positions along the slip planes in the lattice of the metal, and nickel atoms find space among ferrite atoms and form a solution of the ferrite and nickel. This modification of the atomic lattice causes increased resistance to slip by foreign atom blockage (chromium carbide) in the plane and distortion of the uniform planes. Some foreign atoms are larger than their new neighbors and do not slip unless severely stressed. Consequently, when increasing loads are imposed onto the metal the elastic limit increases because the slip planes fail to yield. As the load continues to increase, stress levels also increase to a point where further resistance ceases and cleavage occurs. Such a fracture shows little plastic flow because of the absence of available slip planes. Brittleness in metals then exists due to the presence of numerous foreign atoms in the slip planes of the matrix or the presence of nonmetallic atoms like graphite which will

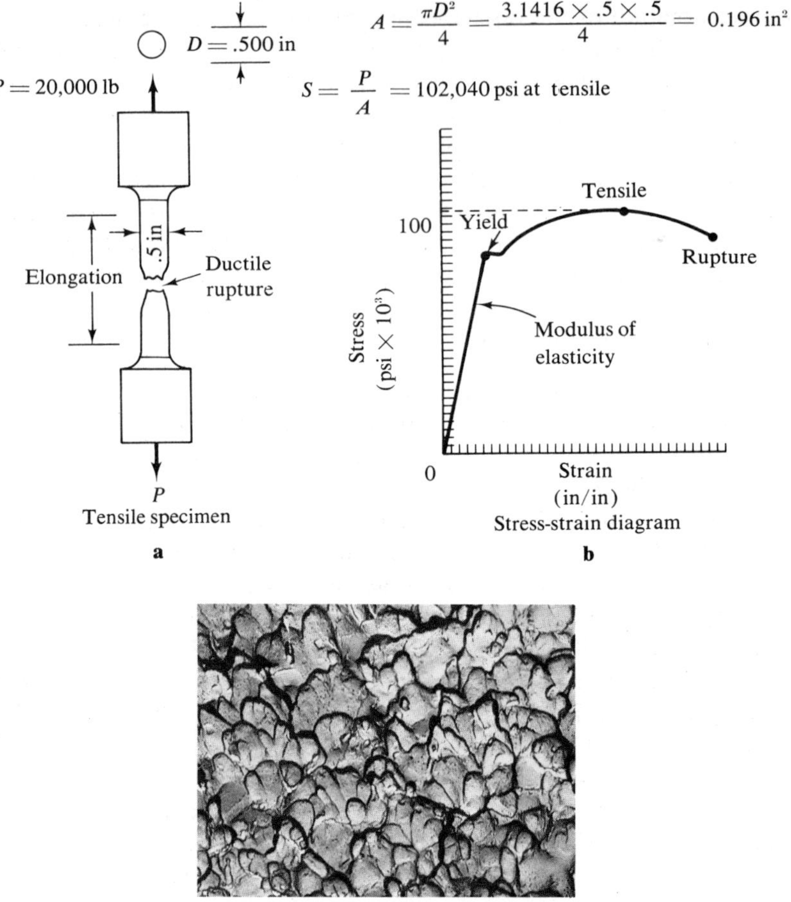

FIGURE 7–14 *a,* A load-to-fracture diagram. *b,* The metal shows a pimple pattern as overload leads to fracture (electron microscope view at 5000×).

not sustain high stresses and allow plastic flow but fracture instead. A simple comparison is illustrated by a ball bearing. As long as all balls are equal in diameter the race turns. If a few slightly larger balls are spaced among the original balls and torque is exerted on the race, no movement will occur. However, if a tremendous overload of force occurs, the race will immediately fracture.

Slip Influences the Strength of the Metal

Because of dislocations and plastic deformation, many metals can be shaped to desired contours and dimensions without tearing. Plastic deformation also provides a means of increasing hardness and strength in a metal. With regard to plastic deformation and fracture of metals, work hardness produced by flowing the metal below its recrystallization point increases the yield and tensile strengths. As the hardness increases, the tensile and yield strengths increase. Severely deformed metal, for example, increases the yield strength closer to its tensile strength because of the reduced ability to flow plastically beyond the yield and because of an increase in the elastic limit. An alloy steel specimen heat treated to a Rockwell hardness of C 43, for example, has a low elongation factor of only 10% because the high tensile strength of 200,000 psi forbids an appreciable plastic flow with its high yield. On the other hand, metals having high elongation factors, which indicate good workability, have larger ranges between their yield and tensile strengths. As the distance between the yield and tensile strengths increases, ductility increases. As the distance between the yield and tensile strengths decreases, the trend is toward brittleness as elongation decreases. Because brittle fracture gives little warning, structural designs such as buildings and bridges must include ductile metals.

MATERIALS FOR HIGH TEMPERATURE OPERATION AND NUCLEAR POWER PLANTS

With regard to materials for use in the nuclear reactor, it becomes a matter of selecting what is available. Uranium is one of the fuels and will be subsequently discussed. Irradiating beams of high energy, accompanied by high temperatures, penetrate the fuel container in the reactor. Walls, fixtures, tubing, and vessels of the container and attached heat exchanger are subjected to high temperatures, irradiating effects of radiation, and corrosion. The closed system contains the hot liquid at high pressures which places the vessels and tubes in tension as if they were expanding balloons. Secondary exchanger tubes are exposed to superhot steam pressures. The tremendously adverse conditions in this environment are not excelled by any other engineering design.

The Need for Nondestructive Testing

Materials for nuclear installations must be selected from those available, but with nuclear grade specifications. Face-centered cubic stainless steel, for example, fulfills much of the need, but not all. Refractory metals, other austenitic steels, and materials like cadmium, zirconium, and graphite have their particular places in the atomic power generation system. Nondestructive inspection of these materials prior to installation is mandatory, and a set of standards for each material and process is established and subsequently used for comparative purposes as time passes.

Several nondestructive tests are performed. For example, a common and reliable test for some of these components is the shear wave ultrasonic which is capable of finding metallic separations in circular parts such as pressure vessels. Radiographs are also produced for studying deep interiors of the metal. Identified discontinuities in large sections of thick metal can be repaired by welding. Such a procedure involves the removal of disturbed metal to form the cavity for filling with sound weld metal deposits and subsequent nondestructive testing of the repair. Such pressure vessel repairs are governed by code repair processes, and the weld is accomplished by a certified welder.

Because time is a critical factor in the life of the metal in this environment, intermittent nondestructive inspections are standard operating procedures. Radiographs done at installation time, for example, are compared with those at some

FIGURE 7–15 Two of the four main steam isolation valves are shown as they are installed in the Baltimore Calvert Cliffs nuclear power plant. Each valve weighs 27 tons, is 8 ft long, and 14 ft 6 in high. If the flow to or from the plant's steam generator must be cut off, the valves are designed to close in six seconds. (Courtesy of Rockwell International, Flow Control Division)

designated later time. If a discontinuity or indication in the material has changed in geometry, or if a discontinuity appears, then evaluation of the material's condition is done and a decision regarding action is quickly made. The quality assurance procedures demand the very best, for example, a main steam valve in the high pressure system of one particular installation is manufactured for a life expectancy of many years under a constant pressure of 1100 psi and at an operating temperature of 580 °F (304 °C) with no significant leakage (Figure 7–15). Stainless steels play an important part in construction of these valves.

All pressure vessels are subject to the formation of discontinuities because of the type and magnitude of the stress they encounter in the adverse environmental conditions. Because a pressure vessel is like an expanding balloon, elastic stress builds up to equilibrium and static conditions. Stresses on one side of the tube or vessel balance with those on the opposite side, and end stresses balance with opposite end stresses. But the stresses are not just tensile along a single axis. They are made up of several multidirectional types of elastic natures at a reasonable and safe level below the metal's yield strength. Under changing pressures and temperatures the several stresses seek a balance, but all stresses are not aligned in the same general direction because the outside surface of the vessel may expand more than the inside due to the thickness of the walls. These radial and longitudinal types of tensile and shear stresses begin to resist each other at angles and place additional elastic stresses within the walls of the vessel. Once a static balance is attained and

FIGURE 7–16 Faulty heat treatment often results in metal failure due to uneven stress levels.

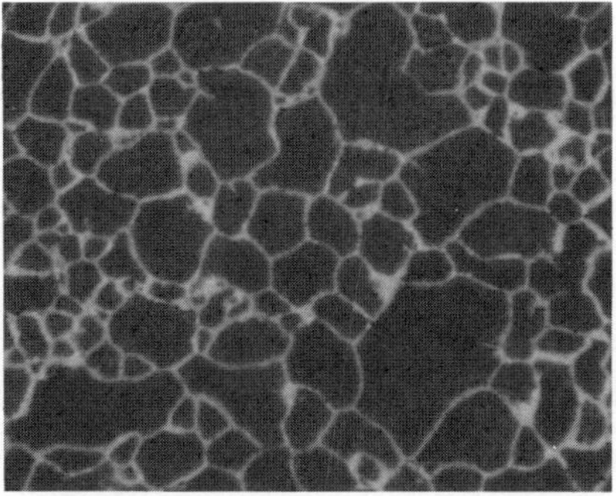

FIGURE 7–17 Intergranular oxidation of AISI 4340 shown at 800×. (Courtesy of Bell Helicopter Company)

FIGURE 7–18 Aluminum alloy (7079-T651) shows stress corrosion at grain boundaries and fracture (electron microscope view at 5000×).

pressure differentials are removed within the vessel, only the projected tendency for the vessel to explode exists, and this is normal.

EXAMPLES OF HARMFUL ENVIRONMENTS

Harsh environments often destroy the strongest metals. Figure 7–16 illustrates the nonuniformity of the grain structure of steel, and Figure 7–17 shows what a chemical attack can do to a once strong alloy. When stress and corrosion combine, stress corrosion often results (Fig. 7–18).

Questions

1. With regard to metals, what is meant by environmental conditions?

2. Define cryogenic temperatures and describe their effects on body-centered and face-centered cubic metals.

3. What is the significance of mechanical properties changing as the temperatures of the metals change?

4. List some metals which are operationally tough at cryogenic temperatures. Explain why these metals are tough.

5. What is the transition temperature in a metal? Why is it important?

6. How does cleavage of a metal occur? Compare this type of failure with plastic failure.

7. Compare the effects of a metal under load at cryogenic temperatures and a metal under load at elevated temperatures.

8. List the metals which can be effectively used above 1600 °F. List several which can be used at cryogenic temperatures.

9. Define creep. What is its significance?

10. What is a refractory material?

11. Compare the slip resistance of solutions and mixtures in metals.

12. Explain the difference between elastic and plastic deformation.

13. Why must public structures such as buildings and bridges be constructed of ductile metals?

14. Explain a slip plane and relate its place in the atomic lattice with unit cells.

15. Differentiate between the ability of body-centered and face-centered cubic metals to slip.

16. How do alloy atoms retard slip and increase the elastic limit of the metal?

17. What does the mean radius of an atom depend on? Why?

18. Describe a lattice imperfection. Explain how a dislocation functions under stress.

19. Differentiate between slip and twinning. Relate these movements to granular rotation under stress.

20. Why are hardness and strength so closely related in metals?

Metallurgy of Ferrous Metals

All metals are divided into two broad categories, ferrous and nonferrous. Such a general classification facilitates the description of each category and provides a means for better understanding of the metals. Basically, ferrous metals are those in which iron is the base material; other elements are included to produce a specific alloy. All ferrous metals contain controlled quantities of carbon, silicon, manganese, sulfur, phosphorus, and iron. These six elements form the alloys called steel and cast iron. However, it must not be concluded that the presence of iron in a metal classifies it as ferrous. Many nonferrous metals contain iron, but the iron is an alloying element and is not the basic metal that exists in the alloy. Three main kinds of ferrous metals are available—steels, cast irons, and wrought irons. All other metals are classified within the nonferrous category.

THE FERROUS RAW MATERIAL

In order that metal working personnel can proceed with the task of producing consumer products, there must first be a means of producing the basic raw materials. Ferrous metals are the product of exact refinement processes resulting from chemical actions and changes occurring in the special types of furnaces especially designed to produce and refine each of these metals. Raw materials such as iron ore, coke (coal), and limestone are initially dug from the earth and subsequently subjected to very high temperatures to melt the constituents and release the impurities into a slag. Removal of the slag leaves a basic ferrous liquid which is subsequently refined into a more usable metal. The chemistry of the liquid is of great importance because it is the resulting solid mass of metal that concerns the study of metals. The solid metal has a history beginning in the liquid, the solid being born from the liquid. Therefore, man's scientific manipulative skills are essential to the production of solid shapes of metals from the liquid.

FROM LIQUID TO SOLID

Ferrous metals originate from the blast furnace. Iron is chemically separated from the ore with the assistance of a flux (limestone, for example, in basic irons) in the presence of very high temperatures which are generated by burning coke and large quantities of air. Control of the initial liquid results in the formation of pig iron which has a high carbon content as well as a higher than desirable content of most other basic elements. However, pig iron is nothing more than the basic raw material for the production of the several types of ferrous metals. The liquid iron is then cast into small castings called pigs. Subsequently, these pigs are melted and delivered to the specific furnaces which produce steel (Fig. 8–1). Cast iron and

FIGURE 8–1   Molten pig iron is poured into a basic oxygen furnace for rapid refinement into steel. Some furnaces produce 200 tons of steel per hour by this process. (Courtesy of United States Steel Corporation)

wrought iron are produced in different furnaces. The steel-making furnaces are different from the blast furnace in that the resulting liquid steel is chemically correct for pouring into castings of consumer products or ingots for further processing (Fig. 8–2). In other words, pig iron is the raw material which is eventually refined and processed into usable ferrous products. In the steel and cast iron production processes, however, a large amount of scrap steel is also used with the pig iron. (Fig. 8–3).

A given shape of metal originates from the liquid. Once the liquid is confirmed as being acceptable according to chemical analyses it is poured into molds to

FIGURE 8–2   Molten steel from an electric furnace is poured into an ingot mold inside a vacuum chamber to produce an exceptionally clean ingot. Subsequent forging or rolling can take place without microcracking of the metal. (Courtesy of Bethlehem Steel Corporation)

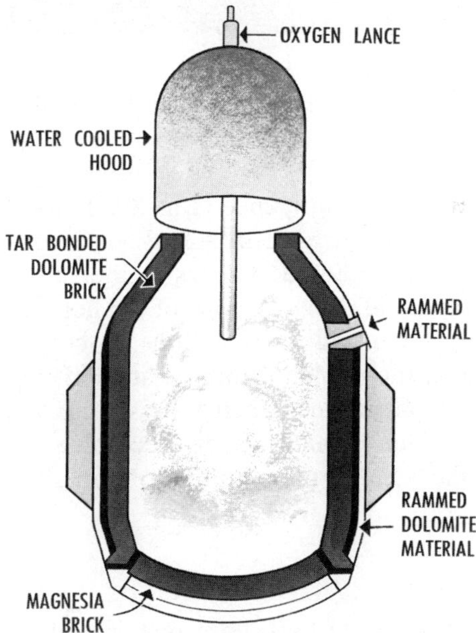

FIGURE 8–3   The basic oxygen furnace is designed expressly to get the best results with oxygen in steelmaking. It can produce steel amazingly fast. In a typical furnace the charge of iron ore, steel scrap, and molten iron is refined into steel by blowing oxygen down from the top through a vertical lance extending to within five feet or so from the bath. During the blow burnt lime and spar are added as fluxing materials. (Courtesy of Bethlehem Steel Corporation)

become steel castings that eventually become usable metal in the final forms of specialized wrought shapes, castings, and metal powders for pressing into a desired shape (Fig. 8–4). The phenomenon of solidification from the liquid is similar in

FIGURE 8–4   The ingot has been rolled into a bloom and, in turn, the bloom will be processed into structural shapes. (Courtesy of Bethlehem Steel Corporation)

most metals; however, the basic difference is in the actions occurring between the pure metals and the alloys. Because a pure metal has a constant freezing and melting temperature (Fig. 8–5) the liquid changes to the solid in a manner dependent on the temperature of the liquid. Inner regions of the hot metal cool more slowly because these regions depend on adjacent regions to cool; this is where the pure metals and alloys differ in their solidification characteristics. The pure metal turns to solid metal when the temperature reaches its solidus point. The alloy contains a mushy stage of metal which is made up of a combination of liquid and freezing metal. Consequently, the solidus temperature of an alloy is different from the liquidus.

Origin of Defects

A common problem applicable to both pure metals and alloys upon solidification is the origin of defects. A pocket of gas, for example, may be trapped in the solid as the result of inability to move from the liquid fast enough. Also, if the freezing metal is not supplied with enough liquid as solidification occurs, holes will appear in the casting. Flaws such as holes and foreign matter inclusions are solidification defects which in many instances remain in the resulting part during its entire lifetime. Figure 8–6 shows the result of metal freezing in the sprue because the pouring temperature was too low.

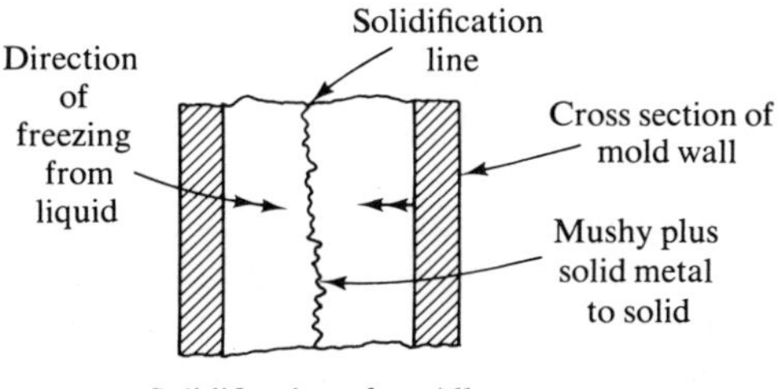

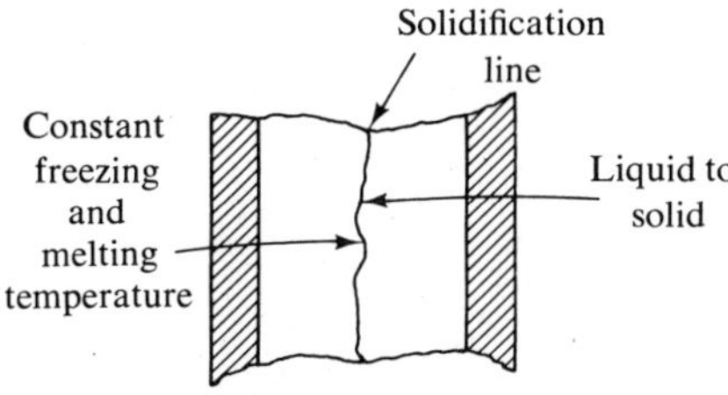

FIGURE 8–5   A pure metal has a constant melting and freezing temperature.

FIGURE 8–6   The result of molten metal freezing in the sprue. Metal was too cool to flow into the cavity at the bottom.

It is not probable that a perfect metal has been produced because perfection would require the absence of foreign materials, the absence of any types of discontinuities, and the absence of nonhomogeneous chemical regions in the metal. But man is striving to better control the chemistry of liquid metals until solidification occurs properly. This physical characteristic of the cooling must ultimately be better controlled to assure homogeneity in the solid mass of metal and freedom from discontinuities, foreign matter, and undesirable segregation. But some degree of nonperfection always occurs. Therefore, it becomes a matter of establishing and using minimum acceptable standards to compare with the quality of the newly produced metal. Cast metals, including the ingot, do not always undergo inspections; the need for inspections is based on the projected use of the metal and the manufacturer's quality control standards. There is a positive trend toward cleaner and more acceptable metallic microstructures. Consequently, many engineers and technicians are planning their production facilities to meet the increasing demands for more perfect metals. Again, it is pointed out that the solid shape of a metal is a product of its history in the liquid (Fig. 8–7), its cooling characteristics from the liquid to the solid, and the many processes which give it its final shape.

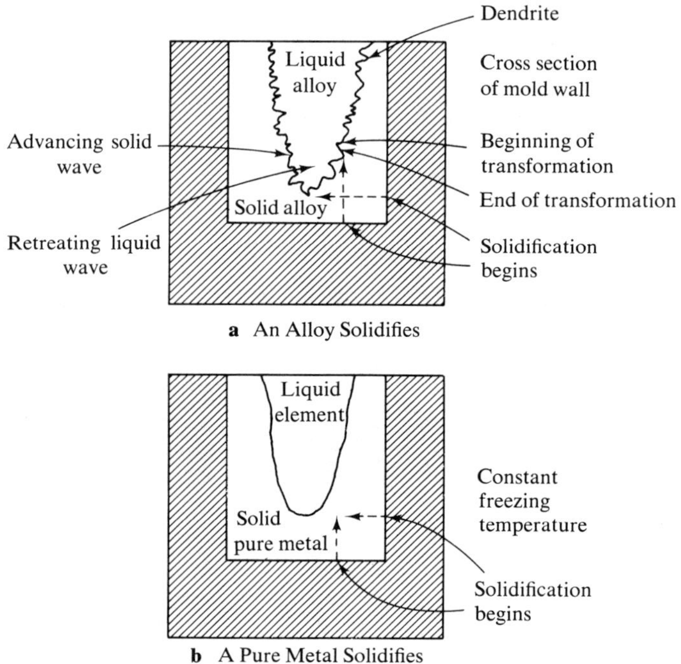

FIGURE 8–7   From liquid to solid in an alloy (*a*) and a pure metal (*b*).

Cooling Characteristics of Ferrous Metals

Because all ferrous metals are alloys their cooling characteristics are somewhat different from those of aluminum, copper, zinc, and other pure metals or elements which have constant freezing temperatures. When the iron-carbon alloys are poured into molds their planned cooling characteristics are partly dependent on their chemical analyses; that is, how much of each of the six or more basic elements is present in each alloy. Because iron or ferrite consumes most of the metal the transformation phases of iron dictate much of the cooling aspects, while the carbon content is considered the most critical quality of the liquid. Together, the amount of carbon and the form of iron existing at any given temperature become important factors when associated with the cooling rate. The descending temperature during cooling causes changes in the metal's microstructure. Atoms shift from weak bonds in the liquid to rigid and strong bonds in the solid. But in ferrous metals an allotropic change occurs in the solid as temperatures continue to decrease toward room temperature (Fig. 8–8). Finally, the resulting mass of metal is a

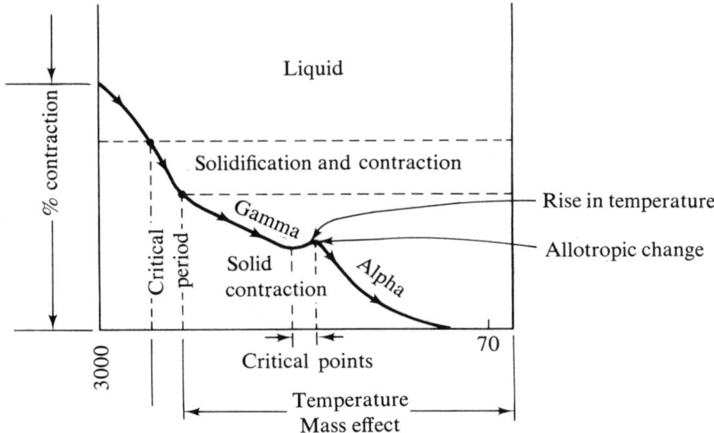

FIGURE 8–8   Allotropic change in a mild carbon steel on cooling from the liquid.

reflection of its mass of constituents and how these constituents are arranged. Of the six basic elements, it is the relationship between the matrix of ferrite and iron carbide (Fig. 3–7) that is considered most important and critical in forming the mechanical characteristics of the metal.

PURPOSES OF THE SIX BASIC ELEMENTS IN STEEL

A summary of the purposes and effects of the basic elements in steel is needed in order to establish a better understanding of the relationship with the alloys which are added as desired. Ferrite (Fe), often wrongfully used interchangeably with pure iron, is a solid solution of *carbon* (C) in alpha *iron* and is the base material of all ferrous alloys. Iron constitutes most of the ferrite, while only 0.008% carbon is in solution. The quantity of other elements in solution is usually small, especially in the plain carbon steels. As for carbon content in the presence of alpha iron, all carbon over 0.008% is mixed with the ferrite as carbide in slowly cooled steels from the red temperature ranges. In steels the carbon content ranges from 0.008% to 1.7%; those alloys having more than this amount (1.7% up to 6.67%) form the series of cast irons. *Manganese* is essential during the manufacturing of these ferrous alloys because it combines with other elements and helps produce a slag whereby impurities are removed. It combines with carbon, iron, and sulfur. Part remains in the ferrite solid solution and part chemically combines with carbon to form manganese carbide, a very hard compound. Manganese is a deoxidizer and helps in the release of iron from its ore. Its content in carbon steels ranges from 0.25–1.65%, depending to a high degree on the carbon content. *Silicon* is also a deoxidizer and helps release the iron from the ore during the manufacturing process and promotes fluidity in the metal. The silicon content range is 0.10–0.30%, the amount again is related to the carbon content. *Sulfur* is not normally added to the molten mass of metal; it is an impurity picked up from the ore and coke. *Phosphorus* is also an impurity picked up from the ore during manufacture. Therefore, sulfur and phosphorus are considered undesirable in most, but not all, ferrous production operations and are maintained at very low amounts, such as 0.050%

sulfur and 0.040% phosphorus in carbon steels. In the steel alloys the phosphorus and sulfur contents are maintained at 0.040%. The above percentages are maximum.

DENDRITE FORMATION

Because most alloys solidify at varying temperatures and because the temperature of the cooling liquid varies in different regions of the melt, density differentials exist throughout the mass of metal. Cooling of the liquid is not completely uniform as solid nuclei form. Parts of these small masses of solid metal are chemically richer in some constituents than are other regions of the liquid (Fig. 8–9). This

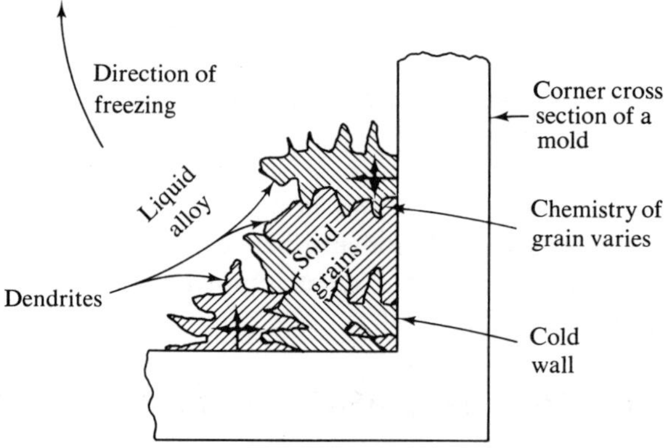

FIGURE 8–9   Dendrites form as the alloy solidifies over a range of temperatures. This illustration shows what happens at the corner of a mold.

means that solid masses develop from the highest melting solution and build granular patterns until all the liquid is transformed to the solid. These many grains, or *dendrites,* are heterogeneous in their chemistry; that is, part of a growing dendrite has a different chemistry than an adjacent part because the dendrite is a buildup of rapidly forming solid metal which varies in its chemistry. Because time is required for the transformation of all liquid to solid an individual grain of metal will reflect a size commensurate with time and with a given shape. The size of dendrites in most castings is large and is nearly always undesirable (Fig. 8–10). Each grain is three-dimensional and contains a given amount of the chemistry of the once molten mass. The solid mass is segregated in chemical content because the cooling liquid is not a homogeneous mass upon solidification as are pure metals. Lower melting constituents of the liquid solidify last. Consequently, the time factor creates a solid metal which is a nonhomogeneous chemical mass of all the grains.

MICROSTRUCTURE OF A METAL

Two major factors exist in essentially all solid metals, whether pure or alloy—a mass of grains having shape and size and some kind of arrangement of the chemical(s) inside the many grains (Fig. 8–11). Together, these two factors make up the microstructure of the metal. The laminations of dark material in the illustration

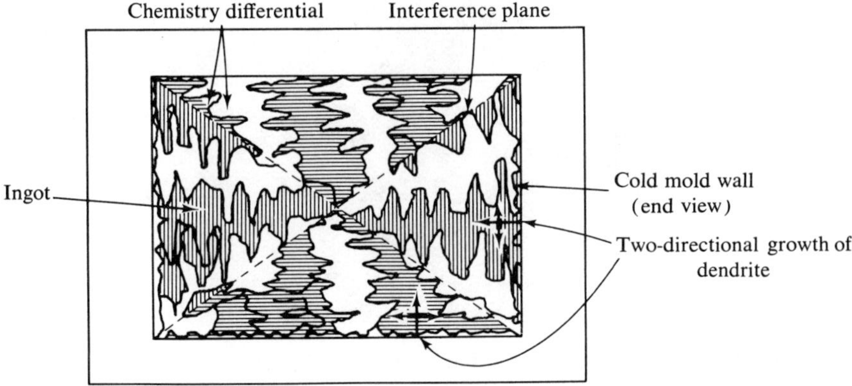

FIGURE 8–10 "As cast" grain size in ingot steel showing exaggerated growth and formation of grains, or dendrites.

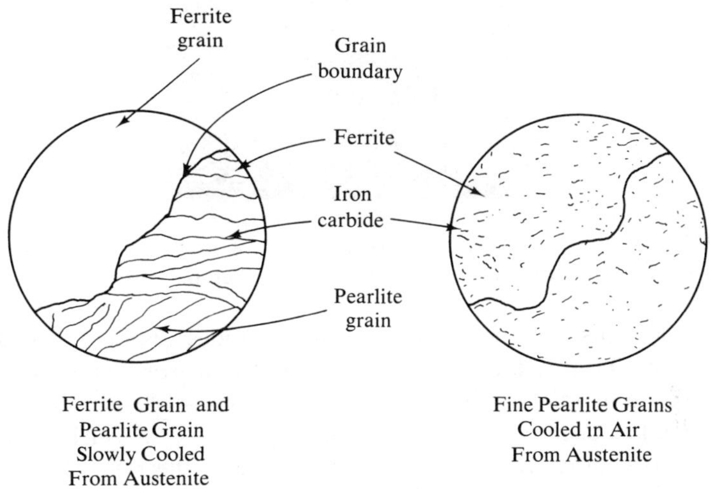

FIGURE 8–11 Chemicals are arranged differently inside the grains of steel as the cooling rate from austenite varies. Slow cooling forms large pieces of carbide, and faster cooling forms smaller pieces.

are iron carbide; the light colored material is ferrite. Combined in the manner shown, the microstructure is called pearlite. A pure metal has only one chemical and one chemical arrangement in its grains (Fig. 2–11), whereas alloys have two or more chemicals with varying arrangements, as pointed out in figure 8–11. If solidification defects such as holes, cracks, inclusions of slag or scale are present, a modified microstructure exists. A crack or inclusion may spread itself across several grains, or the discontinuity may form along the grain boundary.

THREE BASIC TYPES OF FERROUS METALS

Three basic types of ferrous metals are available for manufacturing processes—the many steels, the cast irons, and the several wrought irons. Each metal contains

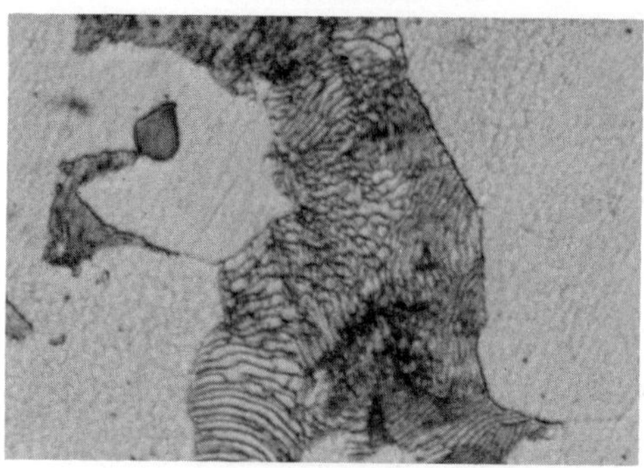

FIGURE 8–12.   Microstructure of an annealed AISI 1040 steel.

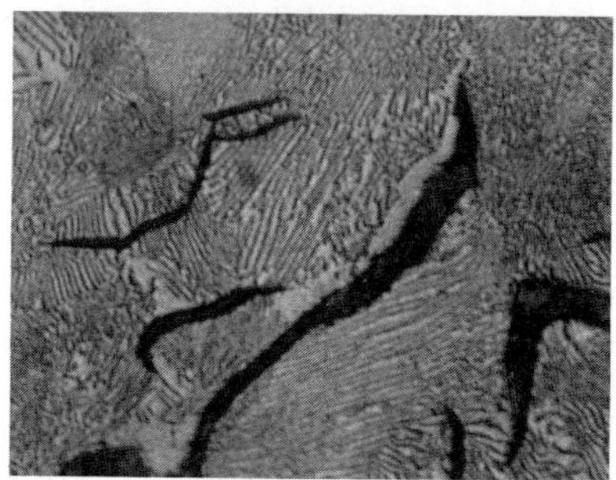

FIGURE 8–13   Gray cast iron showing flakes of graphite in a background
of pearlite. Magnification is 1000× with 2% Nital etch.

the six basic elements, but wrought iron also includes about 2.5% slag with a
carbon content of 0.02%. Steels (Fig. 8–12) and cast irons (Fig. 8–13) solidify
in somewhat the same manner, but wrought iron (Fig. 8–14) is a product of a low
carbon steel that has been poured into a smaller quantity of a much cooler siliceous
liquid slag. It has characteristics of a composite due to the slag which is identified
by the dark stringers. Because of the temperature difference violent turbulences
occur whereby iron and slag are mechanically mixed. While hot and solid the mass

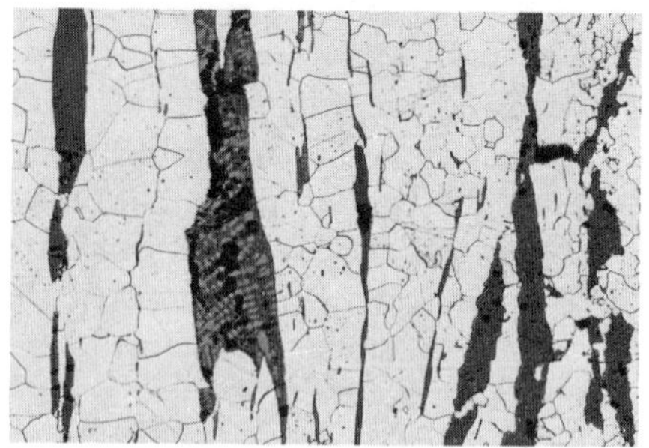

FIGURE 8–14   Photomicrograph of wrought iron, a nail which was made
around 1850. Dark stringers are slag which is stretched out across the ferrite
grains. (Nital etch at 100×.) (Courtesy of Bethlehem Steel Corporation)

is squeezed into smaller solid masses, some being a few tons in weight. Further mechanical or wrought processing results in desired shapes. Each shape has these slag stringers extending through the many grains of ferrite depending on the method of forming. Flakes of dark colored graphite are shown in the photomicrograph of gray cast iron in Figure 8–13. This structure also includes some aspects of the steel structure, but the graphite differentiates it from steel. The term structure refers to the microstructural makeup of the mass of constituents.

Creation of Physical Properties

Physical properties of each of the three ferrous alloys result from the particular chemistry of the metal. Such properties as weight, color, electrical conduction and resistivity, magnetic capability, expansion and contraction coefficients, modulus of elasticity, melting and boiling points, crystal lattice structure, and thermal conductivity are constants for a given analysis and may be assumed to be fixed in relationship to the fixed chemical properties. These physical property data are available in engineering handbooks.

EFFECTS OF CARBON

The primary difference between steel and cast or wrought iron is the carbon content. Wrought irons are inherently weak and have low carbon contents. They are used in applications in the "as produced" condition because heat treating cannot significantly increase their mechanical properties. The wrought irons make excellent transformer cores for the electrical industries and pipe where corrosion is a problem in the soil. Also, due to the fibrous nature of this metal it finds extensive use in the ornamental iron industries and in parts where ductility is required (Fig. 8–14). As for steels, there are many thousands of kinds which serve as the foundation for many engineering and architectural applications (Table 8–1). Carbon contents range from less than 0.10% to 1.7%. The large amount of carbon in cast irons makes most of them suitable for use where ductile steels are not needed. However, some cast irons which are alloyed properly and subsequently heat treated are used for special designs in engineering and effectively withstand high stresses.

Carbon steels, those having only the six basic elements, consume a large percentage of the steel output, but the alloy steels also constitute a large quantity of production. The common low strength steels, the higher strength structural steels, the stainless and heat resisting steels, and the tool steels keep industrial manufacturing processes in operation. On the other hand, the cast irons also account for a very large annual tonnage of pipe, machine bases, grates, manhole covers, machine parts, fixtures, and hundreds of other items. Comparing the ferrous metals, the steels are used for hardness and strength purposes in all kinds of industrial applications, and the common cast irons are used for parts where casting is acceptable and a sufficient strength results from the cast parts. As pointed out previously, alloyed cast irons are competing with steels. Percentage-wise, the wrought irons include only a small amount of production due to their very limited capabilities. The reason why steels are more universally used than cast and wrought irons

TABLE 8-1 CHEMICAL COMPOSITIONS OF TYPICAL STEELS*

AISI	% C	% Mn	% Ni	% Cr	% Mo	% V
1010	0.10	0.45				
1020	0.20	0.45				
1035	0.35	0.75				
1040	0.40	0.75				
1050	0.52	0.75				
1060	0.60	0.75				
1070	0.70	0.75				
1080	0.81	0.75				
1095	0.96	0.40				
1141	0.41	1.50				
1330	0.30	1.75				
3140	0.41	0.80	1.25	0.65		
4012	0.11	0.87			0.20	
4130	0.30	0.50		0.95	0.20	
4140	0.40	0.87		0.95	0.20	
4340	0.40	0.70	1.82	0.80	0.25	
4615	0.15	0.55	1.82		0.25	
4720	0.20	0.60	1.05	0.45	0.20	
4820	0.20	0.60	3.50		0.25	
5120	0.20	0.80		0.80		
5140	0.40	0.80		0.80		
5160	0.60	0.87		0.80		
52100	1.02	0.35		1.45		
6120	0.20	0.80		0.80		0.10
6150	0.50	0.80		0.95		0.15
8620	0.20	0.80	0.55	0.50	0.20	
8720	0.20		0.55	0.50	0.25	

* Average percentages
Si: 0.10–0.30%
P: 0.04%
S: 0.04%; for carbon steels, 0.05%

originates from the microstructure and the desirable atomic networks throughout
the metals.

The Relationship Between Iron and Carbon

Because the relationship between iron and carbon is complex and accounts for
the variable strengths of the ferrous metals there are many possibilities available
for engineering designs. The amount of carbon and its relationship with the iron
is of great importance in predicting hardness and strength of steel and cast iron.
All advanced societies build their technology on metals, and it is mainly the steels
which carry the enormous loads with a high degree of safety. An understanding
of the intimate relationship between carbon and iron atoms in steel enables the
designer to use a minimum amount of metal to safely hold a specified load. He can
forecast the kind of steel required, the condition of the metal in its microstructural
arrangement, and the shape and dimensions needed to accurately do the job. The
choice among wrought iron, steel, or cast iron must be made by someone. Consequently, the choice is based on facts, the facts coming from the combined chemical, physical, and mechanical properties of the metals. The chemistry of the metal
gives the metal its existence. The chemistry also fixes certain properties in the

metal such as weight and melting point, and it is the chemistry that gives the metal its lifelong mechanical property potential.

THE RELATIONSHIP OF UNIT CELLS TO CARBON

A study of the association between iron and carbon atoms points out how variable steel is in its mechanical properties such as softness or hardness. This relationship between elements and compounds gives the steels their tremendous roles in industry. Of the six basic elements in all steels, it is the iron and carbon relationship that primarily establishes softness or hardness. Iron exists in multiple allotropic forms; that is, iron has two main types of cell arrangements, as previously noted.

The unit cell, structured as a lattice, is the smallest mass of matter (atoms) that has the complete symmetry of the particular crystal; several atoms are required to form the submicroscopic size cell. The two most common forms are the body-centered cubic (Fig. 8–15) and the face-centered cubic (Fig. 8–16). Evidence of

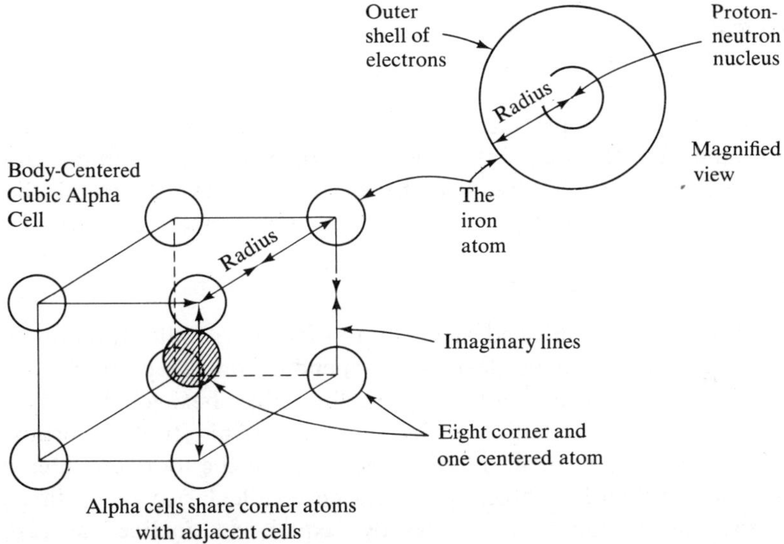

FIGURE 8–15   The alpha iron cell is attracted by a magnet and allows carbon atoms to be mixed with the iron atoms.

the existence of these lattice structures is established when a section of metal is fractured. Observation of the fractured section reveals a crystalline or granular pattern. These grains are readily observable by the eye, but the unit cells cannot be seen, not even with a microscope. The use of X-ray diffraction techniques reveals these space patterns, however. Evidence of these space patterns is also revealed by their actions in causing the allotropic changes from alpha to gamma to alpha arrangements. If an individual grain is theoretically bisected continuously to a point where identity of the metal would be lost at the next bisection, the atom would then be reached. At this time, it would be observed that the infinite number of

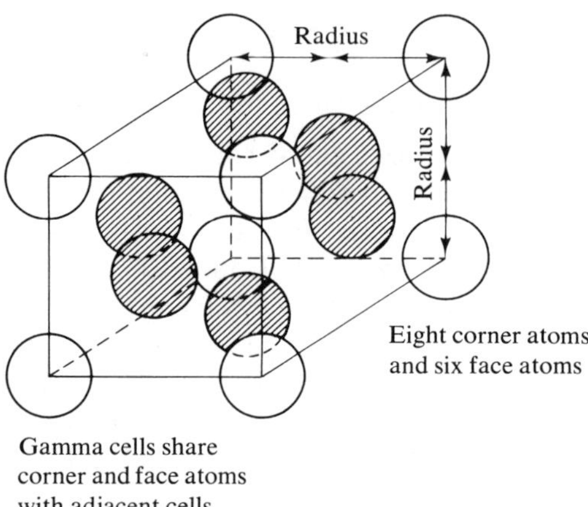

FIGURE 8–16 The gamma iron cell is nonmagnetic and readily seeks and absorbs carbon atoms into solid solution.

atoms would be fairly well organized in rows in three directions and that groups of atoms would take the shape of cubic cells. In metals these atomic groups are called unit cells in a space lattice. It is these cells that make the metal, the sum of trillions of tiny masses of energy laid out in a nearly perfect geometric design.

Origin of the Lattice Structure

Because unit cells are established at the time of metal solidification an organized expansion of each solid nuclei develops. Depending on the particular metal, the lattice will grow into one of fourteen types, the cubic, hexagonal, and tetragonal being most common. Other lattices include the monoclinic, triclinic, orthorhombic, and rhombohedral, arrangements of which account for the fourteen different lattice types. The cubic system has three equal axes perpendicular to each other and one unequal axis. The hexagonal lattice has two equal axes inclined at 120° and a third unequal axis at right angles to their plane. The tetragonal system has two equal axes perpendicular to each other and to a third unequal axis.

Metals having the cubic arrangement include aluminum, chromium, copper, gold, iron, lead, nickel, silver, tungsten, and vanadium. Hexagonal lattices are found in cadmium, zirconium, cobalt, magnesium, titanium, and zinc. Indium and tin have tetragonal lattice systems. The lattice system within a grain determines much of what the metal can do, such as deform along the atomic planes and establish new dimensions under stressed conditions. The cubic system includes by far the most metals.

As a matter of information, the length of the base of these lattices is called the lattice parameter and measures about four Angstrom units. The Angstrom unit is a measure of light waves and is equal to one-hundred-millionth of a centimeter. With reference to a method of measuring the particular space lattice, the two variables of temperature and element material must be first known.

Body-Centered Cubic Cell

According to the physical constants of the elements, iron is cubic in structure and is atomically arranged in the body-centered, or *alpha,* cell structure (Fig. 8–15) when cold. (Austenitic steel is an exception.) Iron is arranged in the face-centered, or *gamma,* cell structure when red hot (Fig. 8–16). Because the structure is organized (Fig. 8–17), cells are placed in a pattern so that the corners of the cells share

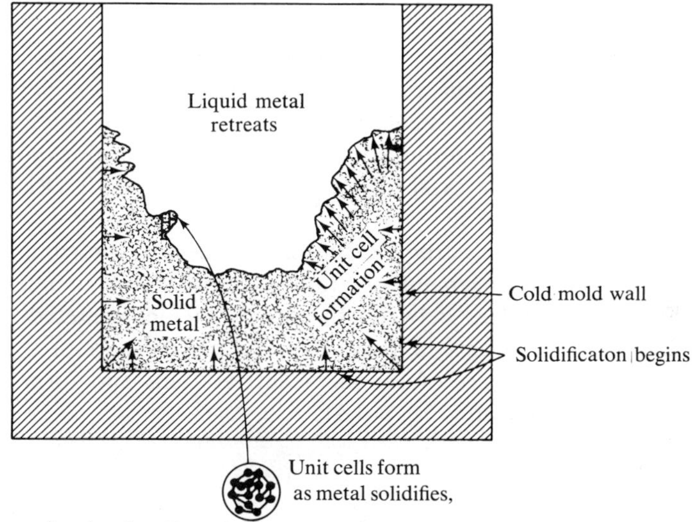

FIGURE 8–17   Solidification of liquid produces unit cells to form the ingot.

atoms with adjoining cells. Consequently, one atom at the corner of the cube is shared by seven adjoining cubes. This adjacent sharing arrangement results in rows of atoms spaced in three dimensions according to their natural radii. The spacing depends greatly on temperature. In order to complete the cell, the alpha cell has an atom located in the center of the group of eight corner atoms. Such a cell is know as body-centered cubic, and it causes the metal to become magnetic in the presence of a magnet. Actually, only two atoms sustain the alpha cell $[\frac{1}{8}(8)+1=2]$. The alpha cell is stable as temperature increases from below room temperature to a critical temperature range of 1333–1670 °F (723–910 °C), depending on the surrounding carbon content of the metal. The critical temperature is where ferrite and carbide begin to form a solution on heating.

Face-Centered Cubic Cell

At elevated temperatures atomic activity is more active. When heated within and beyond the steel's critical temperature range, the alpha cell loses its magnetic attraction and becomes face-centered gamma iron. Reorganization of the mass of iron atoms occurs; the central atoms leave their positions and new locations are established at the center of all faces of each cell in the red-hot metal. Such a shifting along the atomic planes results in an atom at each of the eight corners of the cell and one at the center of each face, or a total of four atoms per cell [⅛ (8) + ½ (6) = 4]. Corner and face atoms share within their respective radii, but a single alpha cell consists of nine atoms and a single gamma cell consists of fourteen. Of course, in nature the single cell is nonexistent; therefore, body-centered cells require the equivalent of two atoms, while the face-centered cell requires four. Other metals in different crystal lattices also have their particular sharing arrangements.

THE SOLID SOLUTION OF CARBON IN IRON

When any ferrous metal is heated, stability of the nine-atom alpha unit cell decreases because atomic vibrations among average radii distances increase. As heating continues in a mild carbon steel, for example, vibrations increase to a point where body-centered cells are no longer stable, and new positions are taken along the face regions at temperatures above approximately 1500 °F (816 °C). Such a shift in atomic plane arrangement produces a loss in magnetic conditions in the new cells and allows adjacent carbon atoms close access to the gamma iron grouping. The face-centered arrangement of atoms is more dense in packing than the body-centered, but the spaces between the atoms in face-centered cells is larger than the spaces between atoms in body-centered cells. Therefore, carbon atoms move quickly to the interstices of the face-centered cells and form an interstitial solid solution of carbon in iron (Fig. 8–18).

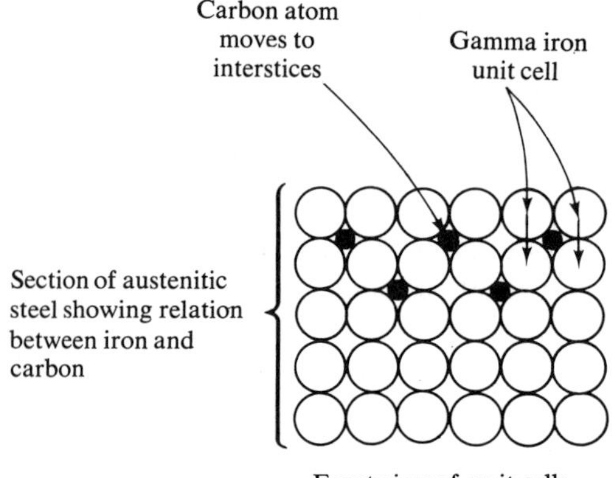

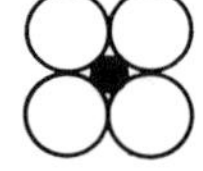

FIGURE 8–18   An interstitial solid solution of carbon in iron. The interstices between the iron atoms are large enough to allow entry of the smaller carbon atoms.

The new arrangement has characteristics of unity of the mass with regard to association between two chemicals, iron and carbon. Solid gamma iron absorbs the solid carbon just as a sponge absorbs water. Such a structure of atoms portrays characteristics of a pure metal in that microscopic examination reveals one material. In steel this material is called *austenite* (Fig. 8–19). The black and white arrange-

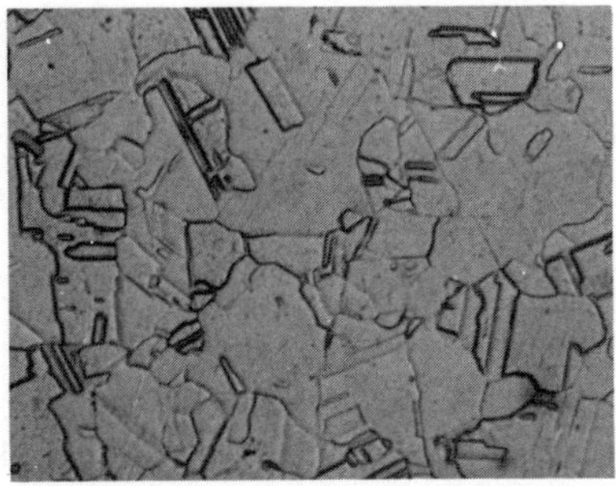

FIGURE 8–19   Austenite at $500\times$ (Viellas etch) in type 321 stainless steel. Lines are grain boundaries.

ment is a material orientation pattern and is not a visible comparison of chemicals. However, grain boundaries are present. Radii in the alpha cell, on the other hand, are too close together to allow entry of any significant number of carbon atoms into the interstices of the iron cells; therefore, on slow cooling to temperatures below approximately 1414 °F (768 °C) to 1333 °F (723 °C), carbide and iron separate, mechanically mix, and become readily observable in the microscope at room temperature (Fig. 8–20). Again, iron responds to magnetic influences (alpha) because it is merely mixed with the carbide.

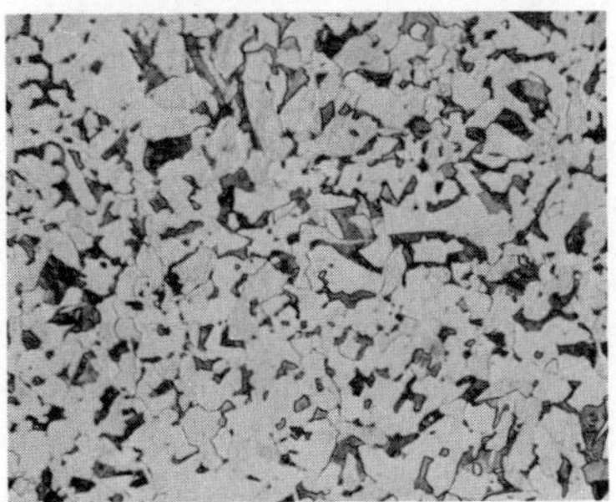

FIGURE 8–20   Microstructure of AISI 1045 steel at $100\times$ with a 2% Nital etch showing about 50% pearlite (dark) and 50% ferrite (white).

THE MECHANICAL MIXTURE OF IRON CARBIDE AND IRON

Stability of the carbon and iron type of austenite exists as long as the proper temperature environment exists, that is, above the critical temperature. As temperatures decrease slowly, atomic vibrations decrease until austenite becomes unstable. Transition to the previous alpha cell arrangement occurs, face atoms move to central density points, corner atoms again take their positions, and strong magnetic properties persist even though the metal is hot. At the same time the solid solution becomes unstable and carbon atoms are squeezed out, just as water is squeezed from a sponge. The body-centered cells cannot form until the carbon atoms are removed; therefore, carbon leaves the face-centered cells at the earliest moment, and that is at the critical point, or in the temperature range from about 1414 °F

(768 °C) to just below 1333 °F (723 °C). Faster cooling causes transformation at lower temperatures. Consequently, the closeness of iron and subsequently ferrite to carbon atoms is readjusted. The two phases, carbide and iron, are again mixed and each maintains a fixed relationship with the other; the carbon immediately combines with iron to form iron carbide, Fe_3C.

Because the solid solution, or oneness, changes on cooling to a mechanical mixture, or plural condition, iron carbide and ferrite form alternate plates and remain in this laminated condition to room temperature on slow cooling. Thus heating of nonaustenitic ferrous metals causes an allotropic change in iron from body-centered cells to face-centered cells, and cooling of nonaustenitic ferrous metals causes a change from face-centered cells back to body-centered. In austenitic steels gamma iron is retained at room temperature in a fairly permanent arrangement of face-centered cells. (This phenomenon is discussed under *Grains of Metal Contain the Constituents*.) The allotropic changes from alpha to gamma to alpha force carbide in and out of solution as atoms shift from one cell arrangement to another. However, extremely fast cooling of the face-centered cell with its captive carbon atoms may result in only a diffusionless change to a body-centered tetragonal condition known as martensite (Fig. 8–21).

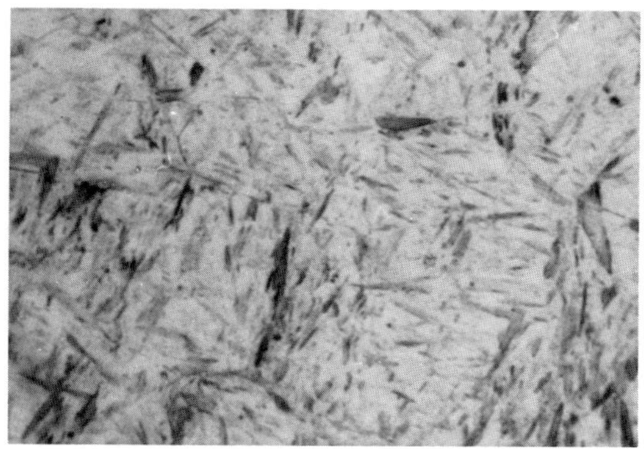

FIGURE 8–21　Martensite is light colored and its spines tend to produce triangular shapes.

FORMATION OF PEARLITE

The chemical or intermetallic compound formed from iron and carbon in ferrous metals is extremely hard and is completely unlike the soft materials from which it is made. Iron carbide, Fe_3C, remains as such during all kinds of mixtures with ferrite. The mixture consists of an extremely hard material (carbide) and a soft metal (ferrite). Once the solid solution of ferrous metal (austenite) cools to a temperature below gamma stability, the relationship between iron carbide and ferrite becomes a function of the rate of cooling from austenite. Slow cooling, such as in furnace cooling, of the metal allows carbon to precipitate from austenite near the critical point of 1333 °F (723 °C). Large pieces of iron carbide form, and at the same time soft ferrite forms between the extremely hard carbides in layer form. This platelike formation of pearlite results because the newly formed alpha cell cannot hold more than 0.008% carbon in solution since there is insufficient space between its atoms for the carbon atoms. Such a sandwich type of microstructure of ferrite and carbide is called pearlite because of the mother-of-pearl or fingerprint appearance of the mixture in three dimensions. The volume of steel at room temperature is composed of the pearlite plus any free ferrite or other material.

Strength Changes as Cooling Rates Change

Steel cooled very slowly from austenite, one having 0.45% carbon, for example, has a tensile strength of about 70,000–80,000 psi and a Rockwell hardness value of about B 84 (Fig. 8–20). This hardness-strength value relates to the microstructure called pearlite or coarse pearlite. As the cooling rate from austenite increases causing gamma iron to transform to alpha at an increased rate, the pieces of carbide have little time to grow and are smaller, more numerous, more scattered, and are closer together in the matrix of ferrite, the total quantity of carbide remaining unchanged. Such a microstructure causes the hardness of the steel to increase above that of the coarse pearlite, a Rockwell C 24, for example, with a tensile strength of approximately 117,000 psi. On cooling the carbide forms quickly as the carbon atoms are ejected from the body-centered cells, allowing the chemical change. The speed of cooling from austenite determines the size of these carbide particles; the faster the cooling rate, the smaller the particles and the harder the metal from a given amount of carbon. A very important point to remember is that the arrangement of the pieces of carbide are wholly within the numerous grains and that the size of the carbide particles has no relationship with the size of the grains which contain them. Grain size is not appreciably influenced by rates of cooling.

A SUMMARY OF THE SOLIDIFICATION PROCESS

When a high carbon steel, AISI 1065, for example, is cast into a mold, cooling progresses fairly uniformly except at the mold's walls where it is faster. Gradually, as cooling continues various regions of the liquid begin to form solid nuclei because of chemistry differences and temperature reductions in the various regions of the liquid. The temperature reduction slows down atomic activity toward a mean radius for each atom so that planes of atoms form in an organized lattice network in the newly forming solid metal. The cooling solid nuclei of hot metal contains the carbon inside its newly formed gamma iron, and millions of unit cells are quickly established as solidification continues.

Random solidification of growing regions of solid metal causes collision and interference with other growing cells in all directions. Consequently, misalignment of planes of cell orientation results at points of interference with adjacent cells. This misaligned orientation of planes of small pieces of solid particles of metal establishes crystals within the misaligned perimeters of organized cells, becomes grain boundaries for the cells, and halts cell growth. As a result, after all the liquid has solidified, the granular structure exists. The grains, or crystals, have three dimensions and form the one section of hot metal.

As cooling slowly continues to points where austenite is not stable, the allotropic change from gamma to alpha occurs, austenite giving up the iron not needed for the trailing eutectoid composition. As the temperature is further decreased the remaining solution of iron and carbon becomes a eutectoid mixture (0.80% C) of iron carbide and ferrite, called pearlite. As cooling continues, the pearlitic and ferritic microstructure is retained to room temperature (Fig. 8–22). When the carbon content is less than 0.8%, free ferrite will form with the pearlite. This phenomenon will be explained later. All of the ferrite and pearlite is contained within the millions of grains which form as solidification occurs. Again, it must be pointed out that the sizes of the carbide particles have no relationship with the sizes of the grains which contain them.

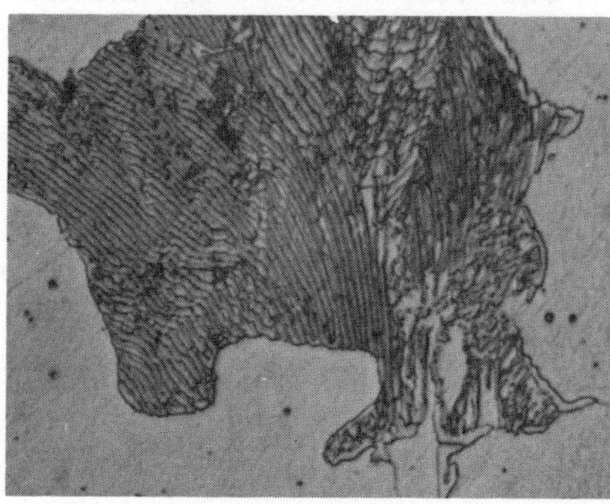

FIGURE 8–22   A slowly cooled AISI 1065 steel showing a microstructure of pearlite and ferrite.

Grains of Metal Contain the Constituents

Basically, a metal consists of individual grains of itself. In steel the chemistry is dispersed throughout the grains in a manner related to cooling rates from austenite and forming processes. The chemical dispersion may result in a mixture of the constituents because of slow cooling of austenite to form pearlite, or the solution may be retained at room temperature as austenite or may be converted to a martensitic solution because of extremely rapid cooling. Solid solutions of constituents such as austenitic steels and yellow brass have mechanical properties similar to the pure metals in that constituent identity is not observable as it is in the mechanical mixtures. Retained austenite often occurs in rapidly cooled high carbon steels, but the austenitic steels are deliberately produced by adding specific amounts of chromium and nickel to the carbon steel's analysis, forcing the retention of gamma iron at room temperature. The solid solution of martensite, however, is an alpha structural solution and is therefore unlike gamma solutions.

IRON CARBIDE AND MECHANICAL PROPERTIES

The alloys of iron and carbon form the bases for the thousands of cast and wrought ferrous metals. As has been pointed out, iron and carbon also establish the hardening and strengthening compound, iron carbide. Its quantity increases with the carbon content, and its relationship with the iron varies from a coarse mixture to an interstitial solid solution, being influenced by wrought operations and by heating and cooling processes. Consequently, a large span of strengths and hardness is often available in a specific steel because the quantity of carbide particles, their sizes and shapes, and their arrangement in the grains are variable, and these factors determine the resulting mechanical properties more than any other factor. With regard to the mechanical properties of hardness and strength, a very close relationship exists between hardness and yield strength; however, the hardness-tensile strength conversion is more commonly used as previously discussed. Mechanical properties are also greatly influenced by the methods used to finish the shapes of metal.

THREE BASIC FORMS OF METALS

Once the liquid metal (ferrous or nonferrous) has solidified, it continues to cool to room temperature. The metal may have been deliberately shaped by the freezing of liquid metal into some particular design which will be used in this condition for the remainder of its life (Fig. 8–23). Metals which are shaped in molds with the

FIGURE 8–23 The right rear corner of a large steel casting which is part of a power shear.

objective of no further shape changes are called *castings*. Much of the present production in the metals industries consists of foundry cast objects. From the metallurgical aspect, the "as cast" casting is the product of its chemistry and time from liquid to solid. Its grain structure is coarse, meaning that its grains are usually large with only equiaxed shapes. These homogeneously shaped grains tend to be spherical in shape with several flat areas in their periphery and offer the minimum in overall strength properties.

Addition of Alloying Elements

When alloys are added to the molten steel, increased strengths usually result. For example, when nickel and chromium are added to liquid steel a much tougher and stronger casting will result. Aluminum acts as a deoxidizer in steel, and vanadium tends to restrict grain growth, resulting in a finer grained steel. Frequently, alloys are added to the molten metal while it is in the ladle just prior to pouring into the sprue of the mold. To further increase the mechanical properties, castings are often heat treated. Heat treatment refines the grain structure and imparts increased strengths into the castings.

Wrought Shapes

Another form of metal shape is the wrought shape. Wrought forms are shaped from a casting (ingot) to some other shape by means of pressure, causing granular deformation (Fig. 8–24). Wrought operations begin with the hot ingot (Fig. 8–4),

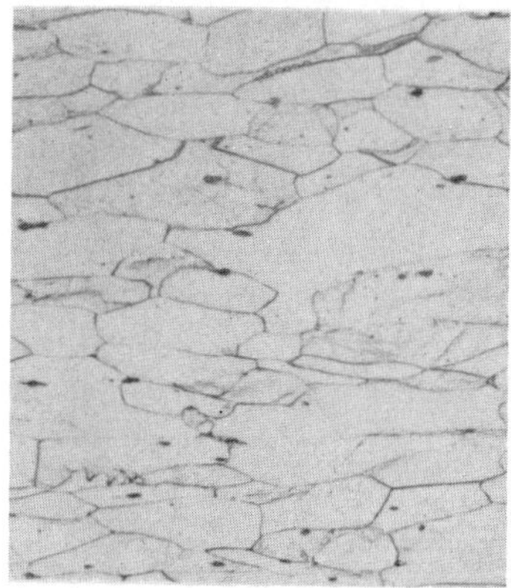

FIGURE 8–24   Cold forming permanently deforms the grain structure of metals.

which is a large casting awaiting further processing. For example, the hot ingot is rolled between powerful rolls (Fig. 8–25) to some particular shape—flat, round, square, or tubular. The wrought process of pressure forming changes the metallurgical condition of the casting by giving the metal directional properties of the grain structure and a finer grain size. Together, these two properties usually provide stronger metals than when the metal is not wrought formed. Then, following a forging operation for example, the metal can often be heat treated to much greater strengths. Forged and heat-treated parts are often superior to other processed parts (Fig. 8–26). Even though wrought shapes have their initial cast type of microstructural arrangements changed, the new grain size and shape still remains with some kind of rearrangement of carbide within the numerous grains. Wrought forming processes provide forgings, extrusions, and rolled, pressed, stamped, and drawn parts. Initially, then, when wrought processes are to follow, the "as cast" ingot is reheated and rolled or forged into smaller shapes for further wrought processing. Subsequent machining operations are equally applicable to the cast and wrought metals.

Powder Metallurgy

The third form of metal is the product of powder metallurgy (Fig. 8–27). In this fast growing process, the different metals are powdered and then pressed into various shapes, called briquettes. After shaping, the briquette is sintered by heating

FIGURE 8–25   A glowing ingot takes on a new shape as it is worked by rolls of a blooming mill. (Courtesy of Bethlehem Steel Corporation)

FIGURE 8–26   This forged and heat-treated part is ready for the finishing operation.

to a temperature below the melting point of the metal or metals and cooled. Parts can be rapidly produced by this process, but limitations make the process the smallest in production of the three basic metal production processes. As in casting and wrought forming, the powdered parts can also be heat treated for increased mechanical properties.

FIGURE 8–27 A sprocket produced from pressed and sintered steel powders.

Ultimately, a metal is formed by some process and its volume consists of the many grains arranged in the original cast condition or refined or deformed into some particular shape by pressure forming. In this respect, details of the numerous and diverse manufacturing processes are comprehensively discussed in another text, *Manufacturing Processes* (Merrill, 1975). With regard to metallic microstructure resulting from forming operations, the three forms of metal shaping have varying influences on the mechanical properties. Usually, the wrought processes produce higher strength metals from a given chemical analysis, but some cast and heat-treated products are equally as acceptable. Consequently, the discussion of metallic microstructure pertains to any form of metal shaping and heat treatment.

IRON-IRON CARBIDE DIAGRAM

The numerous iron carbon alloys can be grouped into an organized arrangement and portrayed as a diagram in which the percent carbon of the alloy is compared to a specific temperature to show near equilibrium relationships. Figure 8–28 is an arrangement of numerous carbon steels plotted with a temperature factor. Along the base or horizontal coordinate are indicated the many steels designated in percent of carbon content. As an example, AISI 1090 is shown as 0.9% carbon, and AISI 1020 is shown as 0.2% carbon. The vertical coordinate indicates the temperature of the particular alloy. A summary of the diagram shows that the phase austenite is stable in the region bounded by the letters *FGHKLEF*. Above the solidus line *FE*, liquid metal, and austenite exist; consequently, all solid steels exist at a temperature below the line *FE*. Temperatures along lines *HKLD* and *JLD* represent critical points in steels and are often thought of as points or lines of demarcation between gamma and alpha iron. Because ferrite carries 0.008% carbon in solution, steels range in carbon content from 0.008% to 1.7%. Figure 8–28 is then a diagram for the carbon steels.

Eutectoid Steel

When an 0.80% carbon steel is slowly cooled from austenite (Fig. 8–28), the solid solution of carbon in gamma iron, a new phase is established as the hot solution cools to below point *L,* or the critical temperature of 1333 °F (723 °C). Because 0.80% carbon is of eutectoid composition a mixture of plates of carbide

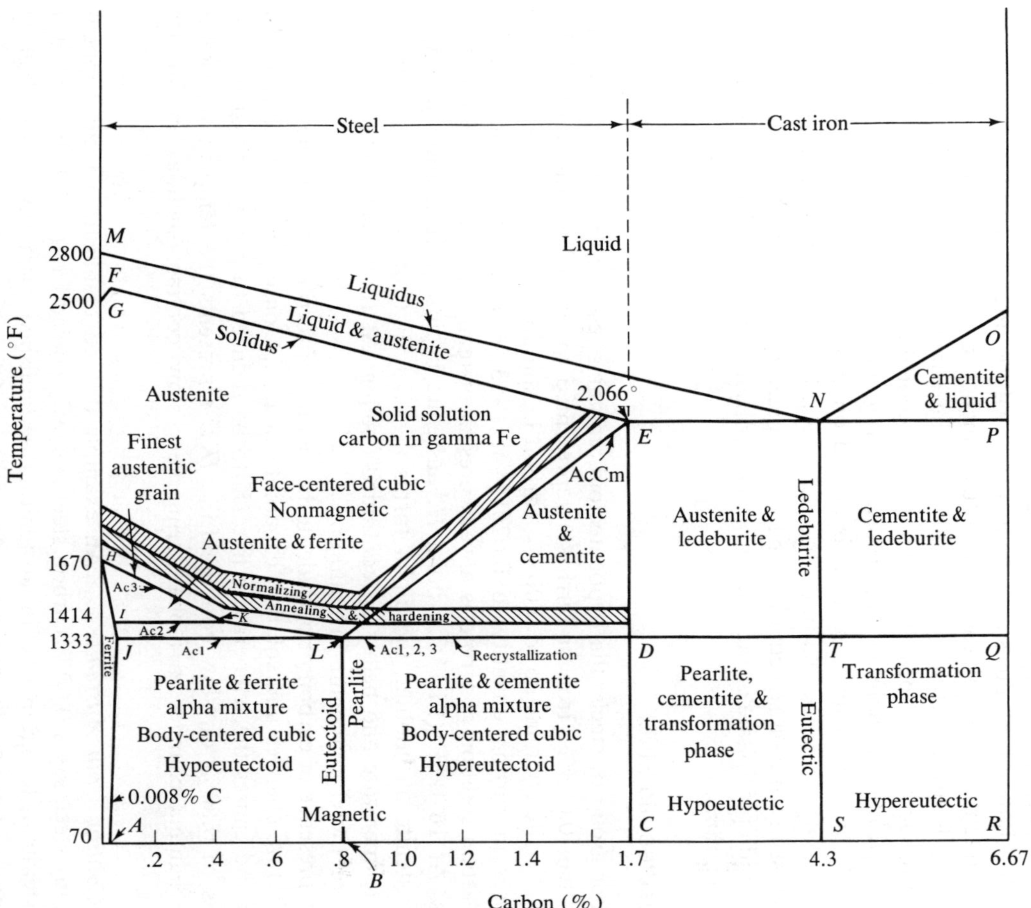

FIGURE 8–28 Iron-iron carbide diagram for steel and cast iron.

and plates of ferrite forms from the cooling austenite within each grain from boundary to boundary in three dimensions. The sizes of the carbide plates or particles vary, but are always large as compared to particles produced by faster cooling rates from austenite. No further change occurs as temperature reaches room temperature. Observation of this steel under the microscope at approximately 500× shows a pearly luster or fingerprint arrangement of carbide and ferrite while further observation shows that the entire granular structure is pearlitic. Mechanical tests of this coarse pearlite have revealed tensile strengths of approximately 85,000 to 100,000 psi; therefore, this slowly cooled steel will require from 85,000 to 100,000 pounds in tension to break each of its square inches in cross-sectional area.

Point L is the eutectoid while point B is the eutectoid composition. Eutectoid steels produce 100% pearlite when slowly cooled to room temperature from austenite. It must be pointed out, however, that gamma iron in austenite transforms to alpha iron in pearlite as point L is passed on slow cooling. The eutectoid then is that chemical composition in a series of alloys that causes the temperature to be the lowest in which a structural change takes place in the solid. At point L ferrite and carbide precipitate at the same time and consume the grain interior with plate-like formations.

Hypoeutectoid Steel

Another group of steels, the hypoeutectoid, contains less than 0.80% carbon and more than 0.008%. Iron or ferrite is represented to the left of the vertical line $AJIH$. Ferrite contains up to 0.008% carbon in solution at room temperature but increases in carbon quantity to 0.025% at 1333 °F (723 °C). Therefore, the iron-carbon solution of ferrite is always present at temperatures below the critical points. At 1670 °F (910 °C), point H, austenite begins the transition to ferrite and austenite on cooling; the alpha forms at 1414 °F (768 °C) and remains to room temperature and below on continued cooling. At room temperature a trace of carbon therefore remains in solution.

The presence of carbon or carbide in the vicinity of ferrite modifies the allotropic temperature changes and solution capability of iron. Consequently, the line $HIJA$ varies and causes the temperature line IK to be reduced to point L as carbon increases to 0.80% on slow cooling. Line IKL represents the temperature points where gamma changes to alpha during very slow cooling, such as in furnace cooling.

Slow Cooling from Austenite   An example is used to illustrate what happens when any steel, say a 0.60% carbon steel is slowly cooled from 1600 °F (871 °C) in austenite. The grain size in austenite is established primarily at the highest temperature reached for any steel. At 1600 ° F (871 °C) the grain size is then established and the metal is nonmagnetic, being a solid solution of 0.60% carbon in gamma iron, and is a stable phase at this temperature. Carbon atoms rest within the interstices of the iron atoms and the solution has characteristics of one material, even when viewed with the microscope, the examination assuming this quantity of austenite is deliberately trapped at room temperature. But as cooling continues, the solution persists until line HKL is reached; that temperature causes proeutectoid ferrite to move from solution into the proximity of the original austenitic grain boundaries. At the same time, the magnetic condition occurs, signifying the presence of the newly formed alpha cell space lattice in the ferrite, the alpha cells having ejected most of the carbon. Also at the same time, line *HKL,* the remaining austenite increases in carbon proportional to the loss of iron. As temperatures continue to decrease more ferrite precipitates from the cooling austenite until the temperature line *JL* is reached. At this point all excess ferrite has been expelled from austenite and the eutectoid composition of austenite is reached and transforms to pearlite among ferrite grains, alternate plates of carbide and ferrite forming concurrently to make the pearlite. No further change occurs to room temperature.

Pearlite and Ferrite   Nearly three-fourths of the resulting microstructure is coarse pearlite grains mixed among the ferrite grains. In eutectoid steels, however, the resulting structure would be all pearlite and no free ferrite. Accordingly, all hypoeutectoid steels produce ferrite grains and pearlite grains on slow cooling from austenite, the amount of pearlite being related to the percent of carbon in the steel. As the carbon content is reduced, such as in AISI 1020 (Fig. 8–29), the resulting

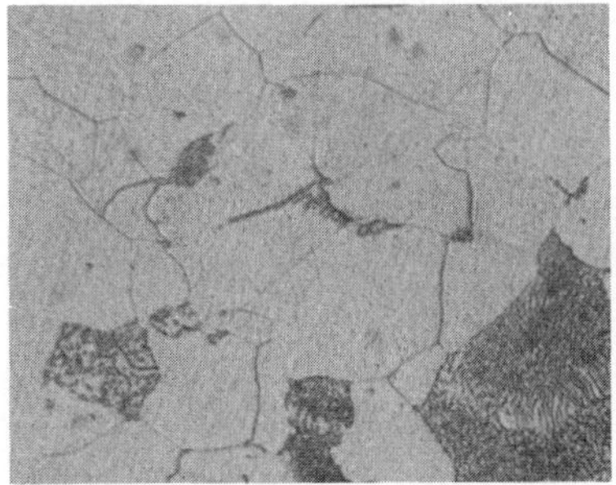

FIGURE 8–29   An annealed AISI 1020 showing about 25% pearlite, the remainder being ferrite in this microstructure.

pearlite becomes negligible while the ferrite quantity consumes most of the microstructure throughout the steel. As the carbon content increases in steels as they are slowly cooled from austenite, the amount of pearlite also increases up to the 0.80% carbon type with 100% pearlite, as in a 1080 steel.

Calculations of Phase Contents   The formation of pearlite from austenite results from extremely slow cooling from austenite. The stable phase of austenite exists above the upper critical points represented by line *HKLD*. The amount of pearlite is dependent on the existing carbon content in the steel. Pearlite increases and ferrite decreases in quantity as the carbon content increases up to 0.80%, at which point 100% pearlite results with no free ferrite. Therefore, if 0.80% carbon produces 100% pearlite, then a fraction can be made to establish a fairly accurate percentage of ferrite and pearlite. Let 8 become a constant in the fraction's denominator, □/8, and let the particular hypoeutectoid steel's carbon content be the numerator, 2/□, for example, denoting an AISI 1020 carbon steel. If $\frac{8}{8}$ (1080) produces 100% pearlite and no free ferrite, then $\frac{2}{8}$ produces 25% pearlite and 75% ferrite. An annealed AISI 1040 steel produces approximately 50% pearlite and 50% ferrite. According to this calculation, 1080 equals $\frac{8}{8}$, or 100%, pearlite and 1040 equals $\frac{4}{8}$, or 50%, pearlite. Then, 100% minus 50% pearlite equals 50% ferrite for the 1040 steel, keeping in mind that the other constituents, silicon, manganese, phosphorus, and sulfur, are included in the ferrite or as compounds.

Hypereutectoid Steels

Hypereutectoid steels can also be analyzed to determine the amount of pearlite and cementite by considering that 0.80% carbon produces 100% pearlite and that cementite, Fe_3C, consists of 6.67% carbon. With reference to Figure 8–28, all austenite exists above the AcCm temperature line, or line *LE,* on heating cementite. On slowly cooling hypereutectoid steels to below the AcCm line, the two phases, austenite and cementite, form below the line *LE.* The amount of cementite forming depends on the carbon content; the first 0.80% carbon is used to form the eutectoid composition in austenite which is to subsequently become pearlite. As cooling continues the excess cementite or iron carbide precipitates along the temperature line *LD* and forms along the original austenitic grain boundaries which existed above the line *LE.* Just below the lower critical point of 1333 °F (723 °C), line *LD,* all of the austenite changes to pearlite due to the transition of gamma iron in austenite to alpha iron which cannot hold the carbon in solution.

Pearlite and Cementite Formation   Consequently, a plate-like structure of carbide and ferrite, called pearlite, fills all of the grains in slowly cooled steels, but many of these grains also contain quantities of the excess carbide along their boundaries. The reason that cementite is called excess carbide is because only 0.80% is required to form the eutectoid and pearlite. Therefore, any steel containing more than 0.80% carbon will include cementite or free iron carbide in some arrangement.

The amount of austenite and cementite in the area *LEDL* can be calculated, and these percentages in the solid material subsequently become pearlite and cementite at room temperature, the austenite transforming to pearlite. The following formulas will then show how much of each phase is present in the slowly cooled steel from austenite and existing in the area *LEDL.* For example, a 1% carbon tool steel at 1400 °F (760 °C) consists of:

$$\% \text{ cementite} = \frac{1.0 \ - 0.8}{6.67 - 0.8} \times 100 = 3.4\%$$

$$\% \text{ austenite} = \frac{6.67 - 1.0}{6.67 - 0.8} \times 100 = 96.6\%$$

Phases are Variable   Because maximum volume is 100%, it is obvious that an increase in one phase or constituent must mean a decrease in the other. Therefore, all slowly cooled hypoeutectoid steels increase in pearlite as carbon content increases and decrease in ferrite up to 100% pearlite at the eutectoid percentage of 0.80. As carbon content increases beyond 0.80% to form the many hypereutectoid steels, the excess phase is cementite, rather than ferrite. Consequently, as free cementite increases in hypereutectoid steels, pearlite must decrease. Verification of these calculations is observable in the microscope at a magnification of not more than 500×.

Coarse Pearlite

Coarse pearlite, when the laminations are clearly observable at 500×, has a tensile strength of approximately 100,000 psi with a Rockwell hardness value of B 94 (Fig. 8–30). Normally, this structure is called pearlite. The hardness value of B 94

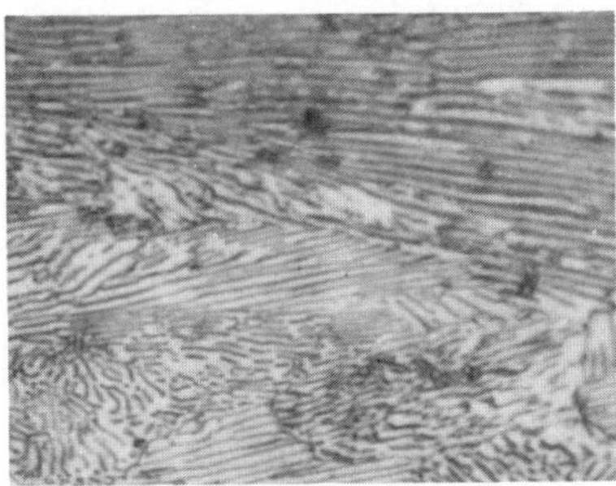

FIGURE 8–30 Pearlite at 500×, sometimes called coarse pearlite to differentiate it from the fine pearlites.

is established as a result of the closeness of carbide plates forming from slowly cooled austenite, such as in a furnace-cooled procedure by which the furnace is turned off while the temperature is in the bright red heat range and the metal is removed the next day. Actually, according to tests the tensile strength range is from 85,000 to 100,000 psi. As slowly cooled austenite transforms to pearlite plus cementite, the solid particles of hard carbide form alternate layers with ferrite. The sandwich, or colony, of pearlite, $Fe + Fe_3C$, is surrounded by other colonies of pearlite, the total amount of pearlite being proportional to the amount of carbon present in the steel. In other words, as the reducing temperature in austenite reaches a point just below line *LD,* the excess phase cementite is slowly expelled from the eutectoid composition (0.80% C) and forms along the already formed grain boundaries which were established in austenite. The transformation is then complete; that is, the gamma to alpha cell arrangement persists because of the newly formed body-centered cubic lattice structure. No further change occurs on cooling to room temperature, either in the size or shape of the grain or in the arrangement of carbide particles inside the many grains. These layerlike formations of hard and brittle pieces of carbide are sandwiched between soft and malleable alpha iron.

Alpha Cells Mix Only with Carbide

Several changes occur on cooling a particular hypoeutectoid steel from austenite. Using an AISI 1020 steel, for example, which is thoroughly soaked at 1650 °F (899 °C), the temperature is reduced by turning off the furnace. At a temperature which corresponds to line *HKL* in Figure 8–28, proeutectoid ferrite precipitates from the cooling austenite in a quantity which is surplus to the needs of eutectoid austenite and settles at the many already formed grain boundaries. The grains take their size and shape mainly from the highest temperature reached and time in austenite. It is at the *HKL* temperature that the solid solution begins its change to the mechanical mixture. Continued cooling brings a magnetic capability to the newly formed alpha cells at about 1414 °F (768 °C); the face-centered cubic iron has allotropically changed to the body-centered cubic lattice structure. As temperature continues to decrease, what was started along the temperature line *HKL* is completed just a few degrees below the temperature line *JL*. This means that the gamma iron solution with carbon has changed to the alpha iron mixture with iron carbide in different layerlike arrangements. The arrangement results from the natural inability of alpha iron to hold any appreciable amount of carbon in solution. At the time of the gamma-alpha transformation an expansion of the atomic lattice occurs and reflects itself immediately into an expansion of the hot metal. The effects of the alpha expansion during the allotropic change remain with the steel even though shrinkage continues to a stabilized dimension at room temperature.

CALCULATING STRENGTH OF MATERIAL

A microscopic examination of the prepared specimen of AISI 1020 reveals a microstructure of a granular network of fairly small, equiaxed ferrite grains mingled among small pearlite grains or colonies (Fig. 8–29). Approximately 25% of the metal is in the pearlitic condition and 75% is in the ferritic condition. If the tensile strength of pearlite in this structural condition, distinctly observable at approximately 500×, is equal to 100,000 psi and if the surrounding ferrite has a tensile strength of about 41,000 psi, then the section of AISI 1020 steel is equal to (25% × 100,000 psi) + (.75% × 41,000 psi), or 55,750 psi in tensile strength. The fractured tensile specimen required a load of 57,202 psi.

Tensile Strength Is Predictable

Another example pertaining to the use and importance of the iron-iron carbide diagram is the prediction of the tensile strength of an annealed high-strength structural steel rod made of AISI 1070. According to the diagram an annealed specimen should contain seven-eighths pearlite and one-eighth ferrite. According to the combined strengths of the two microstructures, pearlite and ferrite, and as pointed out in the microscope, the steel rod should have a tensile strength of approximately 92,625 psi. The calculation is as follows:

$$(\tfrac{7}{8} \times 100{,}000 \text{ psi}) + (\tfrac{1}{8} \times 41{,}000 \text{ psi}) = 92{,}625 \text{ psi}$$

The tensile specimen broke after indicating a tensile stress of 90,105 psi. The calculated and actual strengths are rarely identical because of the numerous variables in the metal, in the testing machine, and in the procedure. Such small differences are acceptable as safe, however, because the gain or loss of actual strength is absorbed in the design's safety factor.

RETAINED AUSTENITE IS NOT INFINITE

The failure of gamma iron to transform to alpha at room temperature occurs in the many austenitic steels. It has been assumed that the failure of an allotropic change to occur as the metal cools assures an infinite life for austenite. But there are circumstances whereby carbide precipitation in some austenitic steels may en-

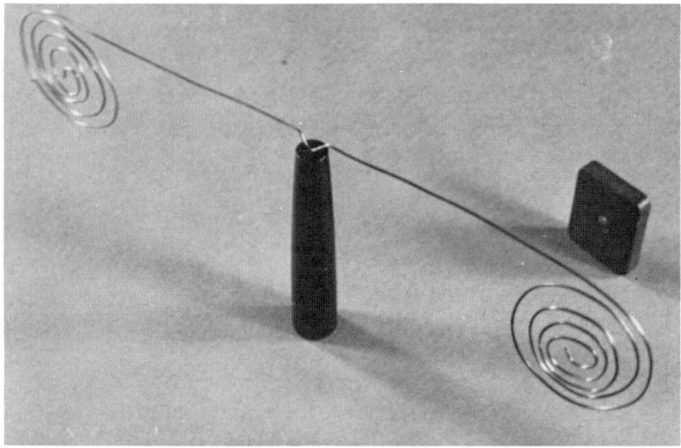

FIGURE 8–31   One coil of wire is attracted to the magnet, but the other coil has very little attraction. The metal is austenitic steel.

courage the formation of alpha cells and allow a dual condition to exist in the metal. An experiment (Fig. 8–31) performed many years ago shows that a slow but continuous movement of atoms from face-centered cells to body-centered has been in existence and continues to occur in only one part of a wire. Evidence of this phenomenon is established by the continually growing magnetic attraction for one end of the wire while the other end remains stable and nonmagnetic. In 1942 the configuration shown in Figure 8–31 was formed from an austenitic steel wire. One end of the wire was furnace cooled from 1900 °F while the other end was not heated. Immediately, the wire was polished and coiled into loops four inches apart, and a balance point was established so that the coils would rotate. During the following year a feeble magnetic attraction for loop *B* appeared while loop *A* remained unaffected in the presence of the magnet. Four years later a significant increase in magnetic attraction occurred in loop *B*. Measurements taken in subsequent years show that loop *B* is more quickly attracted to the same magnet than at the beginning of the experiment. Today the ratio of magnetic attraction between loop *A* and loop *B* has increased to such a magnitude that the wire begins to quickly turn as the magnet approaches loop *B*, but very little attraction is noticed at loop *A*.

Questions

1. Differentiate between ferrous and nonferrous metals.
2. Explain the purpose of the six basic elements in steel and indicate which one is the most critical.
3. How do formation defects occur in metals?
4. What is a discontinuity?
5. What is a metallic microstructure?

6. How does a dendrite form?
7. Describe the three basic types of metals.
8. Differentiate between physical and mechanical properties.
9. What is iron carbide? Describe its relationship with iron.
10. Describe the differences among pearlite, austenite, martensite, cementite, and ferrite.
11. Differntiate between the alpha and gamma unit cells of iron.
12. What is the origin of the granular structure?
13. What is meant by the allotropic forms of iron?
14. Describe the differences between a solid solution and a mechanical mixture.
15. What is the main purpose of the iron-iron carbide diagram?
16. Differentiate among eutectoid, hypoeutectoid, and hypereutectoid steels.
17. How is grain size formed in steels?
18. What determines the hardness and strength of steels from a given chemical analysis?
19. Describe the three basic forms of metals.
20. Describe the relationship between hardness and tensile strength.

Microstructural Effects in Ferrous Metals

In ferrous metals, as in other metals, grains form as the molten liquid solidifies (Fig. 9–1). Once the metal is completely solid and all grains have formed, they

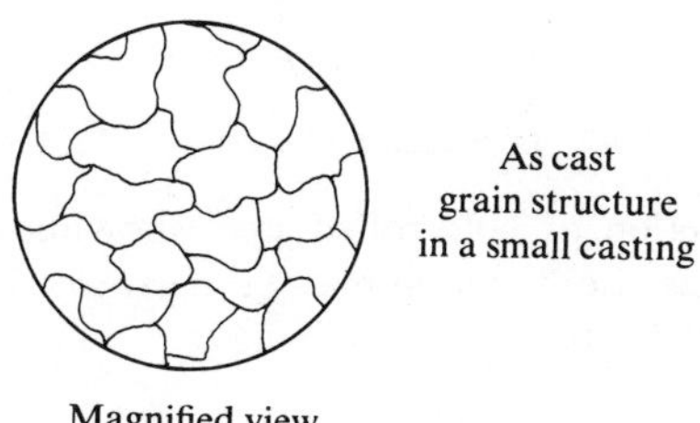

FIGURE 9–1 Solid grains form from the liquid.

remain approximately in this condition as the metal cools to room temperature and throughout the remainder of the life of the metal, unless reheated. If the metal is broken, it will always reveal its granular structure (Fig. 9–2) since metals do not fail because of crystallization.

RECRYSTALLIZATION AND GRAIN SIZE

New grains are formed through recrystallization in non-cold-worked steel only at the critical points on heating (Fig. 8–28), or they may be reshaped by plastic deformation beyond the metal's yield point (Fig. 9–3) on loading. In the latter case, cold-worked grains recrystallize at temperatures lower than the normal critical temperature due to the residual stress build-up in the deformed grains. Steels which are formed at temperatures below the critical temperature of the steel are said to be cold formed, and those formed above this temperature are hot formed. Forming

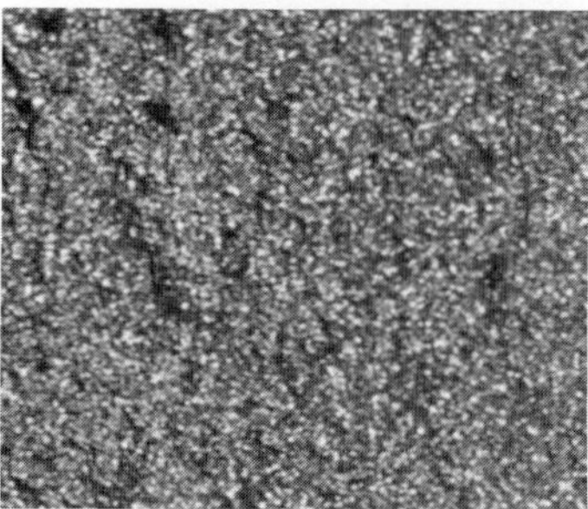

FIGURE 9–2   The natural appearance of a fractured piece of steel is granular. These millions of grains are bound together by the powerful forces of
their makeup, atoms, and when fractured will always reveal this granular
structure.

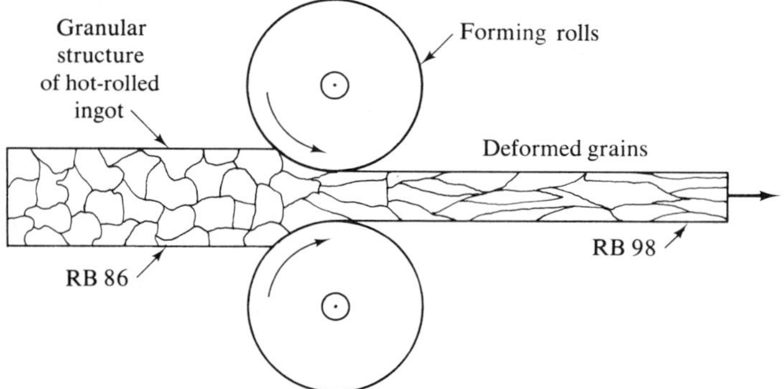

FIGURE 9–3   A section of cold-rolled steel showing directional properties
in the deformed grains. Cold rolling may be performed from room temperature upwards to about 1250 °F (677 °C).

includes rolling, bending, or other pressure forming operations. Only cold working
increases hardness in the metal. After recrystallization occurs in the metal on heating, it is the rate of cooling to room temperature that causes the final arrangement
of compounds and other phases inside the many grains. But it is the grain size in
austenite that predominantly determines the final grain size at room temperature.
On heating an AISI 1080 steel, for example, the grain size remains fairly constant
until the temperature of 1333 °F (Fig. 8–28), line *JLD,* is reached. At this temperature the grain boundary structure begins to collapse and existing grains are
recrystallized to their smallest size (Fig. 9–4). The smallest austenitic grain size
for all steels exists along line *HKLD* (Fig. 8–28), the 0.80% C steel, for example,
recrystallizing at point *L.* Then, when the steel is cooled by any method the approximate same size grain will result at room temperature. The rate of cooling
consequently has no appreciable effect on grain size (Fig. 9–5).

Grains Expand Beyond the Critical Point on Heating

After recrystallization at the critical point and when temperature is increased, the
grains expand to sizes compatible with higher temperatures by enveloping neighbor

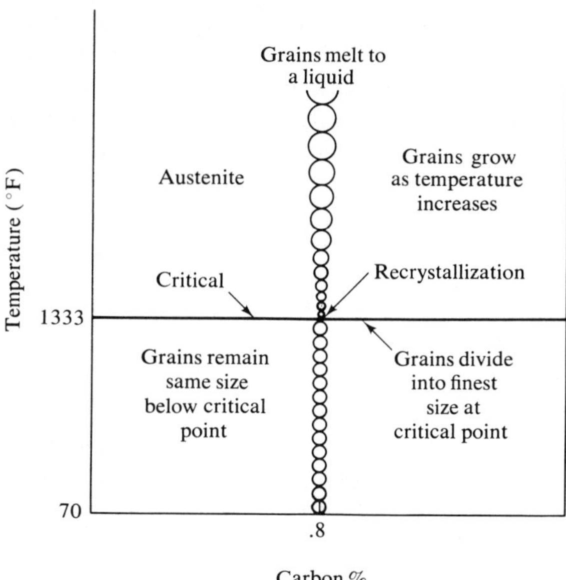

FIGURE 9–4   Recrystallization in an AISI 1080 steel.

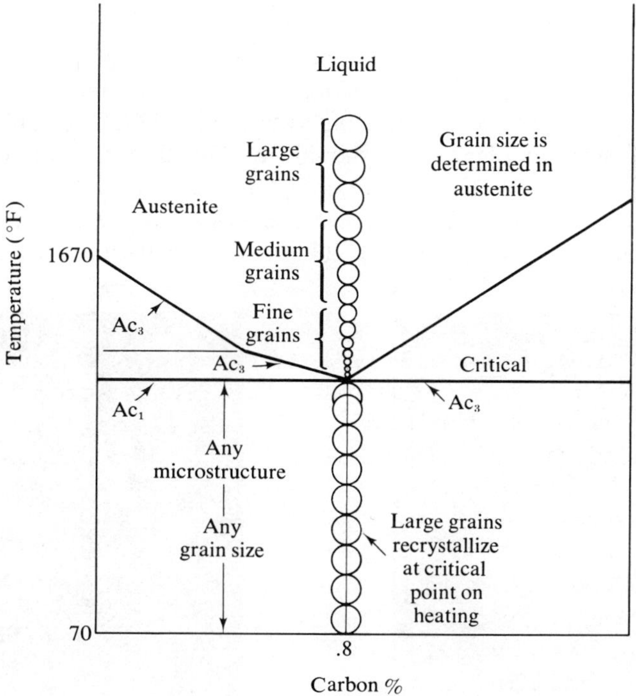

FIGURE 9–5   Grain size in steel is determined in austenite. The cooling rate has no appreciable effect on size at room temperature.

grains. From any temperature in austenite a certain size of grain exists; the boundaries of the grains expand to equilibrium with temperature and time. This austenitic grain size then becomes the room temperature grain size with no significant change, regardless of how the metal is cooled. Again, on reheating, as the metal reaches the critical temperature, its entire grain structure collapses and new grains are formed, the previous grains being destroyed through recrystallization. With regard to critical points, the lower critical temperature line in Figure 8–28 is *JLD* and the upper critical temperature is line *HKLD*. As temperature increases beyond the critical temperature for the specific steel, the grains grow in size and continue to expand until the solidus temperature zone is reached where they begin to again collapse and form a liquid. Accordingly, metals should be heated no higher than necessary to prevent large grains. In summary, grain size in steels is determined by the maximum temperature reached in austenite. Grain shape, however, can be altered by some kind of pressure processing, such as forging with a hammer. The hot and plastic metal easily deforms permanently by flattening and results in elongated grains as compared to equiaxed grains of the casting. Continued heating in austenite, however, can remove hammering effects and also produce equiaxed shaped grains.

Grain Size Varies

Grains are measurable for size in both ferrous and nonferrous metals; these crystals vary from very fine to very coarse. Because grain size greatly influences some mechanical properties, care must be exercised during heat treatment. Basically the smallest grains occur at the critical points of the metal, while the largest grains develop just prior to melting. According to Figure 8–28, the finest austenitic grain exists just above the temperature line *HKLD*. In order to compare grain sizes, the sizes in Figure 9–6 should be studied. Structures existing within the area *HKLJIH* (Fig. 8–28) are not completely austenitic, and their grain sizes are variable due to the temperature effects between lines *JL* and *HKL*. A slowly cooled eutectoid steel from 1400 °F (760 °C) is observed as a fine-grained

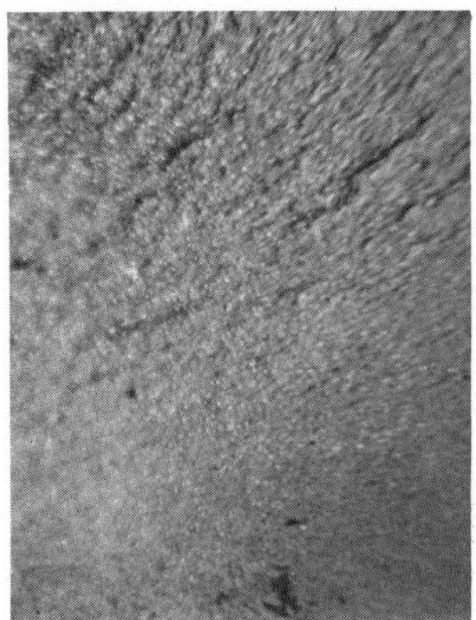 

FIGURE 9–6   Fine (left) and coarse (right) grains of metal observed as fractures.

pearlitic steel because the austenitic temperature is just above the upper critical temperature of 1333 °F (723 °C). Its hardness on slow cooling to room temperature is RB 92. The entire grain structure is filled with pearlite (Fig. 9–7).

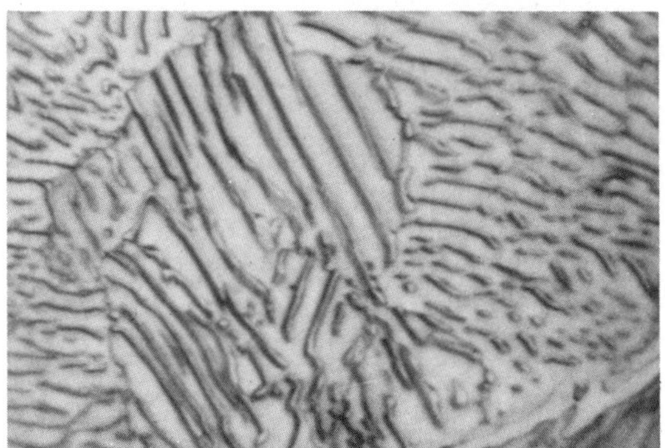

FIGURE 9–7 This eutectoid steel was slowly cooled from austenite, resulting in 100% pearlite.

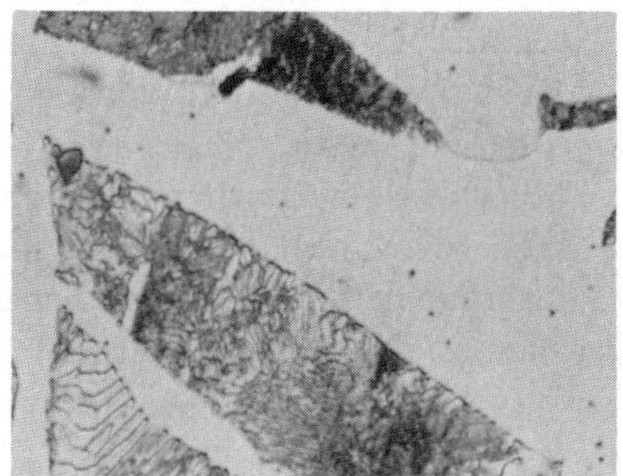

FIGURE 9–8 Microstructure of medium carbon steel cooled slowly from austenite.

A 0.40% carbon steel slowly cooled from 1500 °F (816 °C), just above the upper critical point, is observed as containing about half the grains as fine pearlite and half as fine ferrite (Fig. 9–8). Its hardness is RB 77. On the other hand, the 0.40% carbon steel slowly cooled from 1900 °F (1038 °C) is coarse grained, but it is also about 50% pearlite and 50% ferrite and has a hardness of RB 76. There is no basic difference in hardness even though different austenitic temperatures were used. A 1.5% carbon steel is slowly cooled from 2100 °F (1149 °C) and shows the very large grain structure consisting of pearlite and cementite at the grain boundaries. In this example, the steel was soaked beyond the AcCm temperature line. Its hardness is RB 96 due to the presence of excess carbide, but the steel is still soft. When this same steel is slowly cooled from 1400 °F (760 °C), the grain pattern is fine, but is also pearlitic with a network of cementite at the grain boundaries (Fig. 9–9). Its hardness is RB 95. There is usually no requirement to heat steels above the AcCm line, unless normalizing is necessary to disperse the cementite throughout the grain structure.

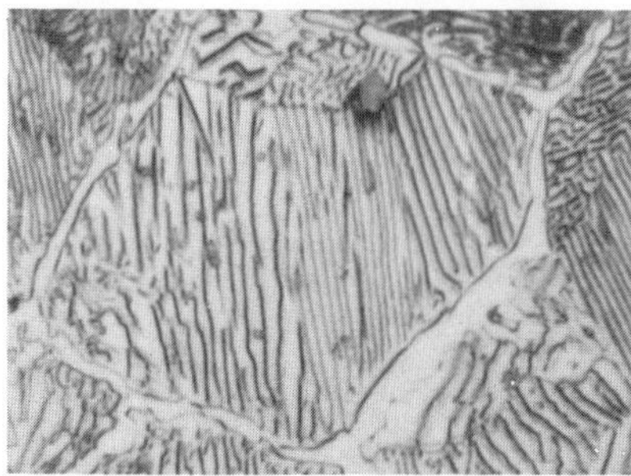

FIGURE 9–9   Annealed 1.5% carbon steel showing a network of cementite surrounding the pearlite grains.

ANISOTROPY

As has been pointed out, grains have size and shape, yet within a single grain certain crystallographic planes cause directionality of mechanical and physical properties. When the properties are measured along equivalent crystallographic planes in the unit cell equal property values exist. A crystal or grain is one independent mass of material because its directional orientation ceases at its boundaries. Adjoining grains are usually oriented in random directions and therefore are not aligned with the crystallographic planes of their neighbors. *Anisotropy* refers to the directional aspects of some particular property, such as mechanical. The property is different when measured in different directions. As an example, the mechanical property of a unit cell which is measured across the face is different from the property which is measured across the diagonal (Fig. 9–10).

EFFECTS OF HEATING AND COOLING STEEL

Heating and cooling of steels have varying effects on the microstructure and, therefore, on the resulting mechanical properties. (The heating and cooling of the wrought irons have no significant effect, but some cast irons are heat treatable.) The heat treatment of steels involves the heating and cooling processes for the purpose of imparting certain mechanical properties into the metals. Some steels are heat treatable for increased strengths while some are not, the chemistry being the determining factor. In order to heat treat a steel for desirable mechanical property characteristics the steel must contain the needed chemicals, mainly enough carbon. Carbon is a hardening element as are chromium and other alloys. For increased benefits to be obtained by heat treating, the metal's potential to be hardened must be known (Fig. 9–11).

THE PURPOSE OF THE IRON-IRON CARBIDE DIAGRAM

A main purpose of the iron-iron carbide diagram for steels (Fig. 8–28) is to determine the metallurgical characteristics of a steel by comparing its carbon

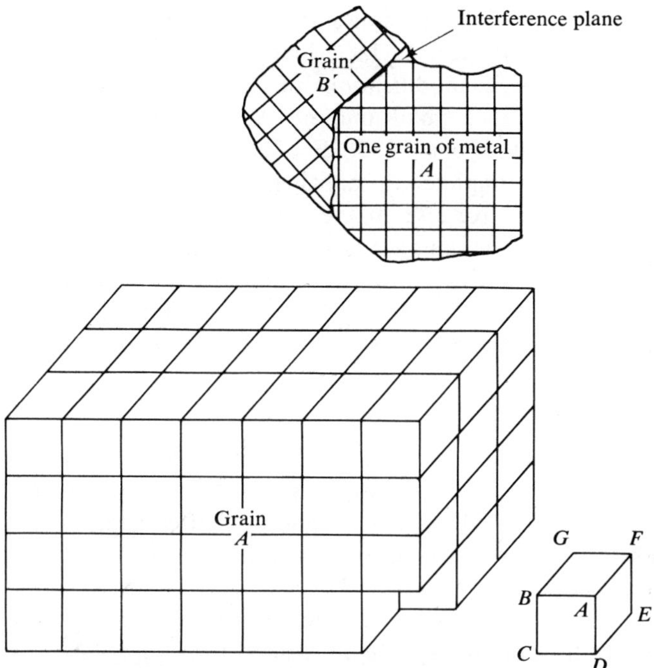

One grain of metal consists of unit cells.
Properties parallel to *BC*, *BA*, and *BG* are equal.

FIGURE 9–10 Anisotropy in a unit cell, illustrating directionality and property values.

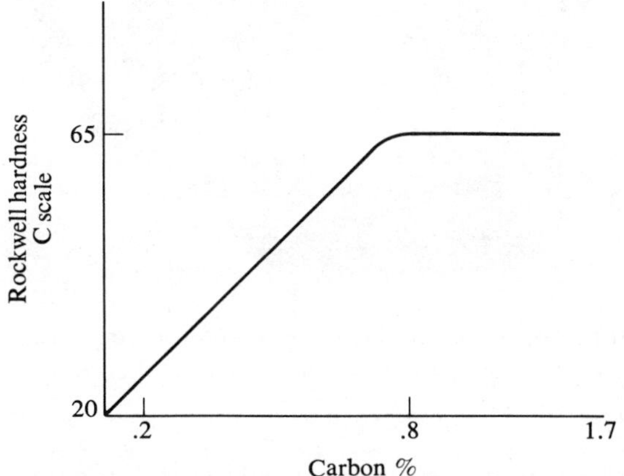

FIGURE 9–11 A graph representing the carbon content and hardenability in carbon steels up to 1 inch in diameter.

content with a potential microstructure. It is the microstructure, grain size and shape, and structural condition inside the grains that determine the resulting mechanical properties of the metals. The diagram also shows the several heat treating temperatures which are available for obtaining these properties. The horizontal coordinate of the diagram shows the numerous carbon-type steels which are available for study and use. In fact, many steels can be superimposed onto the basic diagram with slight modifications of the critical points, the alloys raising or lowering these points by a small amount. As an example, an AISI 1080 steel must be heated to slightly above 1333 °F (723 °C) to be softened or hardened and slightly higher for complete normalizing. The iron-iron carbide diagram is then a multipurpose chart for use in the metallurgical treatments of steels. The temperature line *JLD* on this diagram is known as the Ac_1, or lower critical point, for the particular steel. The line *IKLD* is the Ac_2, the zone in which the allotropic change occurs. The line *HKLD* is the upper critical point, or Ac_3. The AcCm line is basically the solution temperature zone for excess cementite to form all austenite.

Normalizing Steel

An examination of Figure 8–28 reveals a normalizing range, a hardening range, and an annealing range for all carbon-type steels. Critical points and heat-treating ranges for the alloys are usually slightly higher, but this diagram is used. The purpose of normalizing is to remove stress which may be imparted into the metal by fabrication or heat-treating processes. In this respect, stress in steels should be removed prior to further heat treatment to help prevent warping or cracking of the part. An illustration of stress is demonstrated in Figure 9–12. With the use of

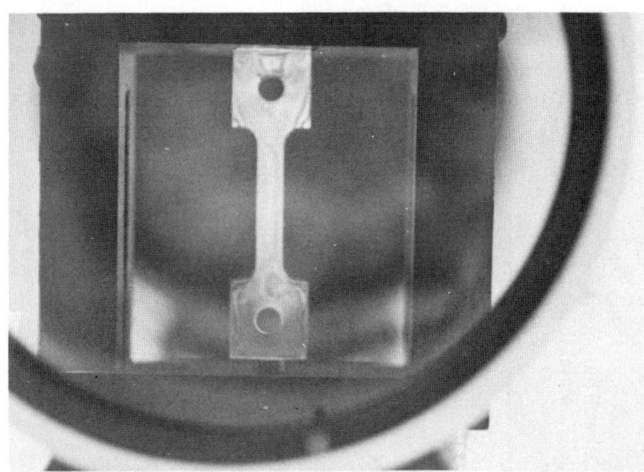

FIGURE 9–12   Stress in a plastic part is observed with polarized light. Stress patterns are shown as dark lines because the column is in compression load.

polarized light stress can be seen in the plastic part, but cannot be seen in opaque objects such as metals. Strain gauges, however, record stress levels in loaded metals.

When normalizing and stress relieving an AISI 1070 steel, preheat at 800 °F (427 °C) and then soak the metal at 1550 °F (843 °C) until the temperature reaches the core and then cool it in still air. Notice that 1550 °F (843 °C) fits into the normalizing temperature range. When cold, the steel part has a microstructure (Fig. 9–13) that shows small pieces of iron carbide thoroughly scattered throughout the metal, the effects of fairly slow cooling from austenite. The result-

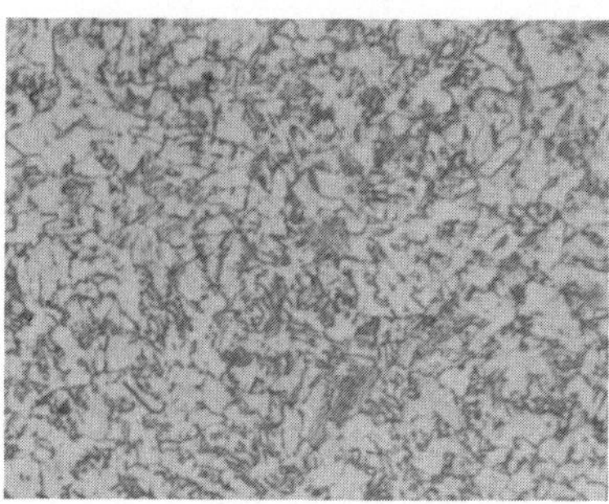

FIGURE 9–13   Fine pearlite resulting from normalizing an AISI 1070 steel.

ing hardness is RC 25 due to air cooling, its one-inch diameter size helping to control the cooling rate. The purpose of preheating is to prevent the exposure of cold steel to elevated temperatures. Preheating should always precede normalizing, annealing, and hardening operations to reduce the possibility of introducing stress into the part because of thermal shock. Normalizing treatments produce stress-free parts.

Because air is the cooling medium in normalizing, the thickness of the cross section of the metal has a great effect on the rate of cooling, thick parts cooling slower than thinner parts. The variance in the cooling rate from austenite causes the size of the carbide particles also to vary and this, in turn, causes the resulting hardness to vary. Consequently, it is the speed of change from gamma iron that determines hardness from a given carbon content. All but extremely fast cooling causes ejection of the carbon atoms. In air cooling, the carbide particles (converted from carbon) have little time to grow; therefore, they are extremely small and well scattered. Also, in normalizing the allotropic change is somewhat faster than in annealing. The microstructure of the RC 25 specimen is fine pearlite, which is unlike coarse or regular pearlite because the carbide particles are much smaller, closer together, and are scattered. The hardness of the specimen before normalizing was RB 89 due to cold rolling its shape. As the cooling rate changes, so does the hardness.

Annealing and Hardening Steel

According to Figure 8–28 the annealing range is just below the normalizing range. When an AISI 1095 is preheated to 800 °F (427 °C) and soaked at 1450 °F (788 °C) until the core reaches temperature and is then allowed to cool in the furnace over night, the platelike structure of pearlite (Fig. 9–14) is observable with a hardness reading of RB 92. Any metal with a hardness value in the RB scale is machinable, one of the purposes of annealing being to soften the metal for fabrication or machining. The microstructure shows all pearlite (coarse pearlite) except a small amount of cementite that the extra 0.15% carbon produced. Now, if the ¾-inch diameter AISI 1095 steel is to be hardened, the same preheating procedure is used and the hardening or austenitic temperature is also the same, 1450 °F (788 °C) (Fig. 8–28). After properly soaking at the austenitic temperature the part is rapidly quenched in water until its temperature is approximately 200 °F (93 °C) and it is then quickly placed in the tempering furnace for

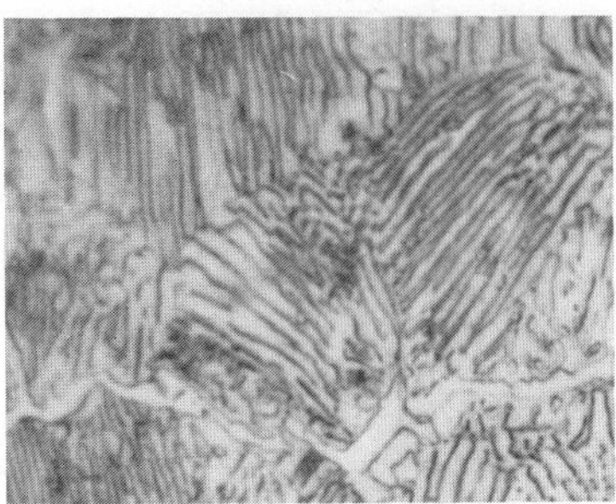

FIGURE 9–14   Annealed AISI 1095 steel showing a small amount of light colored cementite in its microstructure.

an additional treatment called tempering. Here again, the thickness of the part's cross section helps govern its cooling speed, along with the cooling medium, because thick parts cool more slowly than thin parts in the same cooling medium during the hardening operation.

On the other hand, specimens cooled in the furnace are controlled by the furnace. Because this part was cooled very fast a microstructure of martensite is obtained with a hardness of RC 65 (Fig. 9–15). Martensite is completely unlike

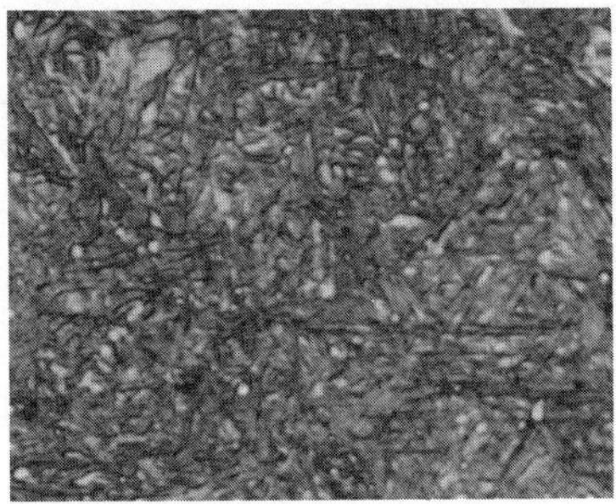

FIGURE 9–15   A form of martensite at 1000×, Nital etch.

the several pearlites; it is a solid solution of carbon in a body-centered tetragonal structure. This means that the carbon atoms in austenite have no opportunity to leave the interstices of the newly forming cells as the cells shear along one dimension to become tetragonal instead of cubic. The carbon atoms in martensite remain trapped between the iron atoms, causing a high degree of stress and hardness and one dimension to be expanded. An important point to remember is that stress is a natural factor and always occurs in a material as a reaction to an external load. Therefore, removal of stress is only desirable prior to imparting a different stress. Further, stress is induced in a metal also by rapid heating and cooling. The resulting hardness is caused by several factors including residual stress and, again, this type of stress is natural. But unless the residual stress is removed before imparting another stress, a crack may result. Martensite is highly stressed because hardness includes stress. Also, because martensite is less dense than austenite the expansion of the part occurs in a diffusionless shearing action of austenite. Martensite is observed as a needlelike, acicular, or triangular microstructure (Fig. 9–16).

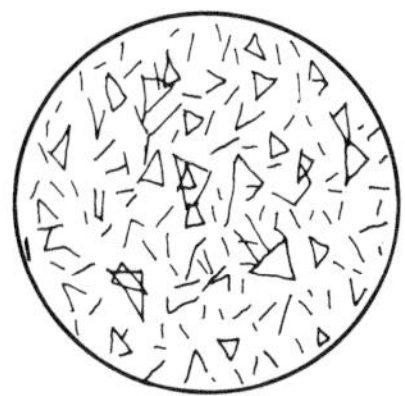

Microscopic view

FIGURE 9–16 Characteristics of martensite include a needlelike or acicular structure which often includes triangles.

The Intermediate Quench

Some high carbon steel parts are often quenched in oil in order to escape the chances of warping or cracking in the water or brine quench. Of course, the oil-quenched high carbon steel part from austenite, 1450 °F (788 °C), only cools at about one-third the speed of water; therefore, its hardness is greatly reduced. The microstructure of this ¾-inch diameter specimen is a very fine pearlite of approximately RC 44 in hardness (Fig. 9–17). As in other cases of rapid cooling,

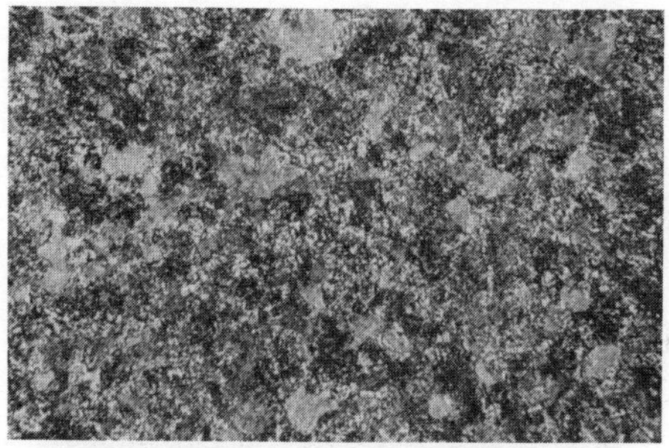

FIGURE 9–17 Very fine pearlite resulting from quenching in oil from austenite. Normally, the microstructure is dark.

the part's diameter influences the cooling rate, which in turn influences the sizes of the carbide particles and, consequently, the resulting hardness. The faster the part cools from austenite, the closer the pieces of carbide get to the limit of their smallness as the hardness increases accordingly. Very fine pearlite is observed as a dark microstructure; the pieces of carbide are so small and so thoroughly arranged throughout the ferrite that magnifications of 2000× are often required to reveal the very small spaces between the carbides. When the same steel is water quenched from austenite with some increase in cooling speed, a higher hardness will result, such as RC 51. In this example, the particles of carbide are much closer together than in the RC 44 example. The microstructure appears darker, but often is partly surrounded with a lighter appearing martensite. The dark colored structure is often bainite (Fig. 9–18) and may require several thousand magnifications to see the tiny plates of well-distributed, organized carbide and ferrite relationships.

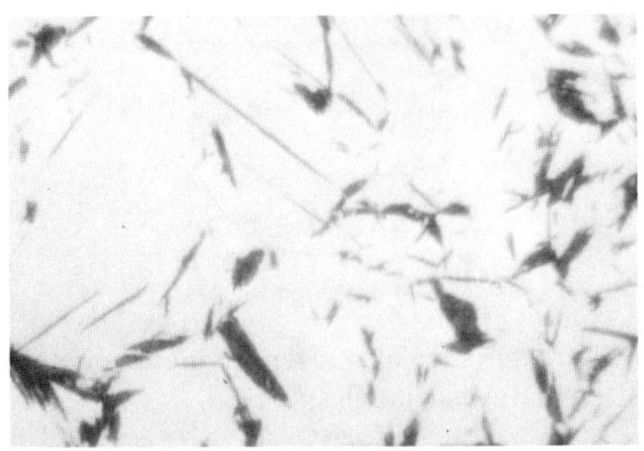

FIGURE 9–18   The dark areas of this microstructure are bainite; the light colored material is martensite.

Tempering

The most critical of the heat treating operations for steel is tempering. This heat treatment gives the metal its exact hardness and strength. Tempering always follows hardening and is always carried out at temperatures below the critical points of the steels. Because tempering follows hardening a rule states that the hardening treatment must always impart the initial hardness in the steel which is equal to or greater than the ultimate hardness required, except in the case of high speed steels. If the "as quenched" microstructure is martensite, then tempering will result in a tempered martensite structure (Fig. 9–19). Temperatures higher than about

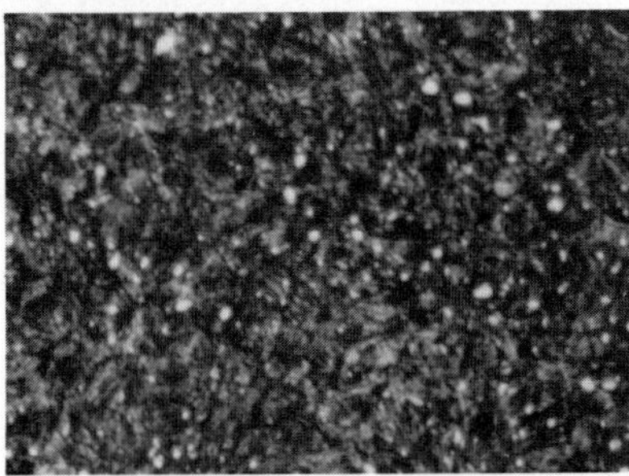

FIGURE 9–19   Tempered martensite in a high carbon steel.

450 °F (232 °C) release the carbon atoms from the tetragonal body-centered structure of martensite and cause the normal alpha body-centered cubic structure to appear. As temperatures increase the particles of carbides grow in size, causing the resulting mixture to become softer due to the appearance of more ferrite. Tempering removes hardness; therefore, if the hardening treatment fails to induce a value equal to or greater than that desired, the tempering treatment is useless. Research has shown that hardened steel exposed to tempering temperatures is reduced in hardness (except high speed steels); the amount of hardness reduction is directly related to the chemistry of the steel and the tempering temperature (Fig. 9–20). Good practice states that best results are obtained when the part is

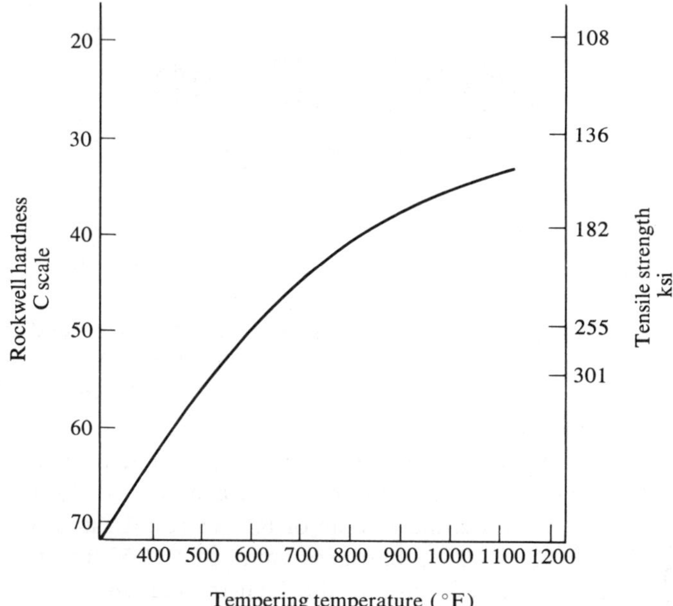

FIGURE 9–20   Tempering temperature vs. hardness in an AISI 1095 steel.

exposed to the tempering temperature for a minimum of one hour, allowing an extra hour for each extra inch of cross-sectional thickness. The part should be air cooled unless stated otherwise.

METALLURGICAL PROCESSING OF A CARBON STEEL PART

With reference to Figure 8–28 an AISI 1090 steel is preheated at 800 °F (427 °C), soaked and annealed at 1450 °F (788 °C) by furnace cooling to at least 800 °F (427 °C), and then cooled in air or in the furnace to room temperature. The steel can then be machined into a tap, for example, since the hardness is only RB 91. After machining, the tap is normalized by preheating again at 800 °F (427 °C), soaked at 1550 °F (843 °C) for the time necessary to heat it to the core, and then air cooled. The hardening treatment follows, again by preheating at about 800 °F (427 °C) and soaking at 1450 °F (788 °C) until the temperature reaches the core. The thicker the cross section, the longer the soaking period. Because a tap has to have a very high hardness and because this tap is made of carbon steel, it must be rapidly quenched in water to trap the carbon atoms in the quickly forming tetragonal cells of martensite. This produces a Rockwell hardness of C 65. This high hardness carries a high brittleness factor; therefore, the tap is immediately tempered at a temperature of 430 °F (221 °C) for one hour and air cooled. Four Rockwell points are lost, resulting in a value of RC 61, which is considered to be the minimum acceptable hardness for cutting tools (Fig. 9–21). The carbon steel tap is then four Rockwell points softer than the high speed steel

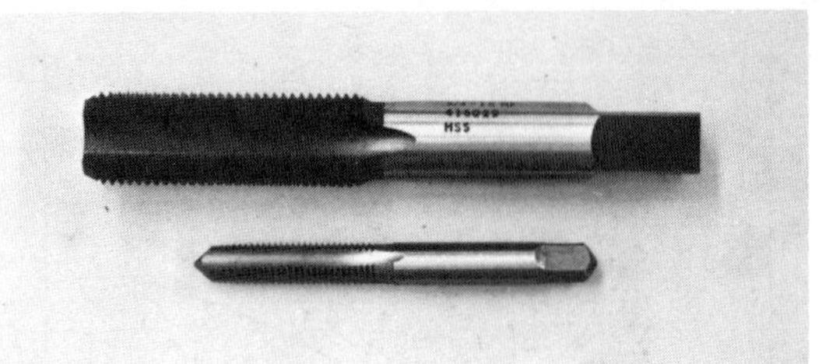

FIGURE 9–21   Taps made of high carbon and high speed steels.

tap which is RC 65. In this respect, the 430 °F (221 °C) is related to an RC 61 and an AISI 1090 when the steel has been properly hardened. If the "as quenched" hardness is C 64, 63, 62, or 61, the resulting hardness is still C 61 after the 430 °F (221 °C) tempering treatment.

As the tempering temperature increases in steels (other than high speed steels) from 200 °F (93 °C) hardness is reduced. When a hardened AISI 1095 is tempered at 750 °F (399 °C) a hardness value of RC 43 results, and when the metal is exposed to 1000 °F (538 °C) a hardness value of RC 32 results which has a tensile strength of about 144,000 psi. If the heat-treated tap described above is accidentally reheated to 430 °F (221 °C), no loss in hardness occurs, but if the temperature increases, the cutting tool is ruined. The microstructure of the finished tap is observed as tempered martensite. As tempering temperatures increase for other parts the martensitic pattern remains.

HIGH SPEED STEEL

Tempering of steels, other than the high speed types, causes a lowering of the "as quenched" hardness and this decrease in hardness has a direct relationship to temperature. On the other hand, steels which are classified as high speed (*HS*) respond in an opposite way to tempering processes. A typical high speed steel includes 5% cobalt, 18% tungsten, 1% vanadium, 4% chromium, 0.75% carbon, and the other five elements common in all steels. Often, molybdenum is used to supplement the tungsten, and this type includes 8% cobalt, 2% vanadium, 2% tungsten, 4% chromium, 8% molybdenum, and 0.90% carbon plus the other five common elements. These specific combinations produce the particular high speed steel. There are numerous other combinations.

After machining and after normalizing, the *HS* part is preheated to 800 °F (427 °C) and again to 1500 °F (816 °C). Then it is quickly placed in the hardening furnace at about 2350 °F (1243 C°) (varies according to analysis) for a few minutes until soaked. The part is quickly quenched in oil, air, or molten salt, and while it is still warm is placed in the tempering furnace at about 1025 °F (552 °C) for two hours and then air cooled. The quenching medium during hardening depends on the particular type of steel. As quenched, some austenite is retained, resulting in a hardness of only RC 60. After tempering, however, five Rockwell points are added because the transformed austenite becomes martensite with an increased hardness value of RC 65 (Fig. 9–22). Secondary hardening increases the hardness value, but the hardness of other steels subjected to the same

FIGURE 9–22   Martensite in the cutting edges of this drill provides cutting actions in the pearlite of the softer metal.

treatment would be severely reduced. With regard to normalizing high speed steels, the process is only performed when cementite must be rearranged by scattering throughout the metal.

Spheroidizing

Sometimes it is essential to perform severe bending of high carbide-type steels, such as high carbon or high speed, whereby small radii result, such as in winding small diameter springs or forming intricate shapes by bending. Too often, fully annealed high carbon steels (RB 94) or high speed steels with their plates of carbide will not deform to the small radius without fracturing. In this situation the steel should be spheroidized; that is, the carbides should be placed in small spheres so that fracturing is not probable during the most severe bending operations.

Cupping operations require the plastic flow of large masses of metal. A steel to be spheroidized so it will withstand such deformation is heated to approximately 1250 °F (677 °C) for ten to eighteen hours and periodically increased and decreased in 50 °F (10 °C) increments. Such a heat treatment causes the carbides to become spherical, the changing temperatures just below the critical point urging the curling movements of the carbides. After forming, tying knots (Fig. 9–23) if desired, the part is normalized, hardened, and tempered. Very small diameter wire, however, is not usually normalized because of possible damage due to the size of the wire.

The spheroidized microstructure (Fig. 9–24) is observable as dark balls of carbides embedded in matrix of light colored ferrite and has a hardness value of RB 72 for high carbon steel. Keep in mind that carbides remain hard even in annealed steels. It is the presence of ferrite which separates the carbides that allows softness to occur. The balls of carbide easily move about in the ferrite during plastic deformation, much like logs move about in mud.

Transformation of Retained Austenite

During rapid quenching of high carbon steels from martensite, plain carbon, or alloys, some austenite is frequently retained as martensite forms. These soft spots

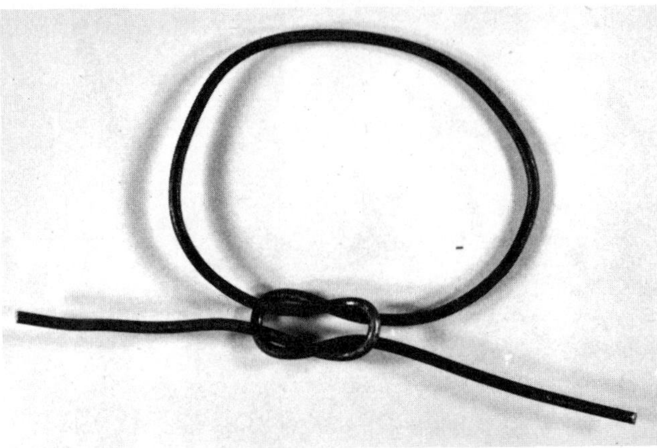

FIGURE 9–23   A high carbon steel wire is tied in a knot to demonstrate the extreme ductility of the metal in the spheroidized condition.

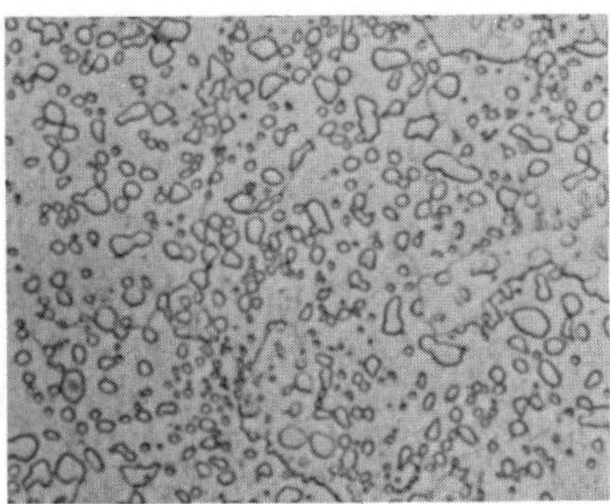

FIGURE 9–24   Spheroidized AISI 1095 steel (1000× with 2% Nital etch).

of austenite are found during hardness testing and they can lead to operational problems. Retained austenite can be transformed into martensite by exposing the steel to subzero temperatures, such as −100 °F (−73 °C). Transformation to all martensite assures structural stability in the metal due to homogeneity of the microstructure. Tempering operations can then follow. However, in order to help prevent cracking of hardened parts, tempering usually follows hardening before the part becomes cold. Therefore, a type of procedure tempers the steel part at about 300 °F (149 °C) to relieve quenching stresses, then exposes the part to −100 °F (−73 °C) for the same amount of time, and then tempers it as required to obtain the needed hardness or strength. A typical 1% carbon tool steel, for example, can retain as much as one-fifth austenite. Potential troubles, such as warpage, cracking, and soft spots, lie ahead unless a complete martensitic structure is obtained.

THE TRANSFORMATION OF AUSTENITE

The hardening and tempering of steels can be accomplished in several ways. The conventional method is the rapid quenching of austenitic steel until the part is cooled to approximately 200 °F (93 °C), followed by tempering at the required temperature. The reason that tempering immediately follows hardening by quenching is to prevent the hot steel from getting cold and possibly cracking. The cracking range is from around 200 °F (93 °C) to room temperature. (A cracked steel part is shown in Figure 9–25.) The hardening procedure is based on the reten-

FIGURE 9–25 A cracked steel punch. The crack resulted from faulty heat treatment.

tion of austenite until the start of the martensite transformation temperature, M_s, begins and then through the M_f, or finishing temperature, by continuous cooling to room temperature. Or, the warm part may be placed in the tempering furnace to help prevent cracking. A hardness value for an AISI 1080, ¾-inch diameter steel will be RC 65 as quenched when austenite becomes martensite and when this procedure is used.

Time-Temperature Transformation

An explanation of austenite transformation is illustrated in Figure 9–26. The coordinates of the diagram are the temperature (vertical) and the time factor (horizontal). Because austenite is stable only above the A_e (equilibrium) temperature line, a transformation curve is plotted to show the beginning and ending of transformation if the critical cooling rate is too slow. The double curve for a carbon steel resembles an S with its nose at approximately 1000 °F (538 °C) and at one second in time. The time-temperature transformation diagram, or S curve, is for one steel only. Consequently, hundreds of diagrams are available. A study of this diagram indicates that maximum hardness is available in the 0.89% carbon steel if austenite can be rushed through the one-second wide gate at the 1000 °F (538 °C) temperature mark without "touching the gate post" or beginning of the transformation line. This means that the temperature must be reduced from approximately 1450 °F (788 °C) to below the 1000 °F (538 °C) zone fast enough to forbid beginning of transformation at 1000 °F (538 °C). The time limit is about one second at the 1000 °F (538 °C) temperature zone. Once austenite is safely through this narrow gate, ample time exists for continuous temperature reduction to the M_s temperature zone where carbon atoms become trapped in a diffusionless transformation from face-centered cubic iron cells to

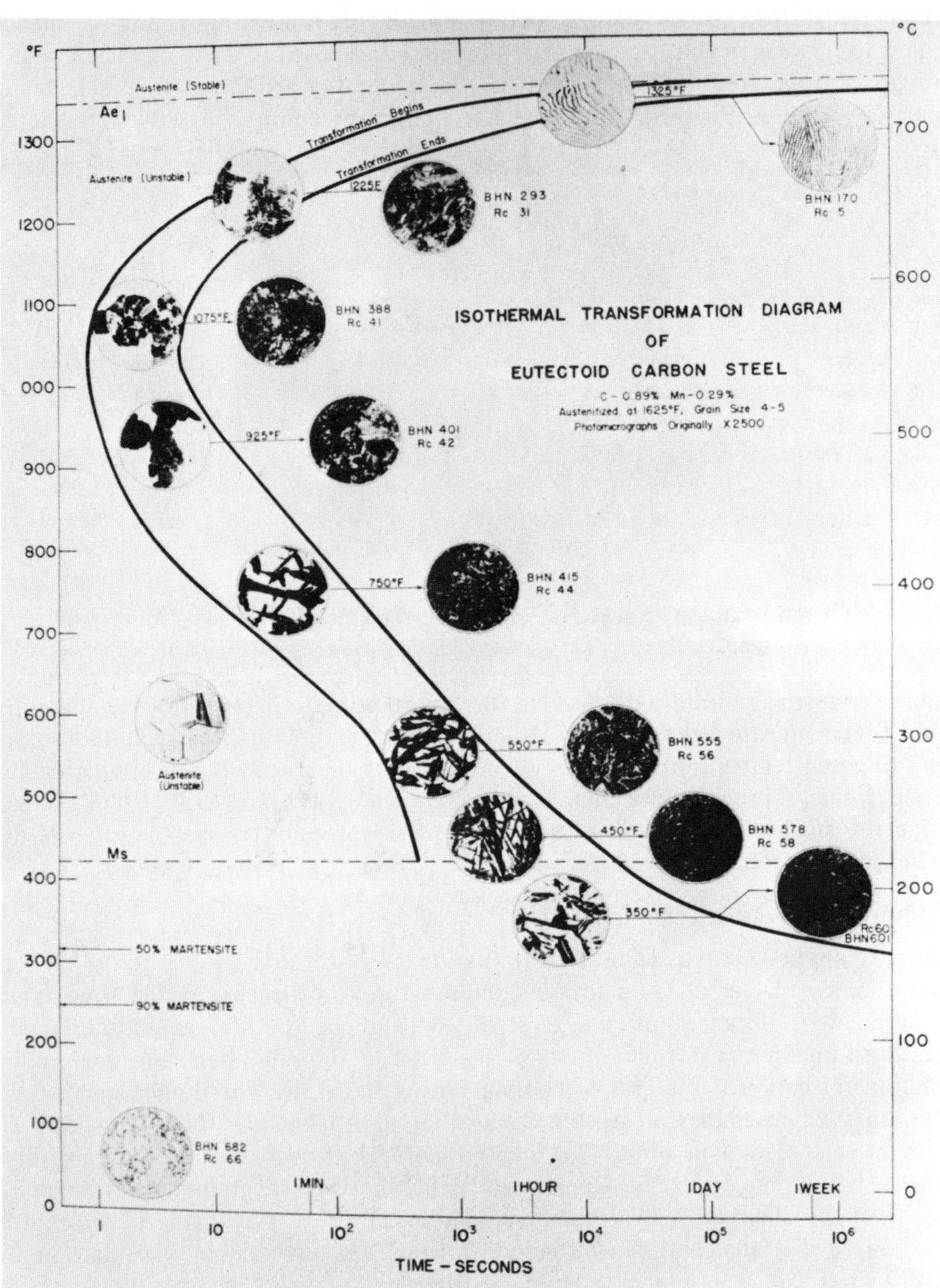

FIGURE 9–26   Isothermal transformation diagram of eutectoid carbon steel. (Copyright 1971 United States Steel Corporation)

body-centered tetragonal cells. The completion of transformation at this low temperature is nearly instantaneous as temperature is continuously reduced and martensite results. The microstructure is observed as martensite with a hardness value of RC 65.

Modified Water Quench

Another typical quenching procedure is performed by water quenching. The water is either hot or is covered with floating oil. As the austenitic high carbon steel part plunges into the liquid rapid cooling occurs, but at a much slower rate than the cold, clean water quench. It is often desirable to quench at a slower cooling rate which is still faster than an oil quench. As cooling continues transformation of austenite begins because the solution fails to reach the exit gate in time. However, due to continuous cooling only a portion of the austenite changes to a fine or very fine pearlite, while most of the hot metal is still retained as austenite. During the next few moments the remaining austenite passes backwards through the beginning of the transformation zone (Fig. 9–26) and again becomes available for transformation to martensite at the M_s and M_f temperature zones which exist at lower temperatures. The resulting mass has a hardness value of RC 45 and the microstructure is observed as approximately 60% martensite and 40% fine pearlite. Another microstructure, upper bainite, is often observable with the martensite, but this too can ultimately be resolved into fine pearlite at higher magnifications with the microscope.

Bainite and Fine Pearlite

Random cooling of austenitic steel parts in water or oil may often result in very fine pearlite or possibly some bainite. As shown in Figure 9–26, the part was cooled fast enough to slide through the time-temperature gate, but steam pockets surrounding the hot part reduced the cooling rate to such an extent that austenite quickly moved into the transformation zone and held the temperature long enough to transform into a type of bainite, often referred to as lower bainite. Its microstructure resembles black feathers surrounded by a light colored martensite and its hardness is about RC 59 (Fig. 9–27).

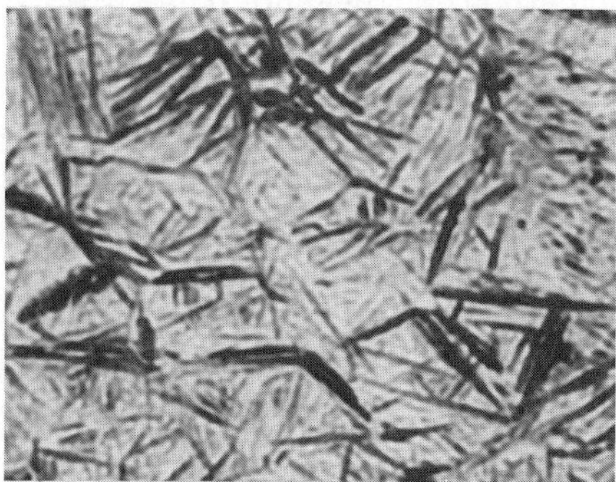

FIGURE 9–27   Dark colored bainite embedded in a matrix of light colored martensite as observed through the microscope.

On the other hand, the section of steel is sometimes either too large to attain the critical cooling rate, the wrong quenching medium is used, or the part is made of the wrong steel. Consequently, as a result of an apparent rapid quench only a hardness range of RC 27 to C 37 is obtained instead of the required RC 63. The hot part fails to make it through the gate, and most of its austenite completely transforms to a fine pearlite with a Rockwell hardness of about C 31. Because fine pearlite sometimes resembles upper bainite it is sometimes difficult to differentiate between the two microstructures. The closeness of the carbide particles in a bainite structure presents a dark feather-type view in the microscope; the fine pearlite appears similar, but does not have the feather design. Air and furnace cooling cause austenite to transform at much higher temperatures; therefore, the metal is much softer due to the presence of pearlite. Austenite transformation at upper temperatures which are below Ac_1 produces the pearlitic structures which are soft. Lower temperature transformation produces the. very fine pearlites, bainite, and martensite which are harder.

Formation of Pearlite and Ferrite

When a high carbon steel part that has been austenitized is allowed to cool extremely slowly as in furnace cooling, alternate layers of carbon and ferrite separate from the solid solution and form pearlite (carbide and ferrite) because no alternative exists for other types of transformations. A Rockwell B 88 is a typical hardness value for annealed AISI 1070 with its accompanying tensile strength of 85,000 psi. If an AISI 1050 steel is annealed, the microstructure reveals about 60% pearlite and 40% ferrite. This particular steel has excellent potentials for high strength mechanical properties due to the pearlitic change to martensite when the hardening treatment is accomplished.

Mass of Metal and Cooling Rate

Many disappointments occur after heat treating because something went wrong. A common conclusion among many heat treaters, for example, is that a surface hardness reading is also a core reading. Actually, as the cross section of the part increases beyond one inch, carbon steels cannot be used when either a high surface or core hardness is desired. Observation of the hardness values from surface to core of a bisected 1-inch round eutectoid carbon steel properly water quenched shows ten points difference in hardness values (Fig. 9–28). The core cannot harden until the surface cools; therefore, mass of metal is limited when a high hardness is needed in a plain carbon steel. Even though water or brine quenching is used Fig. 9–29), mass greater than one inch must consist of an alloy steel, such as one containing chromium. Several alloys such as nickel, molybdenum, and chromium cause the width of the "time gate" in the *TTT* diagram (Fig. 9–26) to widen so that ample time exists for austenite to move safely through the gate at a slower speed with oil quenching and then transform into martensite (Fig. 9–30). Certain percentages of these elements even allow oil quenching of 2-inch diameter parts or larger to obtain hardness values of RC 50 with only four hardness points loss in the core.

Alloys Slow Cooling Rate

Most alloys slow the critical cooling rate and allow more time for austenite to transform into another microstructure. Also, coarse-grained steels harden deeper

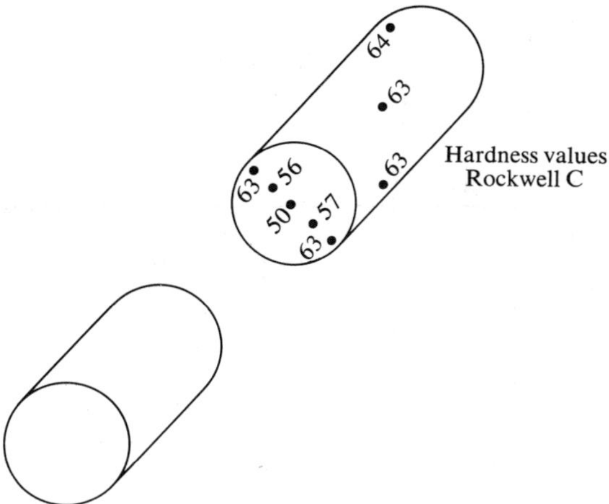

FIGURE 9–28   A bisected AISI 1095 steel showing core and surface hardness when the 1-inch round was water quenched from austenite.

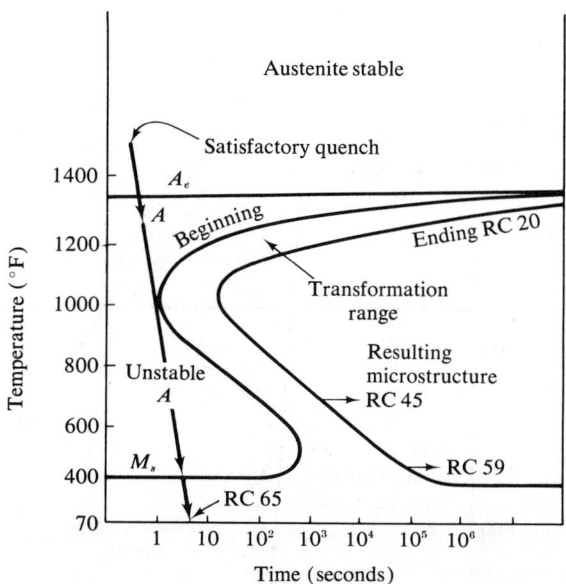

FIGURE 9–29   Time-temperature-transformation (TTT) diagram for a high carbon steel showing austenitic transformation to martensite due to rapid cooling by missing the transformation curve.

than fine-grained, but they are more brittle than fine-grained in impact. Steels are therefore classified as plain carbon or alloy, coarse-grained or fine-grained, and either hypoeutectoid, eutectoid, or hypereutectoid. Another classification separates them as being water hardening, oil hardening, or air hardening. The carbon steels, excluding ones with small diameters, are water hardening up to 1 inch in diameter. Alloy steels are normally oil hardening because more time is available to transform into hard microstructures because of the different chemistry widening the time gate.

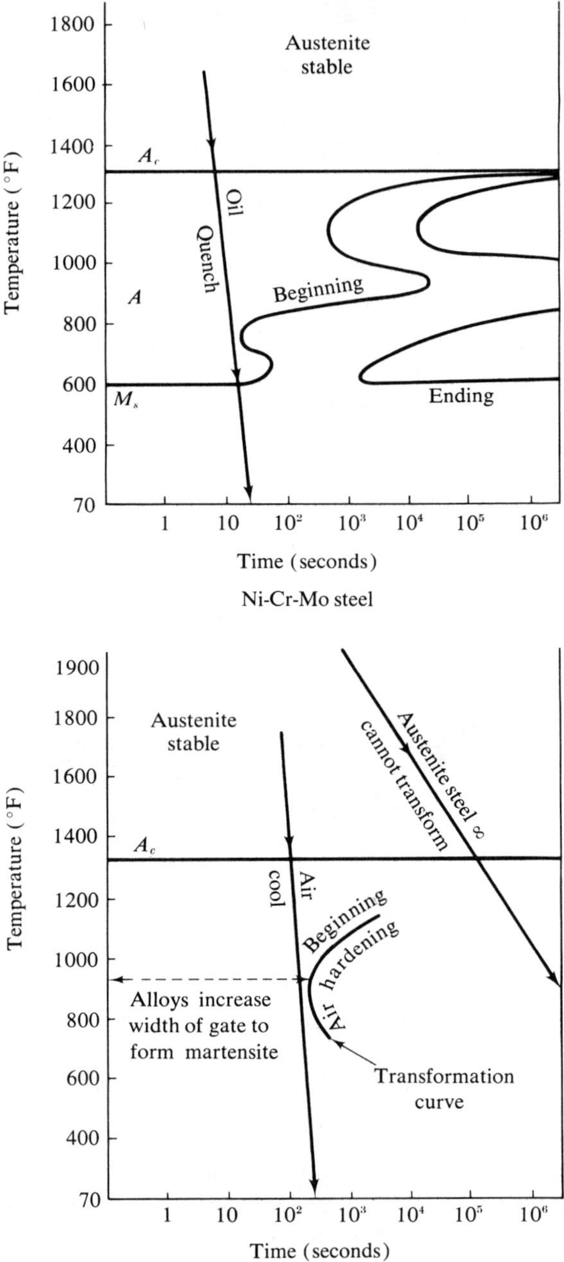

FIGURE 9–30　Time-temperature-transformation diagrams for oil- and air-hardening steels. The austenitic steel has no transformation curve because austenite is stable at room temperature.

However, low alloy contents are sometimes water quenched. Air-hardening steels contain even larger percentages of molybdenum, chromium, vanadium, and manganese to allow air to produce a high hardness such as is needed in intricately shaped tool steels and dies. The time for transformation at the gate is greatly increased (Fig. 9–30). The problem of annealing these steels is solved by furnace cooling.

As more alloys are added to the basic six elements in steels the usual tendency is for the transformation curve to move to the right, thus providing more time to allow hardening to occur. When approximately 18% chromium and 8% nickel are added to the six basic elements, however, the transformation curve is pushed into possible infinity (Fig. 9–30) and austenite has no normal means to transform (Fig. 9–31). Consequently, austenite and gamma iron are retained at room temperature, and when this occurs the steel is known as austenitic steel (Fig. 9–32) and is nonmagnetic. A typical hardness value is RB 89 after annealing, but when cold rolled, hardness values in the low C scale appear. This type of steel cannot be

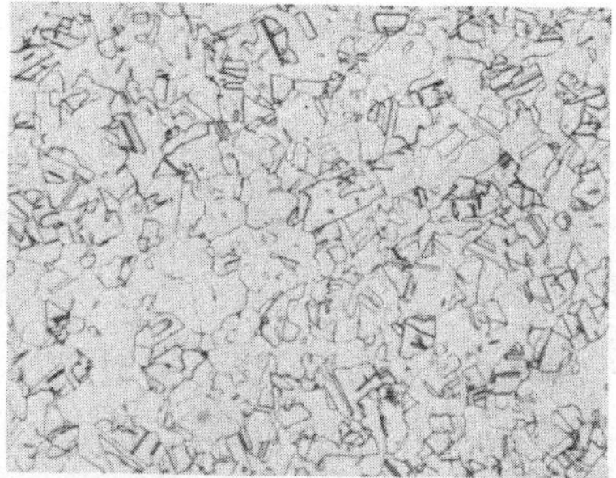

FIGURE 9–31   The 18% chromium and 8% nickel austenitic steel is stable at room temperature. This photomicrograph shows the granular structure of the solid solution along with twin areas (200×, Keller's etch).

FIGURE 9–32   This austenitic steel pressure vessel withstands either high or subzero temperatures due to its face-centered cubic microstructure.

hardened by heat treatment, but can be softened by soaking at 1900 °F (1038 °C) and water quenching. The purpose of the quench is merely to forbid any precipitation of the carbides from the solution and prevent harmful consequences. Austenitic steel has characteristics of a pure metal, therefore, because it is a solid solution, or one homogeneous mass of metal.

Martempering

Several nonconventional hardening processes are available to help assure better uniformity in the steel's mass. A common treatment known as martempering consists of austenitizing the part and quenching it rapidly in a molten salt bath at a constant temperature just above the M_s point, or about 450 °F (232 °C). The part is held at the designated temperature, which is well below the nose of the transformation curve, for a period of time to allow equal temperature throughout the metal and it is then allowed to cool in still air. A martensitic microstructure results throughout the steel which can be conventionally tempered to the correct hardness. With regard to the bath, a typical salt bath consists of sodium nitrite and nitrate along with potassium nitrate (Fig. 9–33).

FIGURE 9–33   A top view of a salt bath furnace used in the heat treatment of aluminum alloy.

Austempering

The heat-treating process of austempering is intended to harden thin sections of steels to hardness values in the middle RC range, such as RC 44, with a bainite microstructure. Such a procedure requires a steel which is capable of moving through the gate at the nose of the transformation curve in time to hold the austenite at temperatures below 1000 °F (538 °C). For example, a medium carbon chromium steel part about 9/16 inch thick is quickly quenched in a sodium nitrite and potassium nitrate combination bath which is held at 600 °F (316 °C). The part is retained in the salt for about 16 minutes to allow complete transformation from austenite to bainite and is then air cooled. A Rockwell C 44 hardness value results with a tensile strength of 207,000 psi. A great advantage of the austempering process is its ability to obtain hardness values other than the low RCs or the upper RCs. Bath temperatures vary in order to vary the resulting hardness. Also, hardening and tempering are accomplished at the same time. Its great disadvantage is that it is limited to fairly small parts.

A word of caution is necessary regarding the use of molten salts for heat treating purposes. Never allow cold or damp parts to enter any type of molten salt because of potential explosion. Never allow nitrates or nitrites to mix with cyanides because an explosion will occur and it can be disastrous. Also, never allow nitrates and nitrites to become hotter than 1100 °F (593 °C) because of possible fire and explosion. If a fire starts, use carbon dioxide fire extinguishers. Further, no form of carbon must enter the nitrate or nitrite bath; consequently, scale from steel must be removed from the salt bath prior to quenching the part.

PURPOSES OF ALLOYS IN STEELS

When a seventh element is added to the six basic elements in steel or when one or more of the six basic elements is increased significantly, an alloy steel results. The purpose of alloy steels is to increase the physical and mechanical capabilities of the metal beyond those possible in the carbon steels. One purpose of adding alloys to steel is to increase the time before transformation of austenite begins so that higher hardness values can be obtained in thick sections of steel and to allow an even hardness between case and core. Nickel, chromium, manganese, and vanadium widen the time gate in the *TTT* diagram. Parts having intricate designs which may crack during water quenching can be made from alloys which harden by air cooling from austenite. Chromium allows depth hardening in thick sections (Fig. 9–34) along with a much slower cooling rate, such as oil or air quenching. Obviously, lower cooling rates decrease the probability of cracking and distortion. Another purpose of alloying steels is to provide extremely hard intermetallic compounds such as the several carbides of chromium, molybdenum, and vanadium.

FIGURE 9–34 A 2½-inch thick high chromium steel plate has been polished and etched to show the fusion zone of the weldment. Chromium adds depth hardness to steels when properly heated and cooled.

Specific Reasons for Using Alloys

Specific reasons for producing steel alloys are numerous. As an example, tungsten causes carbides to form in addition to the iron carbides. Tungsten carbide resists heat better than iron carbide and is therefore used in high speed steel cutting tools which remain sharp even after exposure to 1000 °F (538 °C) (Fig. 9–35). Vanad-

FIGURE 9–35   Tungsten, along with molybdenum and cobalt, is used in the manufacture of high speed steels such as these cutters. A tungsten rod has been placed between the twist drill and the lathe cutting tool.

ium helps to maintain a fine grain structure at austenitic temperatures. Nickel is a toughener in steel as it goes into and remains in solid solution with the ferrite. It decreases the critical point on heating and helps to reduce grain growth at elevated temperatures. Nickel also adds to stainless qualities and promotes an increase in time before austenite begins to transform on cooling. Molybdenum is a hardener and a depth hardener and resists the adverse effects of high temperatures and fatigue failure. Manganese strengthens ferrite, forms carbides for hardness, and is used in nondeforming tool steels such as intricate dies. Its hardenability increases with content up to a specified amount; its work hardening ability is tremendous. As a power shovel rips through soils and rocks, for example, the manganese steel shows its toughness by excellent work hardening and shock resistant characteristics. Bank vaults also contain manganese in their construction.

As pointed out, chromium promotes the formation of hard carbides which not only increases hardness, but allows hardness in depth so that large steel shafts and armor plate can be hardened to required depths. Chromium is often coupled with nickel to produce the great combination of hardness and toughness along with corrosion resistant qualities as in the family of stainless steels. When large amounts of chromium, such as 18%, are alloyed with large amounts of nickel, such as 8%, the regular low carbon-type steel becomes austenitic and remains as such on cooling. These steels are nonmagnetic, are fairly soft (RB 82) with quick work hardening capability, and are stainless and heat resistant. A small amount of the element niobium in a chromium-nickel steel causes the stabilization of austenite in the carbide precipitation temperature zone of 1600–800 °F (871–427 °C) on cooling. Nonstabilized austenitic steels become susceptible to corrosion on slow cooling through this range due to the formation of carbides along the grain boundaries which, in turn, invites future problems.

Silicon in steel strengthens the ferrite and promotes toughness. Boron, in a very small amount, increases depth hardness in steels, while aluminum is a deoxidizer and grain size controller. For nitriding steels(a method of case hardening)alumi-num is added to the analysis in amounts of 0.95–1.30%. Copper dissolves in the ferrite, as does nickel, and strengthens the ferrite. Copper in steel increases resistance to corrosion. Approximately 1% copper in a low carbon steel can provide for precipitation hardening of the part. Titanium inhibits grain growth in steel, is a deoxidizer, and forms extremely hard and heat resistant carbides. The ferrite is also strengthened as part of the titanium moves into solution. Cobalt strengthens the ferrite as it too forms a solution in the ferrite, but its great advantage is high temperature creep resistance of the ferrite when used in heat resistant steels.

There are other elements used in steels; each one imparts its particular benefits. An important point to keep in mind during the study of alloys is that carefully prepared formulas for each steel are made by chemists and metallurgists who have experimented with combinations of the various elements and compounds. An alloy consisting of two elements is known as a *binary* alloy. Three elements produce the *ternary* alloys. Complex alloys provide the benefits of each element added.

CAST IRONS

A simplified version of the iron-iron carbide diagram for steels and cast irons is presented in Figure 8–28. Metal existing above the line *MNO* is liquid, and as the liquid cools along line *FEN* austenite precipitates, leaving the mushy stage. Metal in the region *NOPN* consists of precipitated cementite in the liquid, and solid metal or cast iron is formed below line *ENP*. The eutectic is the iron carbon alloy containing 4.3% carbon, point *S*, which solidifies at 2066 °F (1130 °C), point *N,* and is called ledeburite. The eutectic is a finely divided mixture of cementite and austenite. As cooling continues, austenite, cementite, and ledeburite transform along the line *DTQ* to their new microstructures existing along line *CSR*. In the cast iron portion of the diagram (1.7–6.67% C) slow cooling produces graphite as a transformation product and is observed as such in the microscope. As shown, the cast iron and steel regions of the diagram are combined to form a single diagram so that the relationships of the two metals can be studied.

Gray Cast Iron

Cast irons are classified into several categories and include the gray, white, malleable, and ductile. Both carbon and alloy irons are included in this classification. Gray irons take their names from the gray, or graphite, colored fracture of the broken metal (Fig. 9–36). Gray irons are soft and brittle, an exception to the expected hardness accompanying a brittle condition. Gray irons form as carbon contents increase beyond 1.7% and include the many alloys up to 6.67%. The 4.3% alloy is the eutectic and is a common iron. The gray irons are classified by average tensile strengths and range from below 20,000 psi to more than 100,000 psi in the higher quality groups. The uses of gray irons include the many common castings which require minimum compression strengths. Common cast irons are not normally used in tension because tension stresses impose unpredictable conditions in the metal which lead to sudden failure without warning because there is no plastic flow of the material under increasing loads. Machine bases, housings, grates,

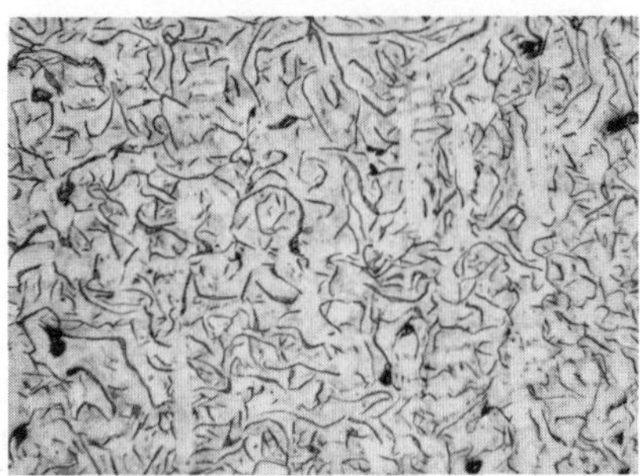

FIGURE 9–36   A photomicrograph of gray cast iron showing the flakes of graphite embedded in the matrix of pearlite and ferrite (Nital etch at 60×).

valves, pipe, and related parts are cast from gray iron. The gray appearance results from the slow cooling of the liquid in the mold, causing graphite to form. A typical gray iron casting is shown in Figure 9–37.

FIGURE 9–37   A typical gray iron casting.

White Cast Iron

White iron has a light colored fracture when broken and is hard and brittle. Much of the structure is cementite. The liquid iron is cast into chilled molds; solidification and cooling are much faster than the cooling of gray iron having the same chemical analysis. Cold spots along the walls of the mold cause faster cooling of the metal. This type of iron is used in machine surfaces which require a fairly high degree of hardness in addition to the casting requirements. Ways on lathes and other parts of machine tables are chilled as the liquid quickly solidifies and hardens.

Malleable Iron

Malleable iron is used when softness and toughness are required in a cast product. The microstructure consists of balls of soft graphite embedded in soft ferrite and is much like spheroidized steel, except the spheres in steel are hard carbides embedded in soft ferrite (Fig. 9–38). However, malleable iron will fracture after a significant deformation because of the graphite. Its great ductility makes it ideal for water pipes and fittings. Its great disadvantage is that it will not withstand freezing

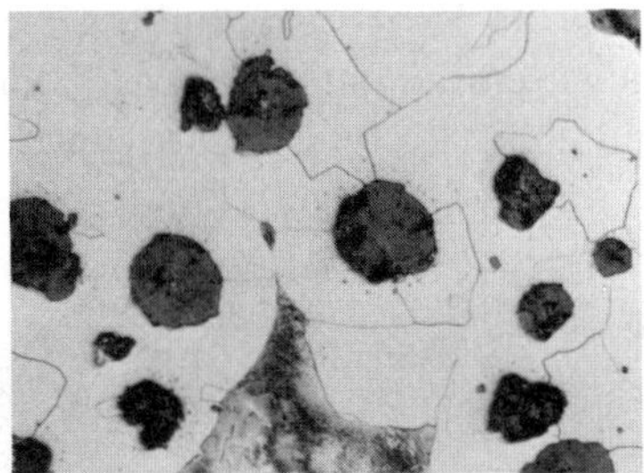

FIGURE 9–38 This photomicrograph of malleable iron shows spheres of graphite and some pearlite embedded in a matrix of ferrite. Grain boundaries are shown as irregularly shaped dark lines (Nital etch at 200×).

temperatures without cracking when filled with water. (All ferrite, by the way, is brittle when cold unless the ferrite is properly alloyed.) The great advantage of malleable iron is the capability to bend and move with shifting soils. Malleable iron is produced from white iron by heating the white iron castings to approximately 1600 °F (871 °C) for five days and allowing the furnace to cool below the critical point of the iron. All of the cementite reacts to the high temperature and forms graphite.

Ductile Iron

A more recent form of cast iron is ductile, or nodular, iron which can be deformed in many of the same ways as low carbon steel. Molten iron in a laddle receives a small quantity of magnesium which immediately causes a turbulence; extensive mixing of the carbon occurs and this results in very small spheres of carbon dispersed throughout the matrix of ferrite. The carbon spheres are smaller than in the malleable castings and the spheres form on cooling from the liquid and do not require heat treatment. Such a dispersion of carbon allows the free ferrite to respond to its inherited mechanical properties of softness with ductility. Thus, brittleness in the iron is excluded and, consequently, a high degree of malleability or ductility is present. Such a combination of engineering skills with regard to casting capability coupled with higher strengths and ductility provides a real competitor to the ductile steels. Another advantage of ductile iron is its ability to respond to heat treatments such as normalizing and hardening. Microstructures of ductile irons verify the ferritic, pearlitic, and martensitic types which result from chemical changes between carbon and iron.

Alloyed Cast Irons

Alloyed cast irons are modifications of the irons already discussed in that specific elements are added to cast iron as they are to steels. When properly alloyed and heat treated some cast irons have tensile strengths of more than 180,000 psi. Again, a particular element is used to provide that property which is needed. Nickel, for instance, induces toughness to the matrix of ferrite and increases the several strengths. Typical high strength cast irons include chromium, nickel, and molybdenum in their chemical analyses. Heat-treated alloyed cast irons are much unlike the common gray irons which are used in the "as cast" condition. A great advantage of cast iron is its fluidity and ability to be cast into small, large, or intricate shapes. However, cast steel also has these advantages (Fig. 9–39).

FIGURE 9–39   This large track shoe for a 200-yard power shovel weighs 6280 pounds. (Courtesy of Marion Power Shovel Company)

INDUSTRIAL CLASSIFICATION SYSTEMS FOR METALS

Since the revision of the metals classification systems following World War II numerous new metals have been produced, both in the ferrous and nonferrous categories. The ferrous system contains so many different steels, more than 50 000, that no one coding procedure will suffice. Consequently, professional organizations and manufacturers have reviewed the massive spectrum and have initiated a series of systems to cover most of the available metals. Coding systems to classify all metals include ASM, ASTM, MIL, FED, AA, AISI, SAE, and numerous manufacturers' codes. In regard to the steels, there are the low strength commercial grades, the structural steels, the high strength structural steels, the pressure vessel types, the aircraft, impact tool, impact cuting, cutting tool, abrasion resistant, railroad and ship, spring, die, armament, pipeline, case hardening, cable, and so on into a near endless quantity. A sample of some of the codes will be discussed. For further information, consult the *Metals Handbook* series published by the American Society for Metals.

Four-Digit Code System

Many of the common types of commercial, structural, and tool steels are included in the AISI numbering system. For purposes of illustration, the four-digit coding system will be explained along with some examples. Many of the common steels, commercial and aircraft quality, are grouped in the four-digit system and include a fair sampling of carbon and alloy steels. Because this is a common and well known coding system, it will be explained. However, the principles of coding in the following system are not transferable to the numerous other systems. The first digits in the four-digit system of AISI are as follows:

1—Carbon
2—Nickel
3—Nickel-chromium
4—Molybdenum
5—Chromium
6—Chromium-vanadium
7—Tungsten
8—Nickel-chromium-molybdenum
9—Silicon-manganese

The first digit on the left in the four-digit code identifies the basic type of steel. For example, a 4130 (read, *forty-one thirty*) is a molybdenum steel with an alloy. The second digit from the left in this example, *1,* indicates the approximate percentage of alloying element and is associated with the first digit. Besides the six basic elements in this steel, there is a chromium content of 0.80–1.10% and a molybdenum content of 0.15–0.25%. In very general terms the alloy chromium constitutes 1%. The alloy 4032, as another example, has a single alloy of 0.20–0.30% molybdenum. The second digit is zero because the *4* represents molybdenum. The last two digits in the first example, *30,* indicate the percentage of 1.00% carbon content, which means that the 4130 contains from 0.28–0.33% carbon, or an average of 0.30%. When the left-hand digit is known, the second digit approximates the alloy content, and the last two digits indicate the carbon content of the alloy in this coding system. Table 8–1 presents chemical analyses of some of the most used steels.

Classification of Tool Steels

There are many steels not included in the four-digit system. Therefore, some of the omitted steels, in this case the tool steels, are discussed along with their present coding systems. The tool steel category is large and diverse. One method codes tool steels by a simple letter-number designation. As an example, the AISI W1 is a high carbon water hardening steel which is used for general purpose high hardness tools such as taps and reamers and is interchangeable with AISI 1095. There are several additional types, such as W2, W3, W4, and so on. The W1 contains a small amount of vanadium and makes a good drill rod. The AISI H11 is a chromium-molybdenum-vanadium air hardening hot work tool steel and is used for punches, mandrels, and similar applications. Several types are available, including H12, H13, H14, and others. The AISI D2 is a high carbon, high chromium air hardening die steel that is used for making dies for shearing and punching and for precision gages. Several other types are available, such as D3 and D6. AISI S7 is a chromium molybdenum air hardening and shock resistant tool steel used for impact chipping and forming tools. Other types include S1, S2, and S5. The AISI

A2 is a medium chromium air hardening die steel used for forming and drawing dies. Type A10 is a graphitic air hardening die steel.

Other tool steels are available in addition to the water and air hardening types. Oil hardening steels constitute a large quantity of steels. AISI O2 is a high manganese oil hardening tool steel used for punches and pressing dies. Other types include O1 and O6. AISI M2 is a molybdenum-tungsten high speed oil or air hardening tool steel used for abrasion and shock resisting purposes and for cutting tools. The M10 type has no tungsten, but has an increase in its molybdenum content. The AISI T4 is a tungsten-cobalt high speed steel which is oil or air hardening and is used for cutting blades and tools. The T1 type is the popular 18-4-1, meaning 18% tungsten, 4% chromium, and 1% vanadium. It is used for general purpose high speed cutting tools. The AISI PPT is a mold steel capable of obtaining a hardness value of RC 37. Several other P type steels are available. The AISI L3 alloy tool steel is capable of producing a roller bearing hardness. It too is available in several other types. The AISI F3 is an oil hardening, wear resistant tool steel used for making gauges and similar measuring instruments. This carbon-tungsten tool steel is also available in several types.

High Strength Steel Castings

In addition to those high strength steel castings and wrought iron products previously mentioned, it must be emphasized that not all castings are inferior to wrought products. When properly alloyed and heat treated, many castings can be used in the high stress applications where both massive design and strength are required. The Crawler transporter (Fig. 9–40) and the armored vehicle (Fig. 9–11) are cases where cast parts and wrought parts are equally acceptable.

A Letter Coding System

The code for these steels does not indicate their chemistry; therefore, a chemical analysis chart must be available to determine their contents and mechanical capabilities. Part of the AISI tool steel classification system of coding is as follows:

Type	Code
Water hardening	W
Hot work	H
Cold work, high carbon, high chromium	D
Cold work, medium alloy, air hardening	A
Cold work, oil hardening	O
Shock resisting	S
High speed, molybdenum	M
High speed, tungsten	T
Special purpose, mold	P
Special purpose, low alloy	L
Special purpose, carbon tungsten	F

Classification of Steels

Another coding system exists for stainless steels other than those listed in the previous system, and it too is large and diverse. These steels are divided into the

FIGURE 9–40 This Crawler-transporter was designed to carry a payload of 6000 tons for the Apollo/Saturn V and holds the load within $\frac{1}{6}°$ of true level even on a 5% grade. Power is supplied by 16 electric motors, two on each crawler track, driven in turn by diesel-electric generators. It is 132 feet long, 114 feet wide, and 20 feet high. (Courtesy of Marion Power Shovel Company)

martensitic, ferritic, and austenitic types. The martensitic stainless steels are heat treatable for high hardness where hardness and corrosion resistance are essential, such as in high quality knives. A typical hardening treatment includes oil or air quenching from 1825 °F (996 °C), followed by tempering at 1125 °F (607 °C). This group is AISI coded into three digits, such as 403, 414, and 420. The 501 and 502 types also fit into this category. The ferritic group is not heat treatable for increased hardness and resists corrosion best when annealed. The 405, and 430, and 446 are typical types. Their main use is where resistance to corrosion is the primary factor. Possibly, one of the most used groups of stainless steels is the austenitic. It includes the 2 and 3 series and is also coded as three digits. Some common types are 202, 301, 310, and 321. The chemistry of this group of steels provides nonmagnetic steels at room temperature which are extremely resistant to corrosion. However, this soft metal in the RB hardness scale work hardens very rapidly during forming operations. For its highest resistance to corrosion it must be annealed and hardness removed. This is accomplished by soaking the metal at 1900 °F (1038 °C) and quenching in water. A small quantity of titanium or niobium in the analysis helps prevent carbide precipitation and corrosion, should prolonged heating occur in the precipitation range. A fourth type of stainless steel in this general code category is the precipitation hardening stainless steels. Several types are available, such as the 17-4PH. This steel is solution-treated by air cooling from 1900 °F (1038 °C) and then caused to precipitate or age by reheating to approximately 950 °F (1038 °C), resulting in an RC 45 hardness value.

FIGURE 9–41 This Command and Reconnaissance armored carrier has high speed durability. Tough alloys are used in its construction to resist battlefield environments. (Courtesy of Cleveland Army Tank-Automotive Plant, General Motors Corporation)

Classification of Gray Iron

Gray iron castings are coded by several systems, such as the ASTM and SAE. Again, no logical means for knowing the chemistry of the metals is shown in the code. The SAE uses the three-digit system, such as a 120 or 122, meaning the minimum tensile strength is 35,000 psi and 45,000 psi, respectively. According to the ASTM system, two digits are used, for example, 30 for 30,000 psi tensile strength and 40 for 40,000 psi tensile strength, and so on. With respect to cast iron, it must again be pointed out that the common irons should normally be used in compression loading due to the inherent brittleness of the gray and white types if used in tension at very high stresses. As an example, the ASTM class 40 gray iron has a compression strength of approximately 140,000 psi which is more than three times stronger than in tensile. During the tensile test for common cast irons, no indication of failure is obvious even as the yield is reached because extremely little plastic flow occurs. Also, the nature of graphite and its mean orientation to the lines of stress make the common irons unpredictable in certain tension situations. Keep in mind that graphite is a nonmetal and cannot retain the higher stresses. On the other hand, as the common iron begins to fail in compression, there are warning signals such as small splinters of the metal popping away from the specimen or a crumbling action which forms a powder. No further loading is possible as complete fracture is eminent.

There are exceptions with regard to the use of cast iron, however. Special alloy types have demonstrated remarkable mechanical properties in tension, compression, shear, and impact. For example, the acicular irons containing molybdenum are used in making some types of crankshafts. One particular acicular iron for a diesel engine crankshaft has developed tensile strengths equal to structural steels and a high degree of toughness as shown by Charpy impact tests. Heat treatment provides higher tensile properties. The needlelike or acicular structure results as the cast iron transforms on cooling. Low transformation temperatures, below 550 °F (288 °C) develop a martensitic pattern while higher transformation temperatures produce the pearlitic types. The acicular microstructure surrounds the graphite flakes or nodules and provides different mechanical properties than the martensitic and pearlitic irons.

Questions

1. How does recrystallization occur in a metal?
2. What is the relationship between grain size and temperature in austenite?
3. Describe pearlite.
4. Describe the relationships between pearlite and ferrite in hypoeutectoid steels.
5. Explain anisotropy.
6. What is the purpose of the iron-iron carbide diagram?
7. What is the purpose of normalizing? When is it done?
8. Describe the relationship between pearlite and cementite in hypereutectoid steels.
9. Why is annealing done? How is it done?
10. Compare the hardening process with annealing.
11. Explain two common classification systems for steels.
12. Differentiate among pearlite, fine pearlite, and martensite.
13. When is tempering performed? What is its purpose?
14. How does the thickness of metal affect the rate of cooling? How does the rate of cooling affect the microstructure?
15. Describe the microstructure of the high speed steel after quenching and then after tempering.
16. What is the purpose of spheroidizing?
17. Explain the time-temperature transformation of austenite in a carbon steel. Compare this transformation with an alloy steel.
18. Differentiate between martempering and austempering.
19. What is the purpose of alloys in steels?
20. Describe the different types of cast irons.

Metallurgy of Nonferrous Metals

The nonferrous group of metals is large; it logically includes all those metals which are not classified as ferrous. Basically, all metals which are not wrought irons, steels, or cast irons are nonferrous. Because most of the elements are metals and because most of the metals are nonferrous, alloys number in the thousands. In addition, the elements antimony, arsenic, and silicon have properties of metals and nonmetals and are commonly alloyed with other metals, further increasing the number of alloys. It has been previously pointed out that an element is the pure material. When two or more elements combine, some kind of alloy or compound forms which is usually completely unlike the materials from which it is made. Some combinations of two or more elements make the numerous compounds, such as iron carbide, and some make the alloys, such as brass. Therefore a long list of intermetallic compounds and alloys is available. When the numerous combinations of the intermetallics and alloys combine, the list of metals is extended. In the ferrous group carbon plus iron provides the compound; when more iron is mixed with the compound, along with silicon, manganese, phosphorus, and sulfur, the new material is called steel. Likewise, in the nonferrous group when copper and aluminum combine chemically, copper aluminide is produced; when more aluminum is added to the copper aluminide an aluminum alloy is produced (Fig. 10–1).

ORIGIN OF NONFERROUS METALS

Nonferrous metals, like the ferrous metals, originate mostly from the earth, but the sea and the atmosphere also furnish several raw materials. The processing of the nonferrous ores is similar to the ferrous; however, nonferrous procedures often include additional processing which is complicated and expensive. Even though the blast furnace and refining furnace are used for some nonferrous metal production, much of the processing involves controlled atmospheric conditions, vacuum en-

FIGURE 10–1   An annealed aluminum alloy microstructure. The dark particles are compounds of copper-aluminide.

vironments, or electrolysis. A detailed explanation of nonferrous metal production and processing is given in another text, *Manufacturing Processes.*

SIMILARITIES OF ALL METALS

All metals, both ferrous and nonferrous, solidify in a granular pattern, and the grains vary in size from metal to metal. These small or large grains can be equiaxed or rounded or they can be elongated due to plastic deformation by force in the forming operation. Also, the inside regions of all these grains have some kind of arrangement. If the metal is an element, there is only one arrangement of the material, but the material can be soft, hard, or somewhere in between due to cold deformation bringing on hardness or high temperatures reducing this hardness. When two or more elements are included in the metal the arrangement within the grains has several possible patterns. Consider the comparison of a room to a grain. The air in the empty room (uniformity) is comparable to the pure metal in a grain. But when ten identical chairs are placed in the room, numerous arrangements of the chairs can be made within the room. And so it is when particles of iron carbide or copper aluminide are dispersed within a grain of metal (Fig. 10–2); the patterns of arrangements of the chemical compounds are many. As an extension of the above example, grain boundaries are comparable to the walls and floors of the room. Many grains make up a metal just as several rooms make up a house.

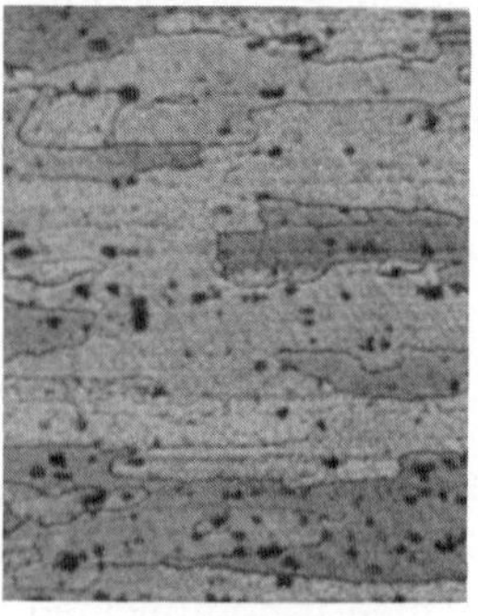

FIGURE 10–2   Heat-treated aluminum alloy showing the solution of copper and aluminum and some $CuAL_2$ after precipitation.

Internal Arrangement of Metals Is Variable

To continue our comparison, when boxes are mixed among the chairs in this room, completely different arrangements are possible among three different things. Likewise, alloys produce different compounds, and when these are mixed with the existing compounds and base materials different patterns of arrangement again result. As the arrangements of the compounds and other phases within the numerous grains change, so do the mechanical properties. Large pieces of copper aluminide, for example, mixed in the base metal allow softness and weakness to exist because more of the softer aluminum solution is exposed (Fig. 10–1). On the other hand, when this same quantity of hard copper aluminide is dispersed more thoroughly throughout the softer metal of the grains into much smaller pieces, the hardness and strength of the grains increase because less of the soft metal is exposed. It is not the size of the intermetallic chemical compound that causes a rise in hardness, but rather the dispersion of the compound. Small pieces are as hard as the larger pieces, but there are more small pieces than large ones.

Another similarity among all metals is the geometric shapes, such as cubes and triangles, which make up their atomic arrangements. The atoms form in groups called cells, and these cells multiply to form grains, and grains form a section of metal. Much of this phenomenon occurs concurrently. The particular arrangements of the atoms within the cells allow movement of the atoms in certain directions better than in other directions. These possible directions of movement are known as slip planes where rows of atoms are organized, even though now and then an atom may be completely missing or the atom may be misaligned. The existence of slip planes has been discussed in a previous chapter. Apparently, then, the main difference between ferrous and nonferrous metals is the material, and the material is, of course, an array of atoms organized within the grains.

Similarities of Microstructures

Much of what has been discussed in ferrous metallurgy is also applicable to nonferrous metallurgy. That is, the microstructure of the metal includes its grain size and shape and the arrangement of constituents inside the many small or large grains. The chemistry of the metal then establishes the physical and potential mechanical properties of the particular metal. One main difference between the ferrous and nonferrous metals, however, is the method of obtaining higher strengths. Most all metals can be work hardened and annealed and many of the nonferrous alloys can be strengthened by heating and cooling processes. It is then a matter of terminology of heat treatments that differentiates between the steels, for example, and the nonferrous alloys. For instance, some steels can be hardened and tempered but aluminum alloy is given the solution and precipitation treatments. Annealing pertains to both ferrous and nonferrous metals. Normalizing and stress relieving apply to the ferrous metals but only stress relieving is applicable to the nonferrous group. Steels are tempered and nonferrous alloys are aged, although not all steels are tempered (some have no capability) and not all nonferrous metals are aged (some have no capability).

Capability or hardenability begins with the chemistry of the metal and this chemistry must be analyzed to establish the physical and mechanical property potentials. The analysis may be simple or complex such as in the following examples. Zinc plus copper produces a soft brass and, excluding cold work and annealing, the alloy remains fixed in its strength properties much like austenitic steel because it is a solid solution. But aluminum plus copper establishes a completely different system of potentials much like the many hardenable steels.

ALUMINUM AND ITS ALLOYS

Among the nonferrous family of metals, aluminum and its alloys are used more than any other metal (Table 10–1). Several reasons account for the tremendous

TABLE 10–1 CHEMICAL COMPOSITION OF ALUMINUM ALLOYS, PERCENT AVERAGE

Alloy	Silicon	Iron	Copper	Manganese	Magnesium	Chromium	Nickel	Zinc	Titanium
1100	1.00	Si+Fe	0.20	0.05	—	—	—	0.10	—
2011	0.40	0.70	5.50	—	—	—	—	0.30	—
2024	0.50	0.50	4.30	0.60	1.50	0.10	—	0.25	—
3003	0.60	0.70	0.20	1.12	—	—	—	0.10	—
4043	5.20	0.80	0.30	0.05	0.05	—	—	0.10	0.20
5052	0.45	Si+Fe	0.10	0.10	2.50	0.25	—	0.10	—
6061	0.60	0.70	0.25	0.15	1.00	0.20	—	0.25	0.15
6063	0.40	0.35	0.10	0.10	0.67	0.10	—	0.10	0.10
7075	0.50	0.70	1.60	0.30	2.50	0.29	—	5.60	0.20
7079	0.30	0.40	0.60	0.17	3.30	0.17	—	4.30	0.10
7178	0.50	0.70	2.00	0.30	2.80	0.29	—	6.80	0.20
8001	0.17	0.58	0.15	—	—	—	1.10	—	—

Courtesy Aluminum Association

tonnage used, including light-weight–high-strength ratio for the alloys, reasonable cost and availability, excellent corrosion resistance factors, and ease of fabrication and machining. The tensile strength of aluminum in the pure condition averages about 10,000 psi with a yield of 5,000 psi. When the copper-magnesium-zinc alloy (7001 T6) is heat treated (Table 10–2) tensile strengths approaching 100,000 psi are available.

The range of tensile strengths shown in Table 10–2 is from 13,000 to 98,000 psi and the yields range from 5000 to 91,000 psi. Elongations reduce rapidly as the yield strengths increase. Because yield and shear properties have acceptable values for aerospace vehicles, many of the modern aircraft include some of these higher strength alloys. In designs involving cyclic loading the fatigue limit is great enough to safely handle numerous situations.

Properties of the Aluminums

Aluminum melts at 1220 °F (660 °C) and weighs 0.0975 lb/in³ while iron weighs 0.2845 lb/in³ and melts at a temperature more than twice that of aluminum. This face-centered cubic metal has a modulus of elasticity of 10 million psi (Table 10–3) while that of steel is approximately 30 million psi. Copper-aluminum and copper-zinc-aluminum alloys weigh approximately one-third as much as steel and have acceptable strengths for medium stressed parts, especially at subzero temperatures, so these alloys are used extensively in the transportation, construction, and

TABLE 10–2 MECHANICAL PROPERTIES OF SOME ALUMINUM ALLOYS

Alloy and Temper	Tensile (psi)	Yield (psi)	Elongation .500 in Dia % 2 in	Hardness (Brinell 500 kg)	Shear (psi)	Fatigue Limit (psi)
1100 O	13,000	5,000	45	23	9,000	5,000
2011 T3	55,000	43,000	15	95	32,000	18,000
2011 T8	59,000	45,000	12	100	35,000	18,000
2024 O	27,000	11,000	22	47	18,000	13,000
2024 T3	70,000	50,000	18*	120	41,000	20,000
2024 T4	68,000	47,000	19	120	41,000	20,000
2024 T361	72,000	57,000	13*	130	42,000	18,000
ALCLAD 2024 T81	65,000	60,000	6*	—	40,000	—
3003 O	16,000	6,000	40	28	11,000	7,000
5052 O	28,000	13,000	30	47	18,000	16,000
6061 O	18,000	8,000	30	30	12,000	9,000
6061 T4	35,000	21,000	25	65	24,000	14,000
6061 T6	45,000	40,000	17	95	30,000	14,000
6063 O	13,000	7,000	—	25	10,000	8,000
6063 T1	22,000	13,000	20*	42	14,000	9,000
6063 T4	25,000	13,000	22	—	—	—
6063 T5	27,000	21,000	12*	60	17,000	10,000
6063 T6	35,000	31,000	12*	73	22,000	10,000
7001 T6	98,000	91,000	9	160	—	22,000
7075 O	33,000	15,000	16	60	22,000	—
7075 T6	83,000	73,000	11	150	48,000	23,000
7079 T6	78,000	68,000	14	145	45,000	23,000
7178 O	33,000	15,000	16	—	—	—
7178 T6	88,000	78,000	11	—	—	—

*1/16 in thick
Courtesy Aluminum Association

TABLE 10–3 MODULUS OF ELASTICITY IN TENSION

Material	E Value (psi)
Structural steel	30,000,000
Wrought iron	27,000,000
Aluminum alloys	10,000,000
Copper-nickel alloy	26,000,000
Brass	12,000,000
Gray cast iron	12,000,000
Tungsten	50,000,000
Nickel	30,000,000
Molybdenum	47,000,000
Cast tin	6,000,000
Titanium	16,000,000
Chromium	36,000,000
Cobalt	30,000,000
Copper	16,000,000
Lead	2,000,000
Magnesium	6,000,000
Manganese	23,000,000
Wood, soft	800,000
Wood, hard	2,000,000
Concrete	4,000,000
Carbon	700,000
Nylon	400,000

FIGURE 10–3   This helicopter, model 204-UH-1, is manufactured from high strength-light weight materials including bonded honeycomb assemblies. (Courtesy of Bell Helicopter Company)

aerospace industries (Fig. 10–3). The aluminum alloy 2024 T86 has about the same tensile strength as common structural steel with only one-third the weight, but it is also only one-third as stiff as steel. Like steels, the aluminums are available in cast and wrought forms and in all standard shapes and sizes. When the aluminums are exposed to temperatures above 400 °F (204 °C) their strengths decline rapidly.

American National Standard Code for Aluminums

Like the ferrous family of metals (AISI), the aluminums are coded by industry and ASTM (ANSI H35.1) (American National Standard Institute) in order to identify many of the various alloys. We will discuss a different coding system, however. The four-digit system differentiates among the eight main types of aluminum alloys. The first number on the left indicates the alloy number or type. The second digit indicates the modification of the original alloy, and the last two digits identify the particular alloy. As in steels, the chemistry of the alloy is found in chemical analysis charts.

CODE FOR ALUMINUM GROUP

Code	Type	Example
1_ _ _	Aluminum	1100 (99% Al)
2_ _ _	Copper	2024 (Al-Cu)
3_ _ _	Manganese	3003 (Al-Mn)
4_ _ _	Silicon	4043 (Al-Si)
5_ _ _	Magnesium	5052 (Al-Mg)
6_ _ _	Magnesium and silicon	6061 (Al-Mg-Si)
7_ _ _	Zinc	7075 (Al-Cu-Mg-Zn)
8_ _ _	Other element	_ _ _ _

Temper Designations for Wrought and Cast Forms

F   As fabricated
O   Annealed (wrought only)
H   Strain hardened (wrought only)
W   Solution treated
T   Heat treated

Temper Subdivisions

H1   Strain hardened only
H2   Strain hardened and partially annealed
H3   Strain hardened and stabilized
T1   Cooled from shaping and naturally aged
T2   Annealed castings
T3   Solution treated and cold worked
T4   Solution treated and naturally aged
T5   Cooled from shaping and artificially aged
T6   Solution treated and artificially aged
T7   Solution treated and stabilized
T8   Solution treated, cold worked, and artificially aged
T9   Solution treated, artificially aged, and cold worked
T10 Cooled from shaping, artificially aged, and cold worked

Additional digits may be added to T1–T10:

T51 Stress relieved by stretching after solution treatment
T52 Stress relieved by compressing after solution treatment

For variations T4 and T6 the following are added:

T42 Solution treated and naturally aged
T62 Solution treated and artificially aged

There are many more temper subdivisions.

Reprinted by permission of the American Society for Testing and Materials from copyright material.

Microstructure of Aluminum and Other Nonferrous Metals

The same basic principles of material relationships which apply to the steels apply to the microstructures of the nonferrous metals. In other words, iron or ferrite is similar to aluminum in that it cannot be hardened by heat treatment but can be cold worked and annealed. Cold working disturbs the grain structure and annealing rearranges the granular condition. The arrangements of the intermetallics, such as iron carbide (Fe_3C) in steel and copper aluminide ($CuAl_2$) in aluminum, strongly determine the resulting strengths of the metals.

The alloy of aluminum and copper, 2024, for example, consists of several elements—iron, silicon, manganese, magnesium, chromium, zinc, and about 4% copper. It is mainly the effects of the copper, however, which give the alloy its outstanding characteristics. Copper and aluminum combine chemically, as has been pointed out, to form the hard intermetallic compound $CuAl_2$. $CuAl_2$ acts in much the same way in the aluminum alloys as Fe_3C does in steels. If the $CuAl_2$ is finely dispersed, the alloy is hard; if the $CuAl_2$ is coarsely dispersed, the alloy is soft. Naturally, from a given amount of the chemical compound, the smaller the size of individual particles, the greater the dispersion and the stronger the metal. Fine dispersions require heating and rapid cooling very similar to steel heat-treating practices, except for the delayed hardening in many of the nonferrous alloys brought on by aging (Fig. 10–2). Coarse dispersions require slow cooling to increase the size of the $CuAl_2$ particles (Fig. 10–1), freeing more of the compound from the matrix and, therefore, making the alloy softer as in the case of pearlite and ferrite.

Differences Between Steels and Aluminums

A main difference between the steels and aluminums is melting points, which greatly influence the heating and cooling processes. Just as steels require different temperatures for heat treating, so do the aluminums. Heat-treating or solution-treating temperatures for the aluminums are much lower than in the steels because of the lower melting temperature of aluminum alloys. When copper, which melts at 1981 °F (1083 °C), is alloyed with aluminum, which melts at 1220 °F (660 °C), in the amount of approximately 5%, the melting point of the resulting alloy is about 950 °F (510 °C). Even though the melting point of copper is much higher than aluminum, the presence of the two elements in certain percentages causes a eutectic to form. The eutectic is the alloy which freezes from the liquid at a lower temperature than any other chemical combination of the alloy. It will also begin to melt before other chemically different regions melt on heating. Therefore, the solution treatment temperature of the alloy must never enter the eutectic temperature zone because of the formation of small liquid pools of the alloy which ruin the metal (Fig. 10–4). The part must be scrapped because the wrought (pressure-

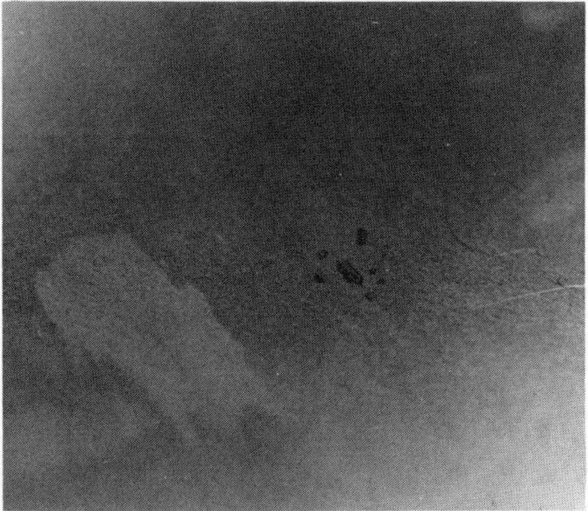

FIGURE 10–4 Eutectic melting of the copper-aluminum solution has occurred on the surface of this solution-treated copper-aluminum alloy. Dark pimples in the center are small castings resulting from overheating the metal.

formed) metal contains small castings and is consequently weaker in the several strengths. Eutectic melting is identified as small blisters on the metal's surface. As for the solution treatment temperature, it has characteristics of austenitic temperatures in steel which provide a solid solution of the constituents. In steel, however, the melting point is much higher than the austenitic or solution temperature, while in many of the nonferrous metals, the solution temperature is often close to the metal's melting point. Therefore, great care must be exercised in the heating and soaking of the nonferrous alloys at solution temperatures.

Aluminum-Copper Diagram

For steels, there is the iron-iron carbide diagram (Fig. 8–28) for use in studying
the system of ferrous alloys; for the aluminums, there are aluminum-copper and
other alloy diagrams. The most used portion of the aluminum-copper system is
illustrated in Figure 10–5. Observe that the percent of copper in aluminum is

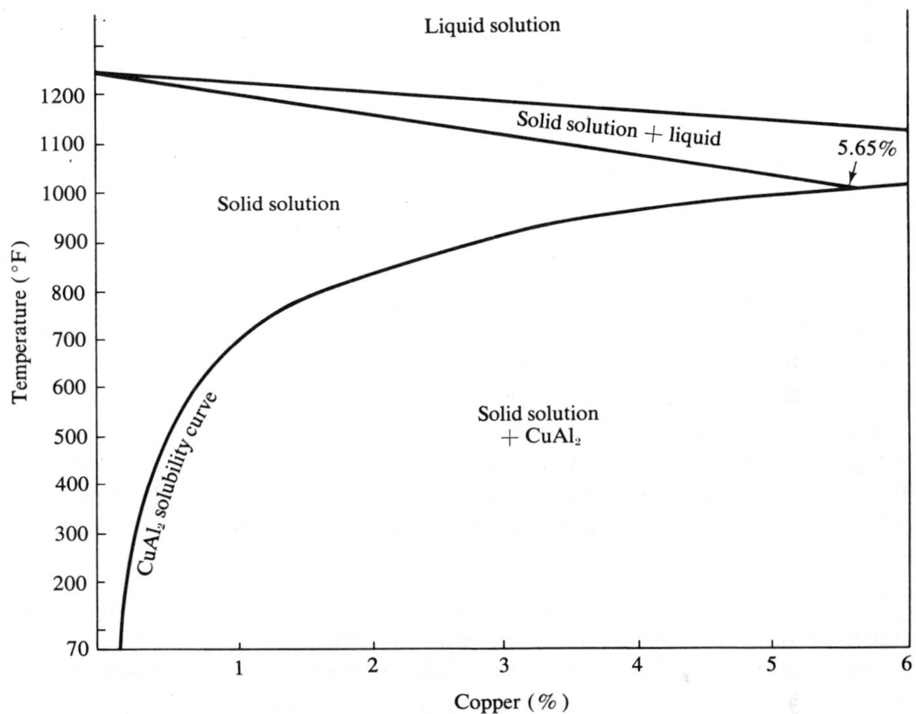

FIGURE 10–5   Aluminum-copper diagram.

plotted against temperature, just as the percent of carbon in steel is plotted
against temperature. A much different solubility temperature exists, however,
which is entirely dependent on a constantly increasing temperature for more of
the solution to form. At 1018 °F (548 °C) the maximum solubility of copper
in aluminum is 5.65%. As the temperature decreases, the solubility also decreases. As a result, some of the copper-aluminum alloys can be heat treated by
heating them to just above the $CuAl_2$ solubility temperature line and quenching
them rapidly in cold water. The cold water quench retains the solid solution at
room temperature, just as martensite does carbon in steel. However, steel hardens
instantly through a unit cell shearing action whereas the aluminum alloys require time to expel their captured intermetallic compound particles. In other
words, the heating and quenching of the aluminum alloy are known as the
solution treatment, and the time and associated temperature factors required to
harden are known as the *precipitation treatment.*

Solution and Precipitation Treatments   The alloy 2024, for example, is solution treated by soaking at 925 °F (496 °C) for a period of time, depending on

its thickness or mass, and quenching in cold water. Quenching must be extremely rapid; the time consumed between opening the furnace door and submersion of the part in the liquid must be a matter of a few seconds. This drastic requirement produces good resistance to corrosion and high strength properties after aging. Specifically, temperature must be reduced very quickly in the range of 750–500 °F (399–260 °C). Agitated water is preferred to assure loss of steam pockets. Then, as quenched, a supersaturated solid solution exists, just as existed at 925 °F (496 °C). However, the alloy, much unlike martensite, is soft, weak, and extremely ductile. In this solution condition, a phenomenon peculiar to the 2024 alloy is its ability to provide its own precipitation treatment by lying idle at room temperature for at least 48 hours. During this supposedly inactive period and because 70 °F (21 °C), its precipitation temperature, happens to be room temperature, an internal microscopic change occurs. Tiny particles of the $CuAl_2$ precipitate from the solution over the long time period and provide intermetallic atoms in the face-centered cubic lattice. Consequently, hardness and strength increase proportionally to the rapidly increasing particles of copper aluminide. After about two days the tensile strength increases from a soft and weak condition to approximately 70,000 psi, only then being ready for structural use. Caution must be exercised, therefore, to assure that the "as quenched" alloy is not immediately called upon for a load reaction; failure will occur if loading is imposed on the part before precipitation is mostly complete. Fabricators take advantage of the very ductile "as quenched" condition and shape the metal. Precipitation is completed as time passes at room temperature and then the alloy is strong enough to use.

Salt Treatment   If molten salt is used instead of air during the heating process, all traces of the salt must be removed after treatment to prevent corrosion. As with any molten salt bath, never allow cold or damp metal to enter the bath (Fig. 10–6) and never allow cyanide, magnesium, or combustibles to enter the nitrate crucible. One mistake may cause an explosion. As for solution treatments of the alloys in air or salt environments, they vary. The 6061 alloy, for example, requires about 980 °F (527 °C) followed by water quenching. As quenched, the alloy is soft and weak and will remain in this condition unless artificially aged at 350 °F (177 °C) for eight hours followed by air cooling. In other words, the 2024 alloy is a naturally aging alloy, and 6061 is an artificially aging alloy. Results of intermetallic precipitation from the solution are the same except that the 2024 alloy is somewhat stronger in tensile strength than the 6061 alloy.

Additional Temperature Treatments   Two additional temperature treatments are applicable to the aluminum alloys. Often it is not feasible to form a sheet of solution-treated alloy within the required 2-hour period. Fortunately, subfreezing temperatures forbid precipitation of the intermetallic compound for an indefinite period of time. Solution-treated alloys can therefore be frozen after the quench and retained in the supersaturated condition until needed. For example, rivets solution-treated today (Fig. 10–7) can be driven while soft next week without splitting if they have been maintained in a frozen condition. Precipitation and strength begin to appear in the rivets or other parts as temperatures reach room temperature

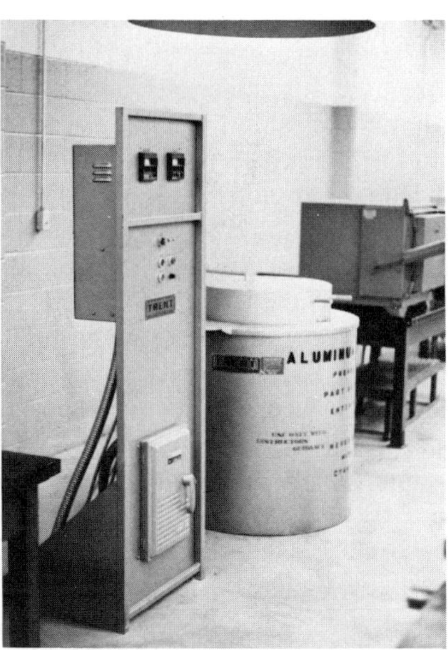

FIGURE 10–6   A typical salt bath furnace for heat treating small aluminum alloy parts. The crucible contains sodium nitrate, a salt; therefore, the part must be thoroughly washed with water after the solution treatment.

FIGURE 10–7   These aluminum alloy rivets have been given the solution treatment and are ready to be placed in refrigeration at about 0 °F (−18 °C) so that precipitation of the strengthening compound will not occur.

and near maximum strength is reached in 48 hours. On the other hand, it is sometimes necessary to soften an alloy for fabrication purposes. Softening or annealing treatments include soaking the alloy at about 800 °F (427 °C), cooling in the furnace to below 500 °F (260 °C), and then air cooling. Such a treatment causes segregation and enlargement of the intermetallic compound (Fig. 10–1), reducing the hardness and strength factors. Quick annealing occurs by soaking at 650 °F (343 °C) and air cooling, the hardness being higher than in regular annealing.

Use of Aluminum and Its Alloys

Aluminum and its alloys are used for numerous items, such as cooking utensils, household appliances, land vehicles, ships, trains, aircraft, sporting goods, toys, space vehicles, buildings, and bridges. The use determines the alloy and the alloy determines the treatment. The least alloyed, or nearly pure metal, types are used in the H, or cold-formed, condition and are not heat treatable for strengths. The highly alloyed aluminums, such as the 2024, 6061, and 7075, are formable in the O condition as well as in the solution-treated condition. Precipitation treatments provide high strengths comparable to some of the steels. With respect to the alclad, or aluminum-coated, alloy, prolonged reheating will diffuse the core into the aluminum skin and decrease corrosion resistance of the skin. An example of several mechanical properties of the aluminums is shown in Table 10–2. It is interesting to note, according to the table, that alloy 2024-T4 has a tensile strength of 68,000 psi with a yield of 47,000 psi, an elongation of 19%, a Brinell (500 kg) hardness of 120, a shear strength of 41,000 psi, and a fatigue limit of 20,000 psi. This fatigue limit is based on 500,000,000 cycles of completely reversed stress.

Such a combination of mechanical properties in the aluminum alloys provides excellent structural applications in many fields. Actually, some heat-treated aluminum alloys compete with the structural steels in several mechanical properties, but not all. As environmental temperatures change, mechanical properties also change. Typical tensile properties at various temperatures are indicated in Table 7–4. In order to correlate mechanical properties with chemistry of several aluminum alloys, their chemistries must be studied. Several of these alloys are indicated in Table 10–1. As expected, the aluminums are readily available in wrought and cast forms similar to the steels. The forging alloy 6061 has structural strength, is light weight, and it has a tensile strength in excess of 45,000 psi, a yield of 40,000 psi, a shear strength of 30,000 psi and an elongation factor of 17%. Proper heat treatments provide these properties. In respect to the 2024-T81, a thermal treatment following the solution treatment is required and consists of soaking at 375 °F (191 °C) for 12 hours and air cooling. Before the last thermal treatment is a cold working operation such as bending and stretching.

Aluminum Provides for Several Systems   The main constitution diagrams of the whole aluminum system are similar to the aluminum-copper system (2______); 2024, for example, is used primarily in high strength structures. The aluminum-manganese alloys (3______) are weldable, as are the several 1______ series designations. Aluminum-silicon alloys (4______) make good castings and are also easily forged. Aluminum-magnesium alloys (5______) are weldable and produce good castings. The aluminum-silicon-magnesium alloys (6______) make fine forgings which respond to artificial aging treatments. Several compounds form in this alloy to give excellent mechanical and physical properties. Aluminum-zinc alloys (7______) produce some of the strongest alloys, a few of which exceed the mild carbon type of structural steel in tensile strength. As for utilization, the aerospace industry uses many of these alloys in its aircraft (Fig. 10–8) and space vehicles (Fig. 10–9) because the weight of the aluminum is only one-third the weight of steel. Titanium alloys are also used

FIGURE 10–8 This F-111 combat aircraft is constructed of exceptionally high strength metals. Its military capability has proven to be one of the greatest weapons systems ever manufactured. (Courtesy of General Dynamics, Fort Worth Division)

FIGURE 10–9 An artist's concept of the future space shuttle *Orbiter* as it jettisons the external fuel tank shortly after reaching earth orbital altitude. The tank, which carries propellants for *Orbiter's* main engines, enters atmosphere and impacts in a preselected ocean area. *Orbiter* returns to Earth in the same manner as aircraft. Heat shielding on its leading edges protects the vehicle from disaster. (Courtesy of Rockwell International)

because they are about one-half the weight of steel. As illustrated in Figure 10–8, the F-111 is one of the most advanced weapons systems in the world. Much of this superior performance is based on the use of high strength materials and the incorporation of new materials in its design in addition to the dependable aluminums. Many of these new materials have been discussed in previous sections of this text.

MAGNESIUM AND ITS ALLOYS

Magnesium is the lightest in weight of all the metals in commercial use. When alloyed and heat treated the alloy is nearly equal to the tensile strength of low carbon steel and is only 25% as heavy. Magnesium is obtained mainly from sea water and its availability is nearly unlimited. One cubic mile of seawater contains hundreds of tons of magnesium. In view of this unlimited supply, magnesium is being used in new and unusual applications such as anodes along pipelines to prevent corrosion of the steel pipe (Fig. 10–10). Several elements

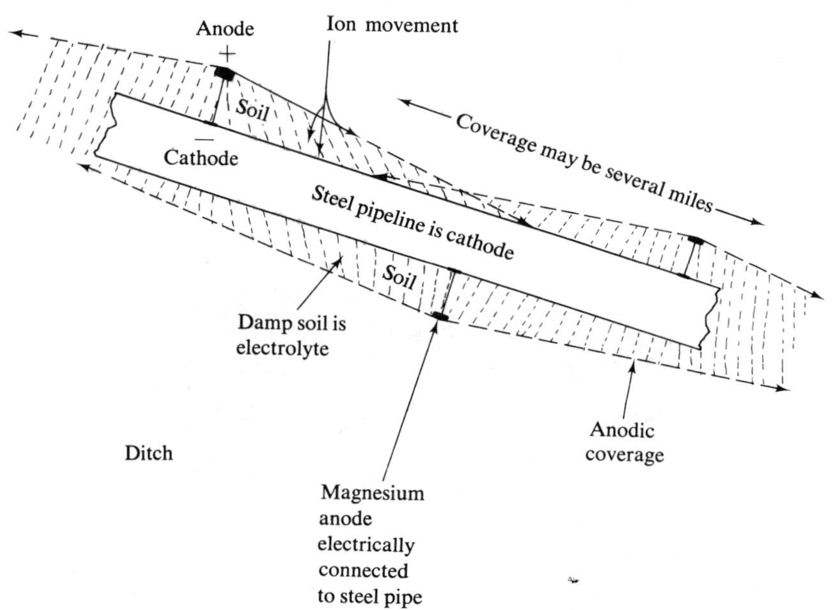

FIGURE 10–10   Magnesium anodes prevent steel pipe from corroding due to loss of anode to cathode and protection of cathode. Magnesium ions move to bare surfaces of pipe, the magnesium eventually needing to be replaced.

combine well with magnesium to form a long list of alloys similar to the numerous aluminum alloys (Table 10–4). In order to identify these alloys the coding system shown in Table 10–5 is helpful. Like the aluminums, the magnesiums are available in both the cast and wrought shapes. Also, the alloys respond to heat treatments in similar ways as the aluminum alloy groups. Magnesium melts at 1202 °F (650 °C), is close-packed hexagonal in lattice structure, and weighs 0.0627 lb/in³. The alloys have different melting temperatures and these vary with the metal's chemistry. When forming operations are required the metal is heated to approximately 500 °F (260 °C) and formed while hot. Cold forming often fractures the metal due to lack of plasticity at room temperature.

TABLE 10–4 CHEMICAL PROPERTIES OF MAGNESIUM ALLOYS

Alloy	Composition				
	Al	Mn	Zn	Th	Zr
AM100A-T6	10.0	0.10	—	—	—
AZ63A-T6	6.0	0.15	3.0	—	—
AX81A-T4	7.6	0.13	0.7	—	—
AZ91C-T6	8.7	0.13	0.7	—	—
AZ92A-T6	9.0	0.10	2.0	—	—
ZK51A-T5	—	—	4.6	—	0.7
ZK61A-T5	—	—	6.0	—	0.7
HK31A-T6	—	—	—	3.3	0.7
HZ32A-T5	—	—	2.1	3.3	0.7
ZH42	—	—	4.0	2.0	0.7
ZH62A-T5	—	—	5.7	1.8	0.7
AZ91A & B	9.0	0.13	0.7	—	—
M1A-F	—	1.20	—	—	—
AZ318-F	3.0	—	1.0	—	—
AZ61A-F	6.5	—	1.0	—	—
AZ80A-T5	8.5	—	0.5	—	—
HM31XA-F	—	1.20	—	3.0	—
ZK21A	—	—	2.3	—	0.45
ZK60A-T5	—	—	5.5	—	0.45
AZ31B-H24	3.0	—	1.0	—	—
HK31A-H24	—	—	—	3.0	0.60
HMZ1A	—	0.60	—	2.0	—
PE	3.3	—	0.7	—	—

By permission, from *Metals Handbook* Volume 1, Copyright American Society for Metals, 1961.

Codification of Light Metals and Alloys (ASTM B-275)

The American Society for Testing and Materials has established a coding system for cast and wrought light metals including the magnesiums. Table 10–5 represents the alloying elements for the particular metal. This coding system is another attempt to simplify the interpretation of the growing mass of alloys of all types. It too is like the AISI coding system for some steels in that only an approximation can be made as to the alloy's contents.

From Table 10–5 a given alloy can be identified by its main contents. For example, AZ91A is an aluminum-zinc alloy. *A* represents aluminum as the alloy used in greatest amount, *Z* represents zinc as the alloy used in second greatest

TABLE 10–5 CODING SYSTEM FOR LIGHT METALS

A	Aluminum	M	Manganese
B	Bismuth	N	Nickel
C	Copper	P	Lead
D	Cadmium	Q	Silver
E	Rare earths	R	Chromium
F	Iron	S	Silicon
G	Magnesium	T	Tin
H	Thorium	Y	Antimony
K	Zirconium	Z	Zinc
L	Lithium		

Courtesy American Society for Testing and Materials

amount, *9* represent the round-off mean of aluminum percentage between 8.6 and 9.4, *1* represents the round-off mean of zinc from 0.6 to 1.4, and *A* is the first alloy assigned under designation AZ91. The letters and numbers indicate the chemical analysis of the particular alloy, but only to the extent of the code number capacity. As shown in Table 10–4, the alloys contain two or three elements and some include the temper designations. Temper designations of the magnesiums (ASTM B-296) follow the same general designations as for the aluminums which have been previously pointed out. For ease of reference, the ASTM standards should be available for those persons working with metals and specifications in both the ferrous and nonferrous groups. These standards are available from the American Society for Testing and Materials in Philadelphia.

Characteristics of the Magnesiums

Some of the characteristics of the magnesiums include the inherited brittleness under bending loads at room temperatures (Fig. 10–11). This brittleness is due

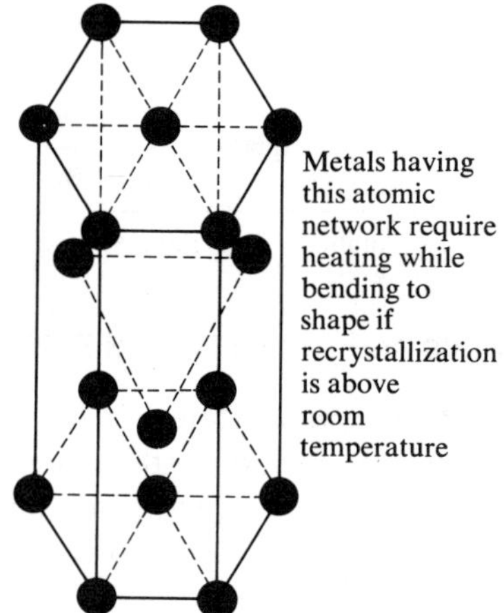

FIGURE 10–11   The hexagonal, close-packed lattice of magnesium. This atomic arrangement of seventeen atoms causes metals like magnesium and titanium to be brittle at room temperature.

to the close-packed hexagonal structure of magnesium, but as temperatures are raised to approximately 500 °F (260 °C), a shifting of atoms to the body-centered cubic arrangement occurs and ductility is induced because of the availability of more slip planes. Therefore, magnesium is hot workable, but no small radii can be produced in the cold condition.

Another very important characteristic of magnesium alloys is their ability to produce and precipitate compounds into the regions of slip planes. Foreign atoms of the intermetallic compounds such as magnesium-aluminum, magnesium-silicon, and others, form along the slip planes during precipitation and increase resistance to shifting of the other atoms. Such a lattice deformation adds rigidity in the metal and, in turn, strength is improved. Several elements form compounds with magnesium, each having its particular beneficial effects. Controlled quantities of aluminum, silicon, zinc, tin, manganese, beryllium, thorium, or zirconium improve the mechanical properties of the base metal. Impurities in magnesiums must be controlled. In steel, phosphorus and sulfur are normally considered as impurities, but in the magnesium systems, iron, nickel, and copper are impurities. Even though the added elements improve several mechanical properties, the modulus of elasticity remains about 6,500,000 psi which makes magnesium ideal for use in aerospace applications (Fig. 10–12), especially with its light weight.

FIGURE 10–12   This intricately shaped aircraft part is cast from magnesium alloy which has an excellent strength-weight ratio.

In all design selections of materials, the material chosen is a compromise among several properties. For example, magnesium and its alloys corrode quickly in corroding atmospheres, especially while the metal is under stressed conditions, so the surface must be protected. Another peculiarity of magnesium is its ability to catch fire and burn like wood, coal, or gasoline. Its burning temperatures reach 3200 °F (1760 °C) (Fig. 10–13), more than enough to melt steel, but it is quickly extinguished by removal of oxygen by smothering with sand or sulfur dioxide. Vertical structures of magnesium, such as the cabin sides of some space-vehicles or earth-type vehicles, which are burning, will not retain the granular materials and can be extinguished with boron trifluoride, a standard fire extinguisher. Never use water on magnesium fires because water contains oxygen and merely increases the hazard. However, *small sections* of burning magnesium can be quickly extinguished in water because the temperature reduction capacity of water is greater than the fire support capacity of small sections of metal.

Heat Treatment of Magnesium Alloys

Magnesium compound-producing alloys are susceptible to precipitation hardening and increased strengths, one of the primary purposes of the alloys. During precipitation of the compounds from the solid solution, hardness and strength increase in an amount relative to the quantity of compound material available under

FIGURE 10–13   Magnesium will catch fire and burn at a temperature of 3200 °F (1760 °C) in the presence of air. Sulfur dioxide or sand will extinguish the fire.

equilibrium conditions. As previously stated, the solid solution of carbon in face-centered cubic iron rapidly precipitates to iron carbide and iron when a carbon steel part is quenched in oil. The presence of the finely dispersed pieces of the compound, Fe_3C, quickly induce hardness during the quench. But, in many nonferrous metals the compound dispersion requires time and controlled temperatures. In the aluminum-copper system, both naturally and artificially precipitated alloys increase strengths through actions of the harder and extremely small pieces of precipitated $CuAl_2$ moving throughout the matrix of the softer base metal. And so it is with other nonferrous metals which have hardening compounds in their chemistries.

Magnesium-Aluminum Diagram

In order to increase strength by heat treatment in the magnesium-aluminum system, the particular alloy is heated to the solution temperature, as indicated in Figure 10–14. The region bounded by letters *ABC* is the solid solution. Below the region *CB* is the eutectic plus the solution. It is the heating of the metal's constituents to above the temperature line *CB* that causes the change from a mixture to a solution to occur. Therefore, the solution treatment includes the soaking of the metal at a temperature above line *CB* but below line *ABE* and quenching rapidly in cold water. As a result, the solution is retained at room temperature with its temporary accompanying weakness. Precipitation and strength occur artificially; therefore, heat must be applied to cause precipitation of the hardening compounds because 70 °F (21 °C) is not sufficient. Remember that failure to accomplish the required precipitation treatment can result in quick material failure under load. According to Figure 10–14, about 2% aluminum is dissolved in magnesium at room temperature while 12% dissolves at 819 °F (437 °C). Consequently, when a temperature of approximately 780 °F (416 °C) is applied to a Mg-Al alloy containing 8.5% aluminum, a solution is formed. Rapid quenching retains the solution. Reheating to about 400 °F (204 °C) for several hours brings

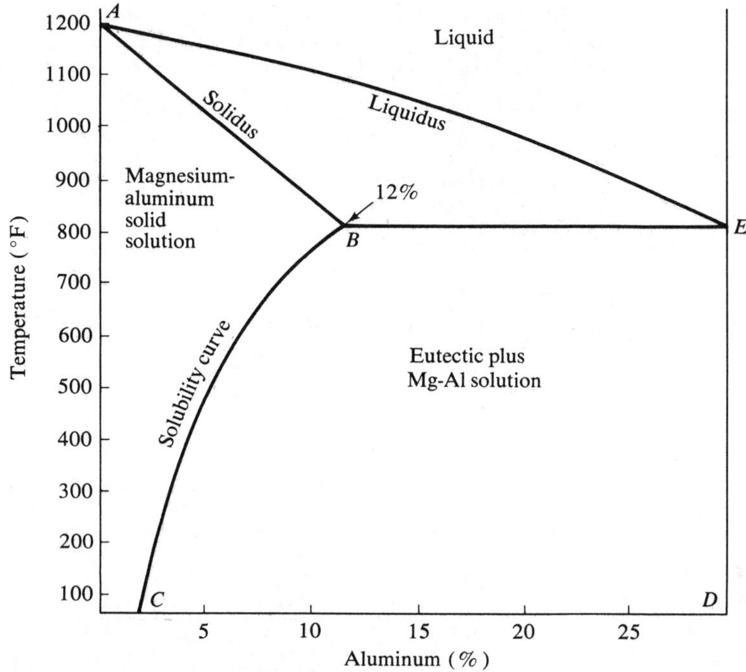

FIGURE 10–14   Magnesium-aluminum diagram.

out the strength through precipitation of the compound. The net strength increase, that is, the difference between loss of strength due to heating and gain in strength due to precipitation of compounds, makes the artificially aged metals important in structural engineering applications. With regard to forging or heat treatment and similar processing, the metal should not be exposed to oxidizing atmospheres because of fire hazards and surface contamination problems. To prevent fire and contamination the atmosphere should be inert or at least 0.3% surfur dioxide should be present.

Use of Magnesium and Its Alloys

Magnesium and its alloys are used in numerous industrial and aerospace assemblies. Automotive and aerospace castings provide good strength characteristics along with light weight. Surface protection treatments include anodic and chemical. Various household appliances and commercial hardware are made from magnesium, as well as optical holders and sporting goods. Aircraft contain many parts made from magnesium, such as instrument cases. Table 10–6 includes typical mechanical properties of several rolled magnesium alloys. Extruded tubing is available with tensile strengths up to 35,000 psi, but rolled products are produced at tensile strengths up to 46,000 psi with an elongation factor of 12%. A typical forged and aged aircraft crankcase alloy containing 8% aluminum, 0.6% zinc, and small amounts of manganese and silicon has a tensile strength of 42,000 psi, a yield strength of 26,000 psi, and an elongation of 5%. Mechanical properties of several rolled shapes are shown in Table 10–7. Products are readily available

TABLE 10–6 MECHANICAL PROPERTIES OF ROLLED MAGNESIUM PLATE

Alloy	Temper*	Tensile (psi)	Yield (psi)	Shear (psi)	Elongation (% in 2 in)
AZ31B	T	35,000	20,000	24,000	17
AZ31B	T	37,000	20,000	26,000	20
AZ31B	T	42,000	28,000	29,000	15
AZ31B	T	42,000	32,000	29,000	15
AZ31B	O	37,000	22,000	26,000	21
HK31A	T	38,000	30,000	26,000	9
HK31A	O	33,000	20,000	24,000	23
HM21A	T	35,000	23,000	19,000	11

*For specific treatment consult manufacturer's bulletin. Courtesy of Dow Chemical U.S.A.

TABLE 10–7 MECHANICAL PROPERTIES OF ROLLED MAGNESIUM SHAPES

Alloy	Temper	Tensile (psi)	Yield (psi)	Shear (psi)	Elongation (% in 2 in)
AZ31B	F	38,000	28,000	19,000	14
AZ31C	F	38,000	28,000	19,000	14
AZ61A	F	46,000	33,000	23,000	17
AZ80A	F	49,000	36,000	22,000	12
AZ80A	T5	55,000	38,000	24,000	8
HM31A	T5	44,000	39,000	27,000	10
ZK60A	F	49,000	38,000	24,000	14
ZK60A	T5	53,000	44,000	26,000	11

Courtesy of Dow Chemical U.S.A.

in the extruded shapes—bars and rods, tubing, forgings, sheet and strip, and special hollow shapes. Many of the alloys are weldable and all are machinable. Annealing is accomplished by soaking the part at approximately 500 °F (260 °C) and air cooling.

COPPER AND ITS ALLOYS

The alloys of copper are more numerous than other nonferrous alloys, mainly because of the many combinations available when different elements are added to copper. Copper, an element, is soft and weak, but it is extremely corrosion resistant. Copper is heavier than aluminum and magnesium and is slightly heavier than steel. It weighs 0.324 lb/in^3 while its alloys weigh less, some as low as 0.302 lb/in^3. It melts at 1981 °F (1083 °C), and as elements are added to form alloys the melting points are reduced. Copper is a very ductile metal, being face-centered cubic in lattice structure. The modulus of elasticity is 16 million psi, approximately one-half the stiffness of steel. However, copper has limited usage because of weak tensile strength. It can only be increased in hardness by cold work, such as hammering or rolling, but it can be annealed by water or air quenching from 1000 °F (538 °C). The rate of cooling from any temperature has no effect on the strength of an element such as copper. However, water quenching loosens the copper oxide scale and leaves a clean surface because of differences in coefficients of contractions between the copper and oxide. With respect to electrical conductivity, power transmission facilities use copper as the medium. Being face-centered in its lattice arrangement, copper can be reduced to subzero temperatures and still conduct several times more electrical current than at room temperature. Consequently, it is used wherever electricity must be transmitted such as in the high voltage power lines where resistance to corrosion is essential.

The Main Copper Alloys

Alloys of copper include the three main alloys—brass, bronze, and the nickel variations (Table 10–8). Brasses are divided into the alpha and beta types, and the

TABLE 10–8 CHEMICAL COMPOSITIONS OF SELECTED COPPER ALLOYS

Alloy	Composition (%)									
	Cu	Zn	Pb	Sn	Fe	Si	Ni	Al	Mn	Other
Red brass	85	15								
Low brass	80	20								
Cartridge brass	70	30								
Yellow brass	65	35								
Muntz metal	60	40								
Leaded brass	65	34	1							
Naval brass	60	39.25		0.75						
Manganese bronze	58.5	39		1	1.4				0.1	
Silicon bronze	96					3				1
Nickel-silver	55	27					18			
Aluminum bronze	82				2.5		5	9.5	1	

alpha is again divided into the yellow and red types. Common yellow alpha brass contains about 70% copper and 30% zinc and has many common uses (Fig. 10–15). Specifically, a brass used for hinges is made from an alloy containing 69% copper, a trace of iron and lead, and the remainder is zinc. Another brass, the naval type, consists of 60% copper, 0.5–1.0% tin, 0.4–1.0% lead, 0.1% iron, and the balance is zinc. Naval brasses have tensile strengths in the annealed condition of 55,000 psi, while in the one-half hard condition (cold rolled) the tensiles are increased to 60,000 psi with yields of 26,000 psi and an elongation of 30% in 2 inches. The common yellow brasses range from tensile strength of 50,000 psi

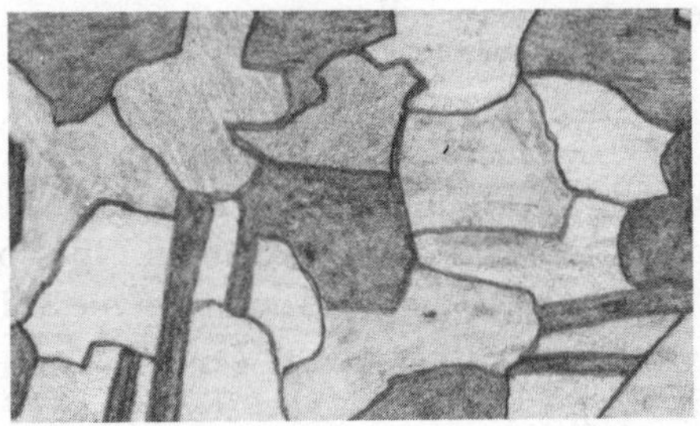

FIGURE 10–15 This microstructure of common yellow brass lends itself very favorably to deforming operations such as bending and cupping because of the plasticity of the solid solution.

to more than 120,000 psi when cold rolled to harder tempers. Cold rolling is performed at temperatures below the metal's recrystallization temperature, which varies according to the kind of metal, while forming loads are above the yield strength and below the tensile strength. Such high stresses often induce season cracking, however, which is either transgranular or intergranular corrosion (Fig. 10–16). A low stress relief (500 °F, or 260 °C) of cold-worked brass will help

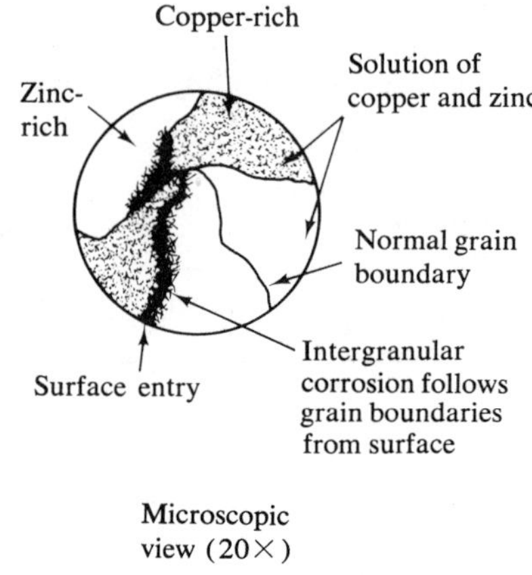

FIGURE 10–16   Intergranular corrosion in yellow brass.

reduce this tendency to crack. Higher temperatures will remove most or all of the stress. After cold rolling and when these brasses are exposed to about 800 °F (427 °C), they soften and lose their strengths which were imparted by the cold-working processes. Brasses and other solutions having no precipitates are not heat treated for increased strength.

The Red Brasses   Another type of brasses, the red, contains less zinc so these metals have a color closer to the brown color of copper. These alloys are less susceptible to season cracking than the yellow brasses. Commercial bronze, a misleading term, contains only 10% zinc, while gilding metal consists of 5% zinc. A bronze may even include a low zinc alloy. Another example of red brass is an alloy which includes 85% copper, 0.06% iron, 0.15% lead, and the remainder zinc. This general purpose alloy has a tensile strength of approximately 39,000 psi. Annealing is carried out as for yellow brass.

Brass Alloys

Even though brass is an alloy, variance in its chemistry enables the metal to provide a wide range of properties. Therefore, two main types of solutions, alpha and beta, are possible with reference to microstructural types. Their main constituent is copper which has a tensile strength of about 25,000 psi as drawn, while zinc has a tensile strength of 15,000 psi. Zinc melts at 787 °F (419 °C) and reduces the copper's melting point of 1981 °F (1083 °C), point *A* on the copper-zinc diagram (Fig. 10–17), to points *BGE*. Only the copper-rich portion of the diagram is shown, which points out that the alpha solid solution persists up to about 38%

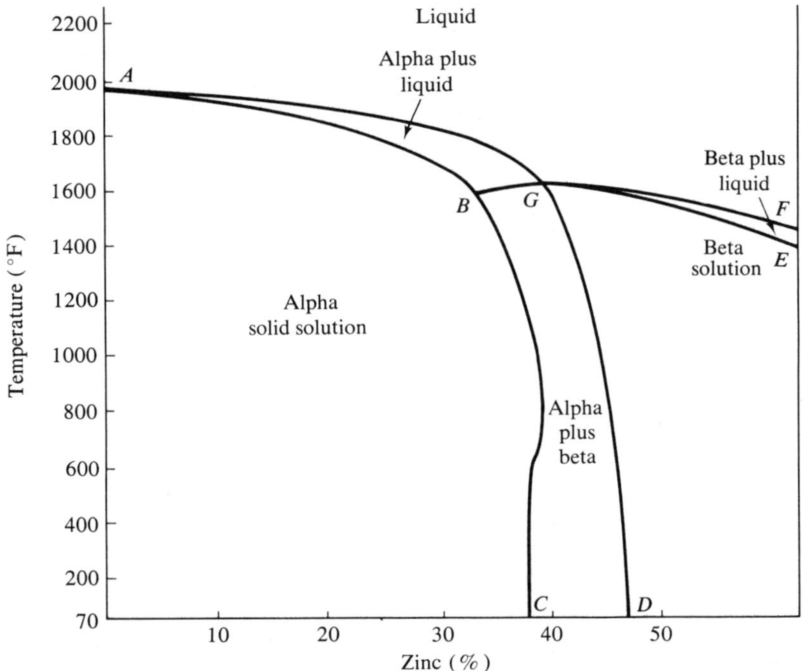

FIGURE 10–17   Copper-zinc diagram.

zinc, point *C*, up to approximately 850 °F (454 °C). The copper-zinc solution phase at room temperature is a face-centered cubic lattice structure of zinc in copper in which the zinc atoms randomly find copper atom vacancies and substitute for the copper atoms (Fig. 10–18). Such a lattice continues to maintain the alpha solution from below room temperature up to the 38% factor, point *C*. Increasing temperatures cause the body-centered cubic lattice to appear with the formation of the beta phase. Increasing temperatures or more zinc provide more beta. Temperatures above the line *ABGE* include the liquid and a solid phase.

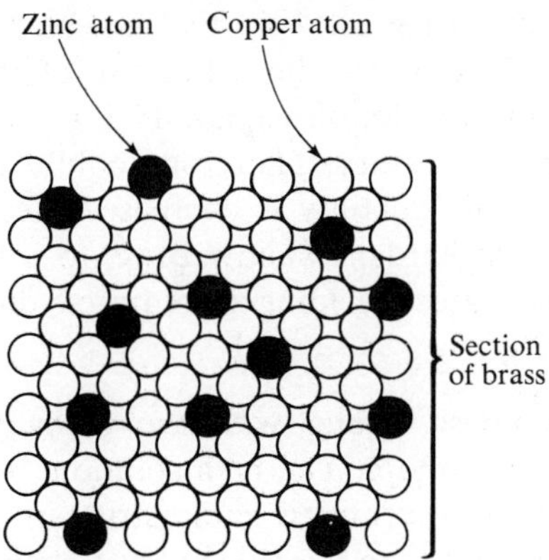

FIGURE 10–18   Substitutional atoms in the copper-zinc system.

Below this temperature-composition line is the solid metal. Above the line *AGF* is all liquid. Alloys having more than 50% zinc are not common. As the per cent of zinc increases the melting points of the alloys are reduced.

The series of alloys are solid solutions and cannot be heat treated for increased strength. But when tin or some other compound-forming element is added to the copper, phases develop and changes allow compound precipitation along with increased hardness and strength. (Table 10–9 points out mechanical properties

TABLE 10–9   MECHANICAL PROPERTIES OF SELECTED COPPER ALLOYS

Alloy	Condition	Tensile (psi)	Yield (psi)	Shear (psi)	Elongation (%)
Red brass	annealed	40,000	12,000	31,000	47
Red brass	half hard	57,000	49,000	37,000	12
Red brass	extra hard	78,000	61,000	44,000	4
Jewelry bronze	annealed	39,000	12,000	31,000	47
Jewelry bronze	spring	74,000	62,000	48,000	4
Commercial bronze	hard	61,000	54,000	38,000	5
Gilding bronze	half hard	48,000	40,000	34,000	12
Copper	annealed	33,000	10,000	22,000	45
Copper	hard	53,000	50,000	29,000	10
Cartridge brass	annealed	49,000	17,000	34,000	57
Cartridge brass	half hard	62,000	52,000	40,000	23
Yellow brass	annealed	46,000	14,000	32,000	65
Yellow brass	spring	91,000	62,000	47,000	3
Manganese bronze	annealed	65,000	30,000	42,000	35
Manganese bronze	hard	75,000	50,000	45,000	30
Phosphor bronze	annealed	50,000	32,000	31,000	48
Phosphor bronze	hard	84,000	83,000	32,000	9
Silicon bronze	annealed	63,000	30,000	45,000	55
Aluminum bronze	hard	85,000	55,000	45,000	30
Aluminum bronze	heat treated	132,000	80,000	—	1.5

By permission, from *Metals Handbook* Volume 1, Copyright American Society for Metals, 1961.

of some copper alloys.) When about 5% zinc is added to copper the tensile strength increases about 10,000 psi, yet the tensile strength of zinc is only 60% of copper. When 30% zinc is added the resulting solution has a tensile strength of 45,000 psi, an increase of 20,000 psi from the element. These alpha brasses have high elongation factors, exceeding 50% in the annealed condition. Extreme ductility results in the copper-zinc alpha alloys because of the resulting face-centered cubic lattice of copper and its solution capability for zinc. But if higher strengths are needed, cold working is necessary. Therefore, cold working increases hardness, strength, and brittleness, while ductility is decreased. Severe cold deformation, as in compression testing, work hardens yellow brass to a brittle condition whereby it shatters like glass. Therefore, shielding is required during compression testing of these types of metals.

Beta Brass   The beta phase, along with the alpha, appears in those alloys containing about 38% zinc or more. The beta solution, being body-centered cubic, reduces the ductility of the face-centered cubic structure of the alpha phase. Beta brass is more difficult to work and cracks easier than alpha. At hot working temperatures, both the alpha and beta phases are plastic, however. An example of beta phase is found in Muntz metal. This 60% copper and 40% zinc metal contains both the alpha and beta phases, which establishes increased strength in the metal.

Bronze

Bronze is primarily an alloy of copper and some element such as tin. The name has also come to include what appears to be the brass family of copper and zinc. Because copper is the base metal, the type of bronze is indicated by the element added to produce the particular kind of bronze. There are several bronze systems, each having its primary effects on the alloys as a result of heating and cooling. The copper-aluminum system, for example, is very similar to the copper-zinc system because both the alpha and beta phases form. However, the copper-aluminum alloy can be heat treated for increased strength in much the same manner as steel.

Copper-Tin Alloys   Other bronze alloys include the copper-tin system. Copper and tin comprise the true bronzes, and these are modified by additions of another element such as phosphorus, silicon, manganese, or aluminum. Tin increases the hardness of copper in the alpha solid solution which results when the alloy is cooled from temperatures above 800 °F (427 °C). As the tin content increases, the solid solution increases in hardness. Phosphor bronze produces sound castings because the phosphorus increases fluidity in the molten metal. When lead is added to molten copper and allowed to cool, a mixture of copper and lead results which provides for good bearing surfaces. Silicon increases the hardness of copper and adds greater corrosion resistance to the bronze.

Other Bronzes   As in most bronzes, the added element, up to a certain amount, forms an alpha solid solution which results in increased hardness and strength. As an example, a 5% silicon bronze in the annealed condition has a tensile strength of 50,000 psi with an elongation of 35%. Manganese and copper produce a beta phase with an alpha phase mixed with the beta. Such a condition with two solid solutions, one mixed in the other, increases hardness and strength. Parts requiring additional toughness have manganese in the two solutions; however, this particular alloy may be considered a brass because of its 40% zinc content. Depending on the degree of cold rolling, tensile strengths of this manganese bronze range from 55,000 psi to 120,000 psi and elongation factors from 15% to 30%. Another manganese-copper alloy contains 0.50% manganese, 0.25% aluminum, 1.0% iron, 0.20% lead, 1.0% tin, 60.0% copper, and the remainder is zinc. This alloy has a tensile strength of 115,000 psi, a yield strength of 68,000 psi, and an elongation of 10% when rolled hard. When annealed this same alloy is reduced to 85,000 psi in tensile, to 45,000 psi in yield, and to an elongation of 20%.

Heat Treatment of Bronze   Unless an additional phase such as beta or gamma can be precipitated from a quenched solid solution, no further hardness can be produced in brass except through work hardening. The heat treatable alloys produce different phases, just as in steels. A 10% copper-aluminum alloy, for example, produces an additional phase. Its phase diagram is similar to the copper-zinc diagram. The base, or matrix, material is the alpha solid solution in which a eutectoid phase is embedded. When the mixture of the eutectoid and alpha solution is soaked in the red temperature range of about 1500 °F (816 °C) and then quickly quenched, a hardened structural condition results. When this structural condition is exposed to about 1000 °F (538 °C), the "as quenched" beta phase transforms to an alpha-gamma combination and further increases hardness and strength. Another heat treatable bronze is the beryllium-copper type. This copper alloy containing 2% beryllium, 0.4% nickel, and the remainder copper will produce more than 200,000 psi in tensile strength after quenching from 1450 °F (788 °C) and aging at 600 °F (316 °C) for 3 hours.

Uses of Copper and Its Alloys

Because copper alloys have excellent resistance to corrosion and fairly good strength properties, they are used for many commercial hardware components and for marine-type fittings. Copper makes excellent wire for electrical power transmission as well as sheet and plate stock for parts requiring corrosion resistance. Brass is used for numerous hardware items such as hinges, turnbuckles, screws, door locks, valves, pumps, tubing, and numerous forgings and castings. Bronze is used where increased corrosion resistance is needed along with higher mechanical properties such as propellers, springs, bearings, gears, valves, tubing, forgings, and castings (Fig. 10–19).

FIGURE 10–19 Typical parts made from bronze include these gears and bearings.

NICKEL AND ITS ALLOYS

The element nickel melts at 2647 °F (1453 °C), weighs 0.322 lb/in³, is face-centered cubic, has a modulus of elasticity of 30 million psi, and readily combines with other elements such as copper, iron, and chromium. Pure nickel is weak in tensile strength, being equivalent to iron (about 45,000 psi), and its yield is low, 9000 psi, due to its very high elongation of approximately 30%. Nickel resists oxidation at high temperatures and resists corroding effects of saltwater and acids. Its use in the alloys consumes more tonnage than any other purpose. When alloyed, the resulting alloys melt at lower temperatures than their constituents. Copper melts at 1981 °F (1083 °C), iron at 2797 °F (1536 °C) and chromium

TABLE 10–10 CHEMICAL COMPOSITION OF SELECTED NICKEL ALLOYS

Alloy	Ni	C	Mn	Fe	Si	Cu	Cr	Ti	Al	Mo	Co
Nickel 211	95	0.1	4.75	0.05	0.05	0.03					
Permanickel	98.6	0.25	0.10	0.10	0.06	0.02		0.50			
Duranickel	94	0.15	0.25	0.15	0.55	0.05		0.50	4.5		
Monel 400	66	0.12	0.90	1.35	0.15	31.50					
Monel K	65	0.15	0.60	1.00	0.15	29.50					
Inconel 600	76	0.04	0.20	7.20	0.20	0.10	15.8				
Inconel 718	52.5	0.04	0.20	18.0	0.20	0.10	19.0	0.80	0.60	3.0	
Hastelloy C	bal.	0.08	1.0	5.5	1.0	—	15.5	—	—	16.0	2.5

at 3407 °F (1875°C). Chemical compositions of some nickel alloys are shown in Table 10–10.

Nickel-Copper Alloys

When an alloy having approximately 66% nickel and most of the remainder copper forms a solid solution, a very good corrosion resistant alloy results. One of the trade names of this metal is Monel. Copper is completely soluble in nickel, and the resulting solid solutions throughout the whole range produce tough and corrosion resistant alloys with mechanical properties greater than the copper-zinc alloys. At cold-working temperatures, the metal is soft, but it work hardens rapidly. Softening and stress relief are accomplished by water quenching from approximately 1600 °F (871 °C) to 1800 °F (982 °C). Again, it is pointed out that the solid solution has characteristics of the pure metal in that increased strength by heat treatment is not possible.

When about 3% aluminum is added to the alloy, however, precipitation hardening is possible. Such a heat treatment requires water quenching from about 1720 °F (938 °C) to provide a homogeneous solid solution and softening. This treatment is followed by reheating to 1100 °F (593 °C) for up to 16 hours, subsequent furnace cooling to 1000 °F (538 °C) for about 5 hours, cooling to 900 °F (482 °C) for 7 hours, followed by air cooling. Cooling should not exceed approximately 15 °F per hour down to 900 °F (482 °C). Alloys of this type which are hot worked can subsequently be age hardened without water quenching because it is the very slow cooling that hardens the alloy, just the reverse of water hardening steel. Tiny chemical compounds move slowly from the solution throughout the matrix and provide the additional phase for hardening and strengthening to occur. Some of these alloys reach tensile strength of more than 165,000 psi after aging. While in the annealed condition, forming can be accomplished as tensile strengths are approximately 70,000 psi with an elongation of 21%.

Nickel-Iron Alloys

Nickel and iron also form solid solutions, and nickel has drastic effects on the gamma-alpha transformation rate. In steels these transformation rates are helpful in producing desired mechanical properties. Because they are solid solutions, however, the nickel-iron alloys are not heat treatable, except those in which an additional phase is obtainable by additions of other elements such as carbon. The ferritic alloys contain up to 6% nickel and readily work harden but do not respond to strength increases by heat treatment. Special alloys with nickel contents within the range of about 6% to 30% are martensitic and are heat treatable to varying strengths. Alloys with nickel contents beyond 30% are nonmagnetic and austenitic and retain their austenitic condition regardless of cooling rates; therefore, these cannot be heat treated for increased strengths. A unique feature of the nickel-iron system is the alloy Invar in which thermal expansion coefficients in the range of room temperature and seasonal temperature fluxuations are nearly negligible. Very little expansion occurs in the range from 32 F° (0 °C) to 212 °F (100 °C). This 36% nickel-iron alloy is water quenched from 1525 °F (829 °C), reheated to 275 °F (79 °C), and then air cooled for expansion stability. It is used where thermal expansion is not allowable such as in static instruments for measuring or in time devices. High nickel content irons approximate the expansion and contraction coefficients of some nonmetals such as glass and are therefore used in fabrication of precision instruments where seals are required.

Nickel-Iron-Chromium Alloys

Many binary alloys of the nickel-chromium series and the ternary alloys of the nickel-chromium-iron series are used in corrosion resistant groups of metals. Also, these alloys are used in numerous applications in the electrical heating element industries where resistance to electricity causes rapid heating and yellow and white temperatures cause no appreciable surface oxidation. Both strength and chemical stability are retained at temperatures up to 2000 °F (1093 °C) and beyond in some alloys. Nichrome, Alumel, and Chromel are trade names for some of these alloys (Fig. 10–20). The combination of 20% chromium and 80% nickel makes excellent heating elements (Fig. 10–21). Often, up to 23% iron is added to the nickel-chromium base metal for use in all kinds of electric irons and toasters as well as heat treating furnaces.

An alloy containing 16% chromium, 8% iron, and 76% nickel is known as Inconel and has found extensive use in the elevated temperature ranges of red

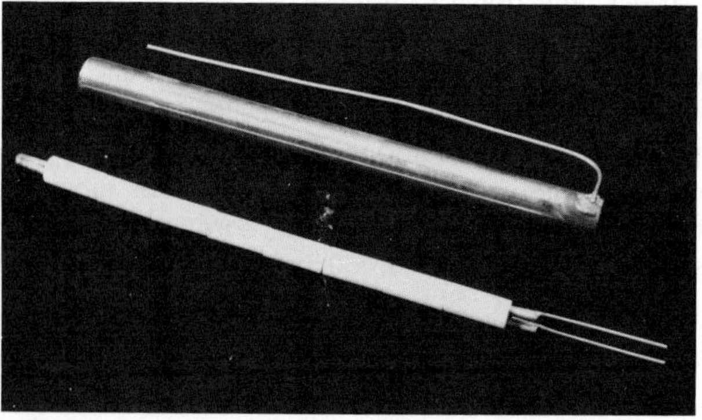

FIGURE 10–20 This thermocouple is made of a Chromel and an Alumel wire twisted together at one end and fused by welding. The open ends are attached to leads and these are connected to the controlling pyrometer. Many thermocouples are housed in a protection tube as shown.

FIGURE 10–21   This melting furnace is heated by elements made of nickel and chromium which surround the crucible. High voltage causes amperes to flow through the metal which resists the flow of current and consequently becomes white hot. (Courtesy of Lindberg Hevi Duty)

and yellow heats. The alloy resists oxidation at these high temperatures and has sufficient strength to perform its function. It is used in food containers for food preparation processes as well as in industrial heat treating processes such as in muffles and boxes for carburizing. When maximum resistance to hot sulfuric acid is needed the nickel-silicon-copper alloy is used. Due to the nickel alloys' capability in the temperature range from around 0 °F (−18 °C) to 1750 °F (954 °C), the several alloys have found extensive universal usage.

Questions

1. Define microstructure and describe its charactistics.
2. Name and describe the two main factors in a metallic microstructure.
3. What is meant by recrystallization of a metal? What are its effects?
4. Differentiate between ferrous and nonferrous metals.
5. How is grain refinement accomplished? Describe two methods of grain refinement.
6. What is an intermetallic compound?
7. Illustrate a substitutional type of solid solution.
8. What are the differences among the alpha, beta, and gamma phases in nonferrous metals?
9. Describe the solution and precipitation treatments which apply to some nonferrous metals.
10. Differentiate between natural and artificial aging.

11. How can aging or precipitation be retarded in some solution-treated metals?

12. Compare the mechanical properties of metals having face-centered cubic, body-centered cubic, and close-packed hexagonal lattice structures.

13. How can magnesium be formed without cracking?

14. When extensive deformation is to be performed on brass what metallurgical treatments are accomplished?

15. What happens during eutectic melting in the process of heat treatment?

16. What is the purpose of a salt bath heat treatment?

17. List some dangers associated with handling molten salts.

18. How does annealing soften a metal?

19. How does the chemistry of a metal influence its physical and mechanical properties?

20. Differentiate between eutectic and eutectoid.

11

Other Nonferrous Metal

In addition to the common structural types of nonferrous metals there are the special types of pure metals and alloys ranging from the low-melting types to the special structural and nuclear fuel types. This vast array of special metals includes the solders and fusible alloys, the precious metals used in coinage and for jewelry, the unique solar cell metals, the refractory metals, lightweight-high strength alloys, nuclear fuel metals and others with unusual characteristics. The availability and use of these metals are greatly responsible for the present state of technology.

TIN, LEAD, BISMUTH, CADMIUM, AND ZINC ALLOYS

The low-melting metals include tin, lead, zinc, bismuth, and cadmium. Recall that when two or more metals combine to form a system of alloys various mechanical and physical properties result; some properties are completely unlike those of the original metals. For example, when two metals are alloyed the resulting alloy often melts at a lower temperature than either of its constituents. The melting points of the low-melting elements are as follows: tin, 449 °F (232 °C); lead, 621 °F (327 °C); zinc, 787 °F (419 °C); bismuth, 520 °F (271 °C) and cadmium, 609 °F (321 °C). An alloy of equal amounts of tin and lead (solder) melts at 421 °F (216 °C). When an alloy consisting of 22.6% lead, 8.3% tin, 5.3% cadmium, and 44.7% bismuth is heated it melts at 117 °F (47 °C). By varying the percentages of these elements a range of melting temperatures can be produced from 117 °F (47 °C) to nearly 500 °F (260 °C).

These low-melting alloys have various uses. One is used to close the nozzle on a fire extinguisher by soldering. The extinguisher system is at a water pressure of 70 psi and is ready to extinguish a fire which melts the fusible alloy of solder when the temperature reaches 160 °F (71 °C). As the solder melts, the water,

TABLE 11-1 CHEMICAL COMPOSITION OF EUTECTIC FUSIBLE ALLOYS

Item	Melting Temp. (°F)	Composition (%)				
		Bi	Pb	Sn	Cd	Other
A	117	44.7	22.60	8.30	5.30	19.10 In
B	136	49.0	18.00	12.00	—	21.00 In
C	158	50.0	26.70	13.30	10.00	—
D	197	51.60	40.20	—	8.20	—
E	203	52.50	32.00	15.50	—	—
F	217	54.00	—	26.00	20.00	—
G	255	55.50	44.50	—	—	—
H	281	58.00	—	42.00	—	—
I	288	—	30.60	51.20	18.20	—
J	291	60.00	—	—	40.00	—
K	351	—	—	67.75	32.25	—
L	362	—	38.14	61.86	—	—
M	390	—	—	91.00	—	9.00 Zn
N	430	—	—	96.50	—	3.50 Ag
O	457	—	79.70	—	17.70	2.60 Sb
P	477	—	87.00	—	—	13.00 Sb

By permission, from *Metals Handbook* Volume 1, Copyright American Society for Metals, 1961.

which is under pressure, puts out the fire according to prearrangements of the nozzles. Table 11–1 indicates the various compositions of several fusible alloys.

Tin

Tin melts at 449 °F (232 °C), is highly corrosion resistant, has a tetragonal lattice structure, and weighs 0.264 lb/in³. It is a basic constituent in the white metals (alloys of the low-melting metals), pewter (Sn 91%, Sb 7%, Cu 2%), solder (Pb 37%, Sn 63%), babbit no. 1 (Sn 90.9%, Sb 4.52%, Cu 4.56%), and fusible alloys. Tin is also used as cladding for steel plate which is used in large quantities in the canned food industry as tin-plate. Another important use of tin is in alloys and for anodes in electroplating processes. Tin has a modulus of elasticity of six million psi. Cast tin has a tensile strength of about 3100 psi, being very ductile due to its elongation factor of 54%. By far the greatest use of tin has been in tin-plate followed by solder. Lesser amounts of tin are used in the numerous alloys.

Lead

Lead melts at 621 °F (327 °C), retards or forbids penetration of waves of X and gamma radiation depending on its thickness, has a face-centered cubic lattice structure, has a modulus of elasticity of 2 million psi, and weighs 0.4097 lb/in³ (steel weighs 0.2839 lb/in³). Lead is a basic constituent in many of those alloys containing tin. It is also used in small amounts to cause steel to machine easier. Some of its greatest uses include plates for wet batteries, ammunition projectiles, sheathing on cables, chemical tank liners, anodes for electroplating, radiation barriers, type metal, solder, and alloys. Lead recrystallizes at a temperature below room temperature; therefore, the folding and bending of the metal does not work harden it. In other words, as fast as permanent deformation occurs the metal recrystallizes and quickly removes any granular deformation or stress. Consequently, lead remains soft because it is hot worked as it is folded and deformed at room temperature.

Lead Alloys   Arsenic is an alloy of lead. This poisonous element is added to certain bearing metals to instill stability of strength at operating temperatures up to about 300 °F (149 °C), for example. Arsenic retards softening tendencies as heat is generated and prolongs fatigue life. From 0.25% to 1.0% arsenic in an alloy helps produce excellent automotive engine bearings; the remainder of the alloy, including 15% antimony and 1% tin, is lead. A lead base type metal alloy (electrotype) made from 2.5% antimony and 2% tin is heat treatable by water quenching the part from approximately 450 °F (232 °C) and then aging at approximately 200 °F (93 °C) for 2 hours. As in all low-melting alloys, tensile strengths are low. For example, a typical arsenical lead has a tensile strength of approximately 2400 psi with an elongation of 40%.

Tensile properties increase rapidly when antimony is alloyed with lead. Basically, as the antimony content increases the tensile strength and hardness also increase up to about 10% antimony. The strengths of these alloys range up to approximately 7500 psi. When tin and lead are alloyed the resulting alloy increases in tensile strength as the tin content increases. There is also the accompanying positive correlation between tensile strength and hardness. Equal combinations of tin and lead produce tensile strengths in the same general strength ranges as the lead-antimony alloys, or about 7200 psi, even though the antimony content is only about 25% of the tin content in the lead-tin alloys.

Zinc

Zinc melts at 787 °F (419 °C), is highly corrosion resistant, has a close-packed hexagonal lattice,and weighs 0.2577 lb/in^3. It, too, is a basic constituent in the low-melting alloys. Possibly the second largest use of zinc is in corrosion prevention as in steel cladding. The familiar spangles on the zinc-coated low carbon steel sheet identifies it as galvanized iron (Fig. 11–1). Zinc is anodic to iron and is therefore

FIGURE 11–1 Spangles on the surface of zinc-coated steel sheet commonly called galvanized iron.

used in the presence of steel where corrosion is a factor. Scratched galvanized sheet or zinc-coated wire fencing does not rust because of cathodic protection given to the steel by the zinc. Ships use slabs of zinc bolted to their sides below the water line to eliminate saltwater rusting of the steel plates. When zinc is used as the anode and steel as the cathode in the presence of water, zinc ions move quickly to bare steel surfaces of the ship's hull for protection purposes. (Cathodic protection is discussed in Chapter 12.) Zinc, aluminum, or magnesium anodes are also used in the soil and are buried at long intervals along pipelines to provide cathodic protection of the steel pipe. Foundations for steel bridges and buildings are also given this protection. Zinc is also used for roofing on cathedrals and other large buildings. Zinc is available in blanks, disks, foils, and ribbons.

Zinc Alloys   Zinc alloys are numerous. Possibly the largest use of zinc alloys has been in the automobile industry where die castings are used (Fig. 11–2). Along with very small quantities of iron, cadmium, and lead, the zinc alloys include varying percentages of titanium, magnesium, aluminum, tin, and copper. These alloys are available in wrought shapes such as sheets, extrusions, rod, and wire. The third largest use of zinc is in the alloy called brass, which is approximately 30–40% zinc. Zinc is also used in paints, electric fuses, ointments, fertilizer, printing plates, teeth, explosives, and insulin. Tensile strengths of zinc and its alloys range from 20,000 psi to more than 47,000 psi. Deep drawing operations require maximum ductility; therefore, pure zinc is more often used for these processes. Additions of lead, cadmium, and copper to zinc increase hardness and strength. Specific combinations of elements raise the recrystallization temperature of the zinc alloy so that increased strengths can be obtained by work hardening.

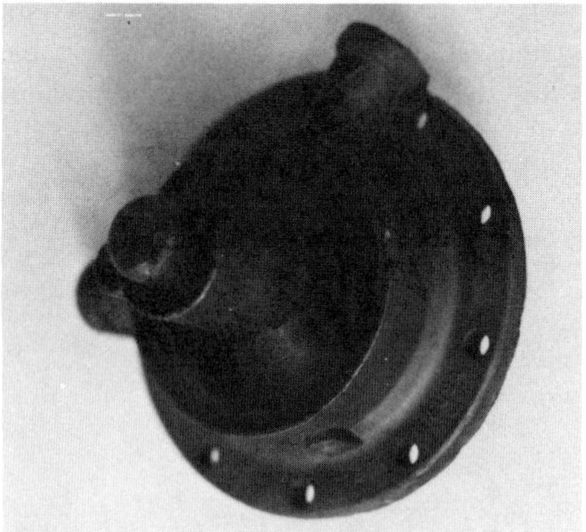

FIGURE 11–2   A die cast part made from zinc alloy.

Bismuth

Bismuth melts at 520 °F (271 °C) and is a constituent in the low-melting alloys. Its lattice structure is rhombohedral and it weighs 0.354 lb/in^3. Approximately one-half of a typical low-melting or fusible alloy includes bismuth and the remainder of the alloy includes tin, lead, and cadmium. A peculiarity of the low-melting alloys is that some expand on cooling and some contract. One particular alloy containing about 58% bismuth expands on solidification while the alloy with 45% bismuth shrinks on cooling, as do most metals. Bismuth expansions on cooling vary up to 0.006 inch per inch and therefore produce fine detail when cast. (Antimony also has this unusual characteristic of expansion on cooling.) The low-melting group of alloys contains varying percentages of cadmium, lead, tin and bismuth and one group contains indium. The term *fusible alloys* has been assigned to many of the low-melting alloys, such as one used in fire extinguisher systems. Tensile strengths of the bismuth alloys vary up to approximately 12,000 psi. Another common use of these low melting alloys is for novelties, such as one that melts when placed in hot water.

Cadmium

The element cadmium is frequently used as an anode (Fig. 11–3) to plate parts for corrosion resistance. Cadmium finds one of its greater uses as an alloy, however, especially in the low-melting group. Cadmium melts at 609 °F (321 °C), weighs 0.313 lb/in^3, is close-packed hexagonal in its lattice structure, and is soft and bluish white in color. Another of its important uses is as a constituent in bearing metals. One of its most recent uses is as a moderator or barrier in atomic fission whereby it helps control the number of available neutrons. Also, one of the compounds of cadmium is used to produce the phosphors in television tubes.

FIGURE 11–3   A used cadmium anode (sphere) showing effects of loss of metal during previous electroplating processes.

THE NOBLE METALS

Gold

Gold is a soft, yellow, corrosion resistant metal that is often called a precious or noble metal. It melts at 1945 °F (1063 °C), weighs 0.698 lb/in^3, is face-centered cubic in lattice structure, is extremely malleable, and work hardens. Because of its combined physical and mechanical properties and limited availability it has been used in the money systems of the world and as a medium of exchange. In addition to its monetary use, gold is used for jewelry, teeth, instruments, and decorations.

It is also used for personal speculation and as a collector's item. In the electronics industry it has numerous applications, such as in transistors and tubes. Because of its exceptionally high degree of ductility and malleability gold can be rolled into foil and leaf so thin that it is often difficult to measure. In fact, measurements have been made which are so small that the foil transmits a light. Foils and leaves of gold are used for decorations and thicker sections are used in delicate instruments, such as suspension wires in galvanometers. When used in electroplating processes, gold finds extensive use in the jewelry and decoration business as well as for those objects and parts requiring extreme resistance to corrosion. But gold is easily contaminated when hot or in liquid form. The atmosphere should be oxidizing when melting the metal. Harmful and unintentional additions include several of the low melting elements. Because gold, silver, nickel, copper, and platinum are all face-centered cubic in lattice structure, numerous soft alloys are available. However, the hardness of the alloy increases as the gold content decreases, such as coining gold which contains a small amount of copper.

Gold Alloys   Pure gold, with reference to the alloy system, is 24 carats; 12 carat gold is 50% gold and 50% other metals. The carat is a measuring system based on parts for precious materials. Twenty-four parts of gold is pure gold, and as the number decreases the percent of gold decreases as the alloying element increases. Gold readily alloys with copper, silver, nickel, zinc, lead, and platinum. As the alloy is added, the metal's color changes along with its physical and mechanical properties. Yellow gold is changed to a green, red, or white color by careful additions of other metals. When copper, zinc, and nickel are added to gold the metal becomes a silver or white color. When silver and copper are added the metal turns to a green or red color. The tensile strength of gold varies as its structure is cold worked and as its alloying content increases. As cold working increases, the hardness and tensile strength, along with the yield and shear values, increase from a low value of less than 20,000 psi to well beyond 150,000 psi. Annealing is performed by applying a few hundred degrees of temperature, often 600 °F (316 °C) being sufficient, and cooling in any manner.

Gold Castings   A 75% gold casting containing 13% silver, 10% copper, and the balance zinc and platinum has a tensile strength of 55,000 psi. But when shaped by cold working its tensile strength increases to more than 100,000 psi. Its modulus of elasticity is slightly more than 11 million psi which gives it sufficient stiffness for use in dental work, such as a gold-covered false tooth attached to an adjoining tooth. Obviously, as pressure is placed on the false tooth a large moment is placed on the connecting gold alloy; therefore, the geometry of the shape is very critical with regard to the allowable bending moment. A *moment* of a force is the product of the force or weight and the length of the straight-line distance from the axis to the acting force. The resulting stress inside the connecting gold must not be greater than the allowable deflection value. Dentists, architects, engineers, and technicians are well aware of stiffness values in materials involving the material and its geometry. As the alloying element is added to the pure gold, the metal changes color from yellow. When less than 70% gold is present the impurity is noticeable. Pure gold is considered to be 99.99% gold and is used as a standard. Normally, not more than .5% impurity is acceptable as a money exchange medium.

Silver

Silver is a soft, white, malleable metal which is also considered a noble metal but is more plentiful than other noble metals. This metal melts at 1761 °F (961 °C), weighs 0.379 lb/in³, has a modulus of elasticity of 11 million psi, is a high conductor of heat, and conducts electricity at a value greater than any other metal. Silver reflects light to a great degree and one of its compounds is so highly sensitive that it finds extensive use in photography. By the way, the silver content of this expended salt used in photography is retrievable. Because of its superior physical properties, acceptable mechanical properties, and availability silver is used extensively in jewelry, electronic components, solder, electrical relay contact points, and money systems. As with gold, the pure metal can be cold worked to a high tensile strength and is easily annealed at approximately 550 °F (288 °C).

Silver Alloys   Silver is face-centered cubic in lattice structure and easily combines with the same elements which alloy with gold. Pure silver is soft and weak in tensile strength, about 18,000 psi. As elements are added its mechanical and physical properties change, as does its monetary value. For example, sterling silver contains at least 92.5% silver, the remainder normally being copper for hardening purposes. This alloy has been used in coining. As copper is added to silver a eutectic is formed at 28.1% copper and at 1435 °F (779 °C). A common use of the silver alloy is for solder. Silver solder is a higher strength metal, like brass and bronze, used in joining two or more metals (Fig. 11–4). The eutectic type is used

FIGURE 11–4   Silver soldering a part to make a high strength joint.

for high strength applications, while common soldering alloys also contain small amounts of cadmium, tin, or zinc. Some of the silver-copper alloys can be heat treated by age hardening after forming at approximately 550 °F (288 °C) for one hour, followed by air cooling. Tensile strengths of the air cooled cast alloys range from 35,000 psi to more than 62,000 psi and are easily workable. A higher strength silver alloy results when small amounts of magnesium and nickel are added; the additional strength of approximately 72,000 psi results from the magnesium oxide precipitate.

Another alloy, a very unusual silver alloy, is called *amalgam* and is used by dentists for filling teeth. Amalgam includes silver, tin, copper, zinc, and mercury. The constituents are available in some physical form such as powder or plastic paste which enables the added mercury to quickly mix and form a solid alloy at room temperature. (Mercury has a great affinity for certain metals and combines quickly with them.) Because the new alloy forms a solid metal quickly, fast and effective amalgamation, rapid mixing and packing, is essential. The resulting physical and mechanical properties of amalgam are compatible with mouth chemicals and temperatures. Because chewing motions cause compression and/or bending forces on the teeth, the resulting alloy is sufficiently strong to adequately withstand these stresses. Even though satisfactory compression strengths are available in this metal, the peculiar alloy is sometimes subjected to unique mouth conditions such as overloading and overheating of the alloy, and these conditions help cause deterioration of the compaction by fracture and crumbling.

Platinum and the Platinum Group

Platinum is a malleable, white metal which melts at 3217 °F (1769 °C), weighs 0.775 lb/in³, is face-centered cubic in structure, and has a modulus of elasticity of 21 million psi. It is extremely resistant to corrosion and is the most noble metal. Production of platinum is small and its cost is high. When annealed its tensile strength is approximately 19,500 psi, being similar to gold and silver, but it can be work hardened to higher strengths. Platinum is face-centered cubic in lattice structure, so its use at cryogenic temperatures is good. Actually, as with several other face-centered metals, tensile strengths increase as temperatures decrease. On the other hand, creep behavior is acceptable at red, yellow, and white temperatures for short durations of time. Because of its exceptional physical properties, platinum is used for high temperature thermocouples, anodes, electronic components, chemical containers, jewelry, alloying elements, and as a catalyst.

Platinum Alloys   Platinum combines with several elements, including iridium, palladium, rhodium, nickel, ruthenium, cobalt, and tungsten. Obviously, each element gives the resulting alloy its peculiar properties; for example, rhodium gives excellent resistance to oxidation at temperatures in the white heat range. Platinum-rhodium thermocouples have acceptably withstood 3400 °F (1871 °C) and molten glass has no significant effect on the alloy. The platinum-palladium alloy is used in electrical contact points to resist arcing pits. It is also used for jewelry because of its luster. When iridium is added to platinum, measuring standards can be produced because of the high combination of unique physical properties. Nickel and platinum produce good electronic components as do ruthenium and platinum. A combination of platinum and tungsten makes excellent spark plug electrodes and metal components in hot electronic apparatus. Many permanent magnets are made from alloys containing cobalt and platinum. Cold-drawn alloys of the platinum group have tensile strengths up to nearly 300,000 psi. Platinum, osmium, and iridium have similar densities, and palladium, rhodium, and ruthenium have similar densities but the densisties are approximately one-half those of the former three elements.

SPECIAL USE METALS

Mercury

Mercury is liquid at room temperature because its melting point is −37 °F (−38 °C). This unusual metal has a rhombohedral lattice structure and weighs

0.489 lb/in³. Its boiling point is 675 °F (357 °C); therefore, the range of temperatures makes it ideal for use in thermometers and other instruments where temperature differences are to be indicated. However, mercury must not be heated except in carefully controlled devices because it is extremely volatile and quickly moves into the air. Mercury is also very poisonous in elemental and compound forms. Being an acceptable conductor of electricity, mercury finds extensive use in electrical apparatus such as silent switches and high intensity lighting systems. The compounds of mercury are many, including the poisons; consequently, its use must be carefully controlled.

Beryllium

Beryllium is a close-packed hexagonal lattice structured metal which melts at 2332 °F (1278 °C) and has a modulus of elasticity of approximately 42 million psi. It weighs only 0.067 lb/in³ and has an exceptionally high stiffness factor, so beryllium finds specialized uses in aerospace vehicles where light weight and stiffness are essential. At room temperature, beryllium is extremely corrosion resistant because of its self-made oxide coating. This lightweight metal has an exceptional capability to resist very high temperatures such as are generated by missiles entering the earth's atmosphere. With its high stiffness value, heat absorption capability, light weight, and ease of fabrication at approximately 1000 °F (538 °C), beryllium finds extensive use as an alloying element in copper for the purpose of producing age hardenable alloys. Some of these alloys attain tensile strengths of 200,000 psi. Beryllium is hazardous to health and must be handled with caution in regard to breathing its dust. No metal should be breathed, but beryllium is exceptionally hazardous.

Iridium

Because of its high melting point of 4449 °F (2454 °C), this face-centered cubic metal is used in the electronics industries where high temperatures and sparking are present, such as in contact points. This metal is about twice as heavy as lead, 0.813 lb/in³, and is extremely corrosion resistant. Iridium has a modulus of elasticity of 76 million psi, more than twice that for steel. All together, these exceptional physical properties provide an excellent capability in the manufacture of spark plug electrodes.

Osmium

Possibly the stiffest of metals (the modulus of elasticity is 81 million psi), osmium melts at approximately 5000 °F (2760 °C) and must be cast into shape and then ground because it is not workable. Its lattice is close-packed hexagonal and its high hardness provides enough wear resistance for it to be used as pivot points for instruments and phonograph needles.

Palladium

Palladium has a face-centered cubic lattice structure, melts at 2826 °F (1552 °C), and weighs 0.434 lb/in³. Its modulus of elasticity is approximately 16 million psi, somewhat less than that of platinum. However, many of its uses are similar to those of platinum, especially in electronic parts. Palladium is widely used in dental, electrical, and jewelry industries where high strength platinum solders are used.

Rhodium

Rhodium has a modulus of elasticity of approximately 42 million psi and melts at 3571 °F (1966 °C). It is a refractory metal as verified by its use as a thermocouple component in high temperature operating furnaces. This face-centered cubic metal weighs 0.447 lb/in³ and is used in jewelry applications and mirrors. One of its main uses is as an alloying element in platinum and as a catalyst in the chemical industries.

Ruthenium

Ruthenium is similar to osmium in lattice structure (close-packed hexagonal), is twice as stiff as steel, melts at 4530 °F (2499 °C), and weighs 0.441 lb/in³. It too has excellent qualities for use in electrical contact points. The tensile strength of ruthenium is equal to that of structural steel, and because of its great stiffness it finds extensive use in control instruments where precision is demanded. Ruthenium's main use is in the alloys, especially with platinum.

Germanium

The element germanium has a diamond cubic lattice structure, melts at 1719 °F (937 °C), and weighs 0.192 lb/in³. It is not easily produced due to its dilution in other materials such as soils, coal, and smokestack soot. Germanium established its need and importance when it provided replacement of the vacuum tube with the electrical transistor; the element has most of the properties and capabilities of the vacuum tube. Besides this major use, germanium is used in optical equipment such as spectroscopes and wide angle camera lenses. Its temperature tolerance is small so it is limited to use where temperatures remain within seasonal fluctuation.

Selenium

Selenium is also a difficult element to produce. It is normally obtained by reclaiming foundry smokestack soot, mud residues from electrolysis, and residues from similar situations where chemical reactions have occurred. The element is hexagonal in lattice structure, is fairly stiff in its elastic modulus (about 8 million psi), melts at 423 °F (217 °C), and weighs 0.174 lb/in³. Selenium has the unique capability to convert alternating electricity into direct current so it is used extensively as a rectifier. It is also used in photoelectric cells because of its photoconductive and photovoltaic capabilities. With the ability to transform light into electricity, selenium finds extensive use in photographic light meters, photocells, and solar cells. It is also used in photocopy processes.

Gallium

Gallium is another hard-to-produce metal. It too is retrievable from places where chemical reactions have occurred, such as in smokestacks. Gallium is orthorhombic in lattice arrangement, melts at 85 °F (29 °C), and weighs 0.213 lb/in³. Fine glass mirrors are wetted with gallium to give their reflective appearance. The element is also used in electronic parts such as solid-state components. Along with mercury, rubidium, and cesium, gallium is liquid in the temperature regions slightly warmer than room temperature. A compound of gallium transforms electricity into light.

Tellurium

This unusual metal is also retrieved from electrolytic residues such as the remaining mud from anodes. Tellurium has a hexagonal lattice structure, melts at 841 °F (449 °C), and is silvery white and brittle. Tellurium catches fire in air and burns with a greenish blue flame. The metal finds unusual uses in thermoelectric mechanisms with the ability to rapidly heat or cool water. One of the main uses of tellurium is in blasting caps for explosives. Like many of these less available metals, it is used as an alloy in other metals for some particular purpose.

Tungsten

Tungsten is body-centered cubic in lattice structure, melts at 6170 °F (3410 °C), has a modulus of elasticity of 50 million psi, and weighs 0.697 lb/in³. This heavy metal has many uses, both as a pure metal and as an alloying element. In steel it raises the temperature utilization for cutting tools up to 1100 °F (593 °C) (Fig. 11–5), and when chemically combined with carbon it produces an extremely hard

FIGURE 11–5 A high speed steel milling cutter made with tungsten and other alloys.

and heat resistant compound, tungsten carbide (Fig. 11–6). Because of its high melting point, tungsten is used as an electrode in electric welding, specifically in tungsten inert gas welding torches. The cold drawing of tungsten is facilitated by its very high recrystallization temperature of about 2000 °F (1093 °C). Based on pounds per square inch of tensile strength, tungsten has been drawn into wire with strengths in excess of 600,000 psi. Most forming and machining is accomplished at red or white temperatures, however.

Tungsten is widely used in the refractory fields and as an alloying element or pure metal in electrical heating elements and contact points. Much of its alloying capability is demonstrated in the high speed steel tools. When powdered and

FIGURE 11–6 A high speed production cutter insert made of tungsten carbide.

pressed it forms the basic ingot for subsequent hot and cold drawing operations in the production of electric lightbulb filaments. Also, many parts which are subject to electric arcing are produced from tungsten, such as those used in vibrators, ignition points, electric razors, and relays. Hard-facing operations frequently employ tungsten since its hardness is not reduced because of high temperature application. In X-ray tubes and shielding lead absorbs the X radiation and tungsten forms the plate that receives the very hot stream of electrons and, in turn, bounces X radiation from the anode through the tube's window for bombardment onto a target.

Tantalum

Tantalum is a body-centered cubic metal that weighs 0.600 lb/in^3 and melts at 5424 °F (2996 °C). Its modulus of elasticity is slightly less than that of steel, but its immunity to human bone and tissue chemicals makes it ideal for use in making artificial body parts such as hip pins and swivel joints. It is one of the highest corrosion resistant metals known. This unusual metal finds specialized uses in the dental, chemical, and electronic industries. In electronics, for example, its gettering ability is high as well as its dielectric and rectifying capabilities. In the vacuum furnace industries, tatalum is used where its physical properties are of great demand. Tantalum is also widely used in the production of electrolytic capacitors.

URANIUM

The metal uranium weighs 0.689 lb/in^3, melts at 2070 °F (1132 °C) and is orthorhombic in lattice structure. The metal U^{238} includes a small amount of the fissionable material U^{235}, and a very small quantity of U^{234} in found in the U^{238} ore. In finely powdered form the metal burns in a manner similar to magnesium and titanium. The tensile strength of uranium is only about the strength of low carbon steel. When cast into ingots and special shapes, at approximately 2500 °F

(1371 °C), either a vacuum is used or an inert atmosphere because of uranium's high reactivity with constituents of air. Uranium is also formed by wrought processes into special shapes, depending on the use. These shapes are weldable in an inert atmosphere.

Fuel for Nuclear Reactors

Uranium isotopes are all radioactive. Natural uranium, however, is partially radioactive; therefore, caution must be used within its effective radius in conjunction with time. Uranium is the highest in order of the naturally occurring elements and may be a decay material from long ago. Its volume in the core and crust of the earth causes a variable amount of heat in the regions near its ores. Natural uranium, U^{238}, is convertible to fissionable plutonium through reactions from U^{238} to U^{239} to Np^{239} to Pu^{239}. Conversion is made possible through the use of the nuclear furnace or reactor where a controlled chain reaction is needed for a continuing heat supply. The fuel or isotope is exposed to neutron bombardment in the presence of a graphite moderator which allows and controls the chain reaction by thermalizing neutrons. Heat results from divisions of the atom, U^{235}, for example, whereby a tremendous energy release occurs in the order of 200 million electron volts per atom. The splitting of the atom causes the chain reaction, and in less than two seconds the fission is operating at the accelerating rate of billions per second. Such a runaway reaction is controlled for peaceful purposes, however, but is allowed to proceed to a mighty explosion when used as a weapon.

Fission Causes Heat

Fission is controlled by cadmium rods which are lowered or elevated in the reactor as required to maintain a continuous and controlled output of atomic energy. Cadmium absorbs neutrons which are the projectiles that penetrate the nuclei of the atoms of the fuel. Splitting of the atoms and energy release in the form of heat occurs within the reactor's container. The container is cooled by a flowing liquid such as sodium or any liquid which will transfer heat from the reactor to the heat exchanger. A core plate (Fig. 11–7) and shroud head (Fig. 11–8) are components of the system. A fuel holder is also part of the core complex. Steam is produced when water flows through numerous coils and tanks within part of the exchanger. Very high steam pressures are generated and the steam moves to the turbine blades at high velocities. As the turbine turns, so does the electric generator; this is the beginning of electric power output. (Fig. 11–9).

The heart of the source of energy is an assembly which contains the core, moderator, and shield. The lower core support is forged from austenitic steel and contains holes in which are placed the nuclear fuel for the reactor. Part of the core also includes components made from zirconium alloys which function at elevated temperatures. The element melts at 3366 °F (1852 °C), therefore, the alloy is a stable support to the core complex. Within the reactor complex, fission occurs whereby the fuel, U^{235}, for example, is converted into heat, gas, and radiation. Control of the fission is a function of the moderator which surrounds the

FIGURE 11–7 A core plate being completed for a nuclear reactor. Austenitic steel finds top priority in fabrication of nuclear reactor components.

FIGURE 11–8 Austenitic stainless steel is also used in fabricating this shroud head for a nuclear reactor.

core. To shield the outside regions the shield is fixed around the assembly. Pipes, tubes, and pressure vessels are part of the total assembly. Radiation is trapped within the shield, heat from fission is directed at pressure vessels containing fluid whereby steam is generated, by-products are removed for other uses, and cheap electricity is produced.

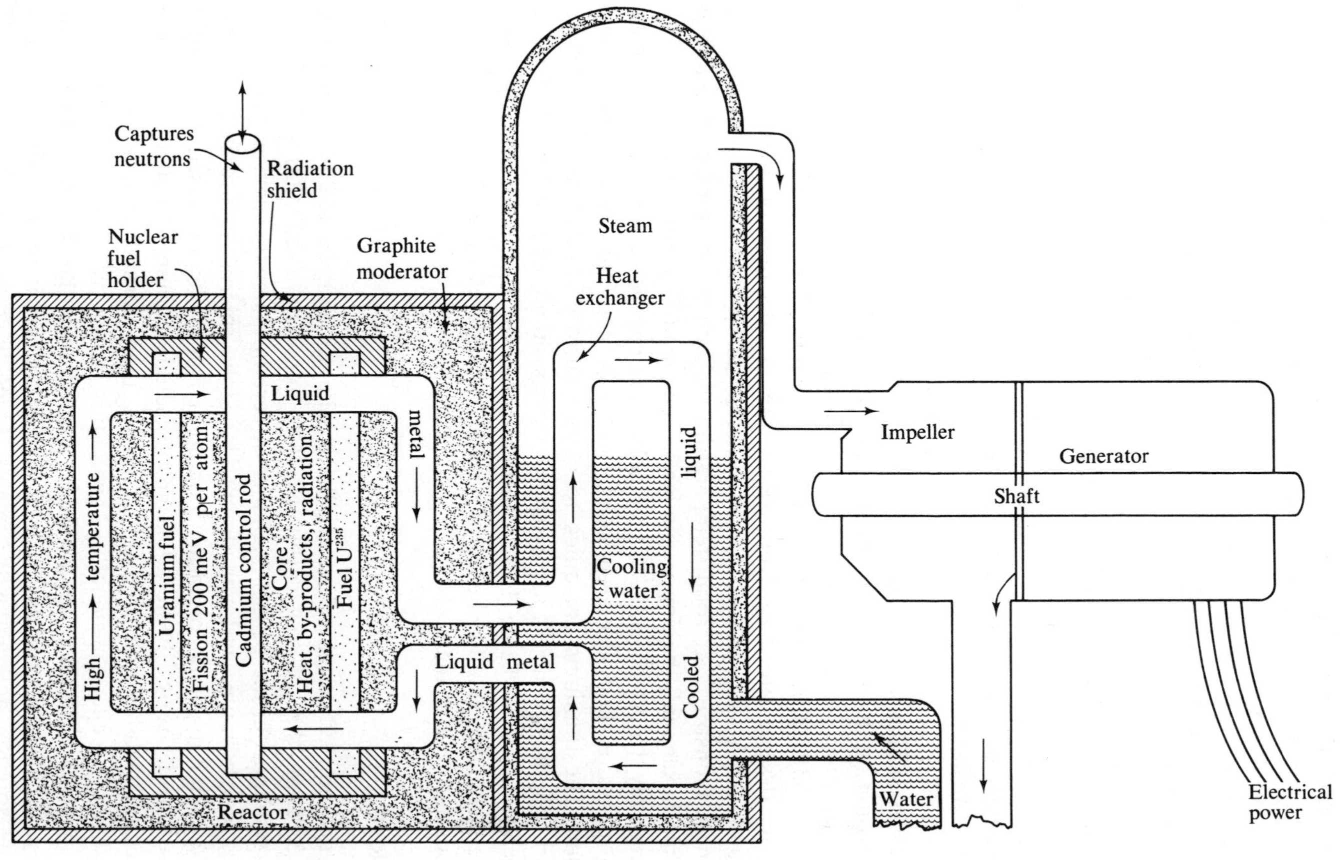

FIGURE 11–9 Schematic drawing of a nuclear power-generating station which uses uranium fuel to make steam. The steam turns the impeller assembly which generates electricity.

Approximately 10 pounds of fissioned uranium is comparable to 15,000 tons of coal. With a natural source of water, therefore, only the fissionable material is needed to sustain a continuous 24-hour electrical output for many years. But the nuclear reactor produces two products, heat for conversion into tremendous quantities of cheap electricity and a residue of radioactive material. Millions of kilowatts of electricity are available for distribution across the land to feed the society and its industries with the required energy, and the dangerous residue is contained.

TITANIUM AND ITS ALLOYS

Titanium has characteristics so different from most metals that its structural alloys are nearly unique. Some of its characteristics are similar to steel, however. The element melts at 3035 °F (1668 °C), has a modulus of elasticity approximately one-half that of steel, and weighs 0.1628 lb/in³. Some of its alloys are heat treatables similar to steel and aluminum alloys, and these alloys resist salt-water corrosion to a very high degree. Titanium forms four main classes of microstructures, and its alloys have an operational range of –400 °F (–240 °C) to 1000 °F (538 °C). This metal has an endurance limit higher than most metals if its designs are free from stress raisers. Because of their excellent mechanical and physical properties the titanium alloys are used extensively in the aerospace industries. At room temperature the lattice structure of titanium is close-packed hexagonal, making it similar to magnesium in brittleness. However, at approximately 1625 °F (885 °C), its allotropic change is from alpha to beta. The beta phase is body-centered cubic and this condition allows permanent deformation. The titaniums are quick to absorb surface contaminants; therefore, atmospheric handling controls are essential. Compositions and typical mechanical properties are listed in Table 11–2. With reference to the indicated heat treatments, the technician should check the specific procedures because of the critical time periods and cooling methods.

Alpha Alloys

With the addition of certain elements the tensile strength capability of titanium is increased from that of low carbon steel to more than 200,000 psi. Such a condition results from the proper blending of certain elements such as aluminum, chromium, molybdenum, and vanadium. The alpha microstructure is hexagonal close-packed and this type of lattice condition cannot be increased in strength by heat treatment. Alloys in the alpha group include aluminum and often tin for microstructural stability and strength at higher than normal temperatures (Fig. 11–10). Their excellent weldability gives this group of titanium alloys top priorities in many aerospace applications. Even though the Ti-8Al-1Mo-V is an alpha alloy, it can be increased in strength by production of a small beta phase through the solution and aging treatments. Quenching in water from 1800 °F (982 °C) and aging at 1100 °F (593 °C) for 8 hours produces the increased strength. The pure metal, however, is weak in tensile strength but high in corrosion resistance. On heating to 1625 °F (885 °C), the lattice structure changes

TABLE 11-2 COMPOSITION AND MECHANICAL PROPERTIES OF SELECTED TITANIUM ALLOYS

Alloy	Condition	Type	Tensile (psi)	Yield (psi)	Elongation (%)	Treatment*
99.0 Ti	Annealed	—	95,000	80,000	25	1300 °F and air cool
Ti-5A1-2.5 Sn	Annealed	Alpha	125,000	120,000	18	1450 °F and air cool
Ti-3A1-13V-11 Cr	Heat treated	Beta	135,000	130,000	16	1450 °F and air cool
Ti-3A1-13V-11 Cr	Heat treated	Beta	180,000	170,000	6	1400 °F to air; and 900 °F and air cool
Ti-2Fe-2Cr-2Mo	Annealed	Alpha Beta	137,000	125,000	18	1250 °F and air cool
Ti-2Fe-2Cr-2Mo	Heat treated	Alpha Beta	179,000	171,000	13	1480 °F. to water; 900 °F and air cool
Ti-8Mn	Annealed	Alpha Beta	138,000	125,000	15	1300 °F to 1000 °F and air cool
Ti-4A1-4Mn	Annealed	Alpha Beta	148,000	133,000	16	1400 °F to 1000 °F and air cool
Ti-4A1-4Mn	Heat treated	Alpha Beta	162,000	140,000	9	1450 °F to water; 900 °F for 24 hr and air cool
Ti-4A1-3Mo-1V	Heat treated	Alpha Beta	195,000	167,000	6	1625 °F to water; 925 °F for 12 hr and air cool
Ti-5A1-1.5Fe-1.4Cr-12Mo	Heat treated	Alpha Beta	195,000	184,000	9	1600 °F to water; 1000 °F for 24 hr and air cool
Ti-6A1-4V	Heat treated	Alpha Beta	170,000	150,000	7	1700 °F to water; 975 °F for 6 hr and air cool
Ti-7A1-4Mo	Heat treated	Alpha Beta	190,000	175,000	12	1650 °F to water; 900 °F for 6 hr and air cool
Ti-3A1-13V-11Cr	Heat treated	Alpha Beta	180,000	170,000	6	1400 °F to air; 900 °F and air cool

*Refer to specific heat treatment data.
By permission, from *Metals Handbook* Volume 1, Copyright American Society for Metals, 1961.

FIGURE 11–10   Photomicrograph of titanium alloy (Ti-6AL-4V) showing alpha phase at grain boundary and platelets (light) and beta phase (dark). Hydrochloric-hydrofluoric etch at 500×. (Courtesy of Bell Helicopter Company)

to body-centered cubic and, in turn, hot working and permanent deformation are obtainable.

Alpha-Beta Alloys

With the addition of beta-forming alloys titanium beomes capable of heat treatment for higher strengths by a combination of the alpha-beta phases (Fig. 11–11). The solution and aging treatments apply to this group of titanium alloys.

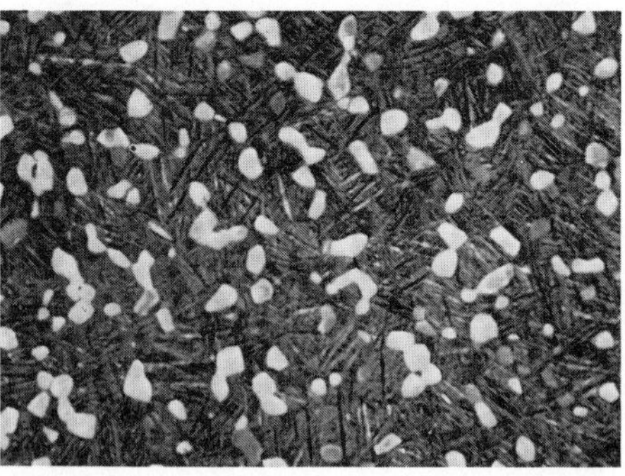

FIGURE 11–11   Microstructure of Ti-6Al-4V solution treated at 1750 °F (954 °C) for 1 hour and water quenched and aged for 24 hours at 1000 °F (538 °C). Light areas are primary alpha, and the dark acicular matrix is aged alpha prime and beta (500×). (Courtesy of Bell Helicopter Company)

Alloys such as Ti-6Al-4V are available in the bar, rod, forging, and sheet forms and respond to the alpha-beta strengthening treatments. Elements such as vanadium, chromium, and molybdenum tend to stabilize a beta phase as the alloy cools from the red heat range. Such chemical changes result in a mixed lattice structure, alpha close-packed hexagonal and beta body-centered cubic.

Working of the metal is therefore possible. When these alloys are subsequently heated to approximately 1700 °F (926 °C) and rapidly water quenched the beta phase is suppressed enough to form a partially stable structure. When subsequent aging at 950 °F (510 °C) for 4 hours is accomplished the tensile strength increases because of the fine dispersion of the alpha constituent throughout the prior beta phase. The purpose of quenching is to retain a supersaturated beta phase which yields to precipitation during the proper temperature environment. Air cooling is accomplished from the aging temperature. Because heat treating drastically rearranges the microstructure, evidence of this change is very pronounced as magnifications of the microstructure increase. Figure 11–12 is a structure of titanium alloy as viewed through the electron microscope.

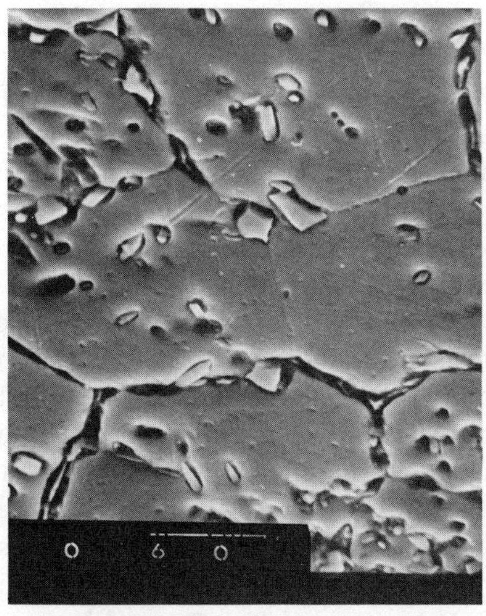

FIGURE 11–12   An electron microscope view (3000×) of niobium segregation in 17-4 PH stainless steel. Light particles are niobium-rich compounds. (Courtesy Bell Helicopter Company)

Requirement for Surface Protection

Conditions dangerous to titanium exist in the red and yellow heat ranges when oxygen and hydrogen are present. Both scale with embrittlement characteristics and a thin-skinned alpha phase are created at all surfaces of the metal unless a vacuum is available while heating or a conventional salt bath or inert atmosphere is used. Consequently, the facility for heat treating the titaniums is more expensive than facilities for many of the steels and aluminums. With regard to surface protection, however, the surface of steels would also be better protected if heat treated in inert or controlled atmospheres. Solution temperatures are high, often higher than most austenitic temperatures for alloy steels (except high speed), and chemical disturbances at the surface of the metals are easily induced. Annealing temperatures are approximately 350 °F (177 °C) lower than solution temperatures, but the extremely slow cooling to about 1000 °F (538 °C) provides the possibility of surface contamination. Like steels, titanium can be removed from the furnace at about 1000 °F (538 °C) and air cooled for satisfactory softening.

Characteristics of Titanium Alloys

The strength-to-weight ratio of titanium is ideal for specialized applications such as parts for spacecraft. Titanium alloys are equal in strength but lighter in weight than steel. Fatigue properties of titanium are also excellent, as are strength properties up to 1200 °F (648 °C). Titanium alloys have excellent buckling resistance strengths and provide good rigidity values. On the other hand, titaniums must be welded with inert gas. Machinability is good when sharp tools are used along with slow speeds and heavy feeds.

Uses of Titanium Alloys

This unusual metal, when properly alloyed, has excellent strength properties at cryogenic temperatures, room temperatures, and at temperatures up to 1200 °F (648 °C). However, like magnesium, it will burn violently if ignited and must not be allowed to contact flammable situations when in very small pieces or in the powder state. G-1 powder will extinguish the fire, but water or carbon dioxide will increase the burning capacity. Because the weights of titanium alloys are about one-half that of steel, their nonmagnetic conditions are like austenitic steel, their corrosion resistance under several kinds of corrosive conditions is excellent, and they are easily formed and heat treated, the titanium alloys are used extensively in many types of applications. Specific uses are found in aircraft, including helicopters, space vehicles, missiles, and rocket nozzles, and in compressor cases, rotor hubs, pressure vessels, aircraft wheels, engine housings and auxiliary components, fuel and oxidizer tanks, valves, pipes and tubes, compressor blades, and in high strength hardware such as bolts, exhaust systems, chemical containers, and numerous structural shapes.

When welded structures and parts are subject to oxidation at temperatures up to 1000 °F (538 °C) the alpha alloys are used because of satisfactory strengths. Short-term temperatures up to 2000 °F (1093 °C) do not destroy the basic strength of several titaniums. The alpha-beta alloys are not as easily formed but have higher strength capabilities. Sections as thick as 4 inches are heat treatable such as the Ti-7Al-4Mo. One of the most used general purpose alloys is the Ti-6Al-4V. The Ti-13V-11Cr-3Al is a beta alloy and is formable at room temperatures, is weldable, and can be solution treated, formed, and aged. Also, several titanium alloys have proven exceptional under loads up to 1000 °F (538 °C) and at −423 °F (−253 °C), making them safely usable for space travel and planetary landings (Fig. 11–13).

The use of titanium alloys is rapidly growing. For example, an AISI 4130 aircraft landing gear part is heat treated for 160,000 psi tensile strength. At about one-half the weight and if a less stiff modulus of elasticity is acceptable, the Ti-7Al-4Mo titanium alloy can be used since it is also heat treatable to 160,000 psi tensile strength. With reasonable safety factors the titanium alloys can do the job at one-half the weight of steel, particularly in space vehicles. Consequently, much of the structure of current space vehicles is fabricated from titanium alloy.

FIGURE 11–13   An artist's concept of the space shuttle liftoff. With main engines and solid rocket motors roaring, the space shuttle lifts off from the launch pad. This is the first reusable space transportation system. The shuttle will take off like a rocket and return to earth, landing similar to a jet transport. Only a limited number of materials will withstand the environment of the exhaust system, subzero space, and the heat of reentry. (Couresty of Rockwell International)

Moon landings have been guaranteed with the use of titanium as well as austenitic steel, aluminum, nickel, and copper. Inner skins of the space vehicles have been made from thin sheets of spot- and seam-welded titanium. Spacecraft skin, bulkheads, stringers, longerons, brackets, rivets, bolts, fasteners, rocket forgings, pressure tanks, and landing gears have been produced from the titaniums. Jet engines include large quantities of titanium alloy. Keel beams in fighter aircraft are frequently produced from the titaniums as well as wing spars, channel rings, and wing flaps. Titanium is a unique metal because of its package of mechanical and physical properties. It has answered the need for structural stability in space where attacks from heat, cold, and radiation occur.

MOLYBDENUM

Molybdenum melts at 4730 °F (2610 °C), weighs 0.369 lb/in³, has a modulus of elasticity of 47 million psi, and a body-centered cubic lattice structure. This element is used normally as an alloy to some other metal such as in the tougher steels (Fig. 11–14). When molybdenum is included in the analysis of structural steels such as AISI 4340, the steels' tensile and yield strengths are higher and resistance to impact loading increases. When molybdenum steels are heat treated they have a greater hardenability than the carbon steels because the time-temperature-transformation factors allow deeper hardening at a slower cooling rate. When vibration, impact loading, cyclic loading, or creep possibilities are present, the molybdenum steels and alloys rank among the very best.

FIGURE 11–14   This aircraft landing gear is manufactured from nickel-chromium-molybdenum steel and heat treated to effectively withstand the forces of takeoffs and landings.

In steel a small quantity of molybdenum is usually sufficient to provide the intended property. For example, the strong and tough AISI 4340 steel contains only an average of 0.25% molybdenum. But as the need changes from high tensile strengths to high compression strengths at temperatures above normal (70 °F or 21 °C) increased amounts of molybdenum are required to provide the unusual property of red hardness or resistance to softening at 1000 °F (538 °C). The molybdenum type of high speed steel, M15, for example, contains 3.5% molybdenum along with 6.5% tungsten, 5% cobalt, 5% vanadium, 4% chromium, and a big 1.5% carbon. This amount of carbon, which forms the complex carbides with most of the above elements, provides a very high resistance to abrasion; therefore, the use of M15 as a cutting tool is well known. On the other hand, when a small amount of molybdenum is included in the structural steels, machinability is usually improved because of better shearing action of the cutting edge of the tool.

Molybdenum steels offer free scaling surfaces during forging and rolling operations; consequently, less chance of surface inclusions exists. After forging and heat treatment, parts made from molybdenum steels have high resistance to fatigue, partly because of excellent hardenability potentials and resistance to attack by corrosion. When case-hardened parts, which are processed by the carburizing or nitriding methods, are placed in use the presence of molybdenum increases the operating life of the part by promoting fatigue resistance in parts undergoing variable loading.

Special Purpose Alloys

When metals are used at subzero temperatures, as pipes, tubes, and pressure vessels, for example, molybdenum is often part of the analysis. This is especially true in the steels. The elevated temperature stabilizing capability of molybdenum is also well known. The austenitic stainless steel 316, for example, contains around 2% molybdenum, and when annealed has a tensile strength of about 85,000 psi with an elongation in 2 inches of 55%. Formability is good providing the operation is carried quickly to completion because of work hardening of the metal. When a special type of Inconel, for example, one containing 4.5% molybdenum, is heated to 1200 °F (648 °C) its yield strength remains at approximately 90,000 psi. Of course, other elements such as chromium, titanium, aluminum, iron, and niobium are also included in this alloy. Boilers and pressure vessels having molybdenum in their analyses function satisfactorily at temperatures up to 1000 °F (538 °C) with yield strengths more than three times greater than the carbon steels. Exhaust valves, springs, and turbine wheels, all operating at elevated temperatures, usually contain molybdenum. It must be pointed out that the metal molybdenum is usually not the only element which establishes these excellent mechanical properties. It is frequently used in conjunction with chromium or nickel and often accompanies tungsten in some high quality tool steels.

Molybdenum in Cast Iron

The use of special types of cast iron is increasing because tough and ductile irons are available for use in designs previously employing steels. Mechanical properties of the iron, such as higher strengths, are greatly improved both at room and elevated temperatures. Hardenability of the iron along with increased uniformity in microstructure improves in the several types of constituent arrangements when molybdenum is present. However, the relationship of hardness to tensile strength in the cast irons may not be as close as that in the ductile steels. The presence of graphite (a nonmetal) will not convey the same stress as the metal; therefore, hardness values of the irons are relative among the irons but are not convertible to correlations with the nonbrittle steels. On the other hand, molybdenum improves most microstructures with regard to strength.

Questions

1. Differentiate between the formation of a eutectoid and a eutectic.
2. Why can lead be continually folded and deformed without work hardening?
3. Name several uses of lead.
4. Name three important uses of zinc.
5. What metal conducts electricity at the greatest value?
6. List the noble metals.
7. Name several elements which rely heavily on the retrieval of manufacturing wastes for their production.
8. Why is tungsten considered to be such a very important metal?
9. Name the metal which is frequently used in the human body for engineering designs.

10. Name a radioactive metal. What does "radioactivity" mean?

11. Why is titanium an important structural metal?

12. Explain the effective operational temperature range of some titanium alloys.

13. If the same geometric shapes of steel and titanium become mixed, list several ways to separate them.

14. Explain several advantages of adding molybdenum to a metal.

15. How does molybdenum increase hardenability?

16. Why is molybdenum included in those steels which are subjected to flexible loading?

17. Differentiate between interstitial and substitutional types of solid solutions.

18. Differentiate between binary and ternary alloys.

19. What property does lead have which makes it a barrier to X and gamma radiation?

20. Name a refractory metal which is used in thermocouples.

12

Fatigue and Corrosion

When metals are placed in service some are used in static conditions where service loads are constant, or nearly so, and others are subjected to fluctuating loads under various circumstances. These are the metals which are exposed to cyclic loading over an extended period of time where tension stresses are involved. They become subjected to possible early failure even though the stresses never exceed the static yield strength of the metal. With respect to yield strength in fatigue environments the yield is only a comparison for static conditions ond is not applicable in limiting elastic stress in fatigue. Fatigue is the term given to materials exposed to cyclic loading. Static stress and cyclic stress can be compared to a person remaining still or running. Running brings on fatigue and exhaustion, but sitting in a chair is quite different. Fatigue is a phenomenon that incorporates so many variables that it is only partially understood. It is known, however, that certain cyclic stressing will bring quick failure to some metals and to some designs, but if the stresses are lowered somewhat or the design changed slightly, the metal and the assembly will function indefinitely. Then the question is asked, What is fatigue failure?

FAILURE WITHOUT WARNING

Failure of a material or of an assembly often occurs without warning because the process of fatigue involves normal functioning of the material over a long period of time. Fatigue failure occurs more than any other type of failure. Poor designs with stress raisers, such as notches, holes, and grooves, and improperly hardened metal lead to fatigue failure over a period of operation. Another source of fatigue failure commences with imperfections in metals under cyclic loading. Submicroscopic size defects appear in the material, some are already present, and in the presence of flexible loading these irregularities may signal the beginning of the end of the design. Continued operation under normal conditions usually extends into months or years and then, without warning, the fatal fracture occurs. On the other hand, some discontinuities may never elongate into the primary stress pattern but remain dormant. In metals the specific causes of fatigue failure are numerous, mainly resulting from improper designs and imperfections in the metals. No metal is perfect; therefore, something is wrong somewhere and this imperfection is frequently focused on the atomic lattice and the atoms. This is a justifiable statement because a metal is merely a pile of atoms layed out in an organized manner with some exceptions. These exceptions, such as dislocations and voids, are numerous and it is very often the exception that leads the acceptable design to progressive failure. Possibly the quickest way to accelerate this type of failure is to create stress concentration points in the part's design.

IMPROPER DESIGN LEADS TO FAILURE

Even though a part or assembly has been inspected and certified as being acceptable, many fail after only a few months of operation. A study of one of the failures, a machine spindle in this example, showed areas where excessive stress occurred and the alternating stress concentration quickly signaled the beginning of destruction of the spindle (Fig. 12–1). From the designer's point of view the spindle's geometry was simple and was drawn to existing standards and concepts. But an analysis of a particular area in the design points out mistakes which simply cannot be repeated. For example, the bottoms of the splines were milled with square corners. The length of the spline reduced the amount of metal between the end of the spline and the surface of the steel. Because a high shear stress existed in the surface of the one-inch diameter metal and a high stress was generated at the corner of the spline, the metal between these two points was doomed to overload. Subsequently, atomic slip provided the fracture route. Because the load was increased and decreased in rhythmic sequence, a prime condition for fatigue failure was generated. Even though the maximum operating stress was substantially lower than the metal's yield, that is, a reasonable safety factor existed, observation shows that lines of stress became too crowded in area A. This area then became a prime point for overloading and eventual metallic separation. A modified design would provide fillets in the splines and shorten the length of the splines while maintaining a reasonable safety factor. On loading, stress response would then flow smoothly and no excessive concentration would occur. Basically, draftsmen and designers must avoid directional changes without generous fillets. In regions where stress becomes excessively concentrated, more metal must be provided. The elimination of stress concentration can extend the service life of the part into infinity.

PROCESS OF FATIGUE FAILURE

The process of metal fatigue and potential failure involves a load on the metal; the load must vary, the load must include tension, and the process must operate over a period of time. Such a situation can involve a complete cycle of loading from tension to compression and back to tension, a variable load in tension to

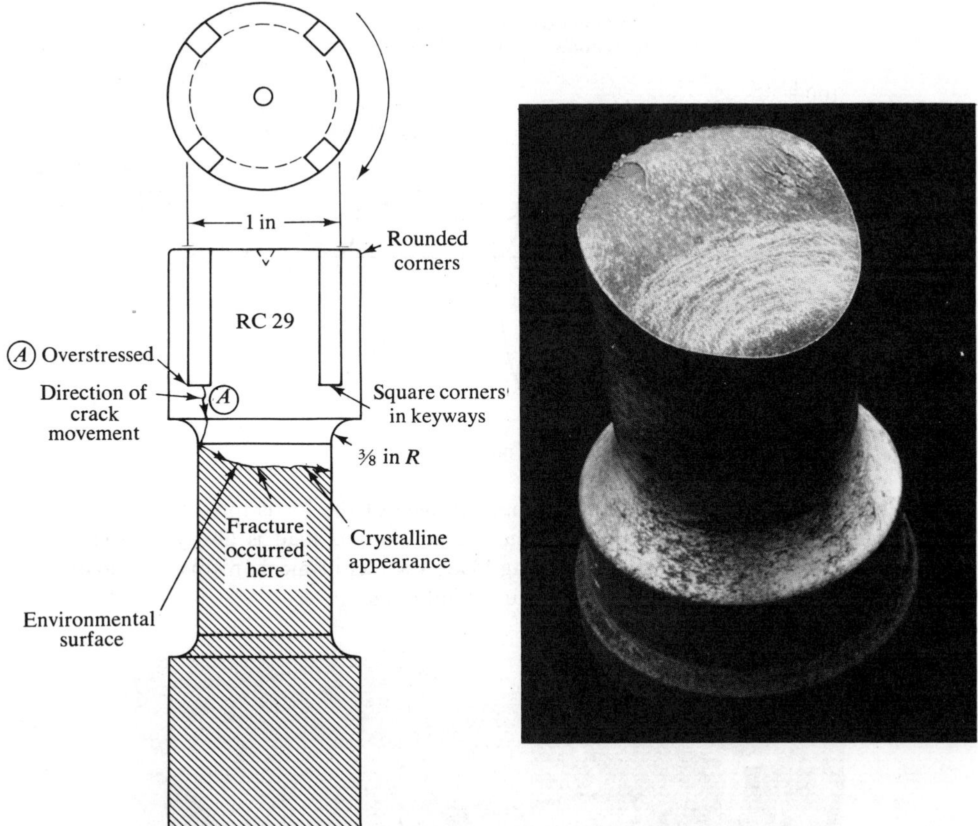

FIGURE 12–1 A machine spindle that fractured under fatigue conditions. The clamshell markings show evidence of surface grinding and the dark fracture is crystalline.

zero and back to tension, or a variable load in tension only where a minimum load is always present (Fig. 12–2). All of these conditions are common. Operating situations where these stressing conditions can exist include bending, torsional, axial, and shear loading. A flexual loading device is illustrated in Figure 12–3.

Types of Loads

Bending loads induce tension and compression loads into a part at the same time; maximum tension stress occurs on the surface of one side of the part as the metal elastically stretches and maximum compression occurs on the opposite surface. A fluctuating bridge beam is an example where tension loads are alternately applied and removed on only one side of the beam causing the cycle (Fig. 12–4). Incidentally, these compression and tension stresses are often equal because of the symmetrical shape of the beam. Torsional loads are twisting loads which establish shearing stresses in the part, the stress varying from maximum at the surface to zero in its center (Fig. 12–5). Shafting is an example whereby the load (engine) is placed on one end of the shaft which drives a mechanism on the other end such as an automobile rear axle.

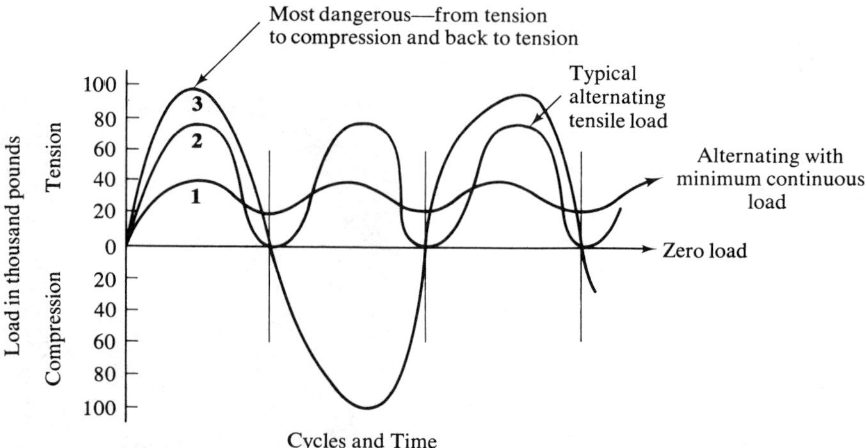

FIGURE 12–2   Types of fatigue loading. One of the loads must be tension. Load *1* is found in some designs where a continuous load is always present. Load *2* is found in alternating pulling loads such as in mechanisms, and load *3* is found in rotating beams and flexual conditions.

FIGURE 12–3   These turning shafts are exposed to flexual loading tension occurs in one side while compression stress occurs on the opposite side. If the weights do not stress the shafts beyond their endurance limits, the shafts will operate infinitely.

Another type of load is axial and this occurs in pulling and pushing situations such as are found in piston rods in certain types of engines. Also, bolts are often used in tensile designs where they hold heavy loads. In this example the stresses are spread evenly across the cross section of the part so central stresses are equal to surface stresses. If the line of stress slips from the centroid axis in this situation, eccentric loading will occur and uneven stresses will be distributed across the cross section.

Shear stresses occur from several types of loading situations such as torsional conditions and certain bending operations where both tension and compression

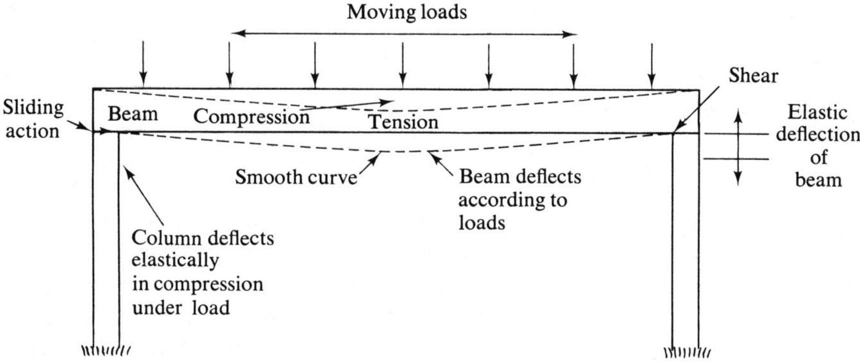

FIGURE 12–4 Bridge beam deflects as loads increase. Beam deflection returns to zero as loads move away. The beam elastically stretches in alternating cycles as loads appear and disappear.

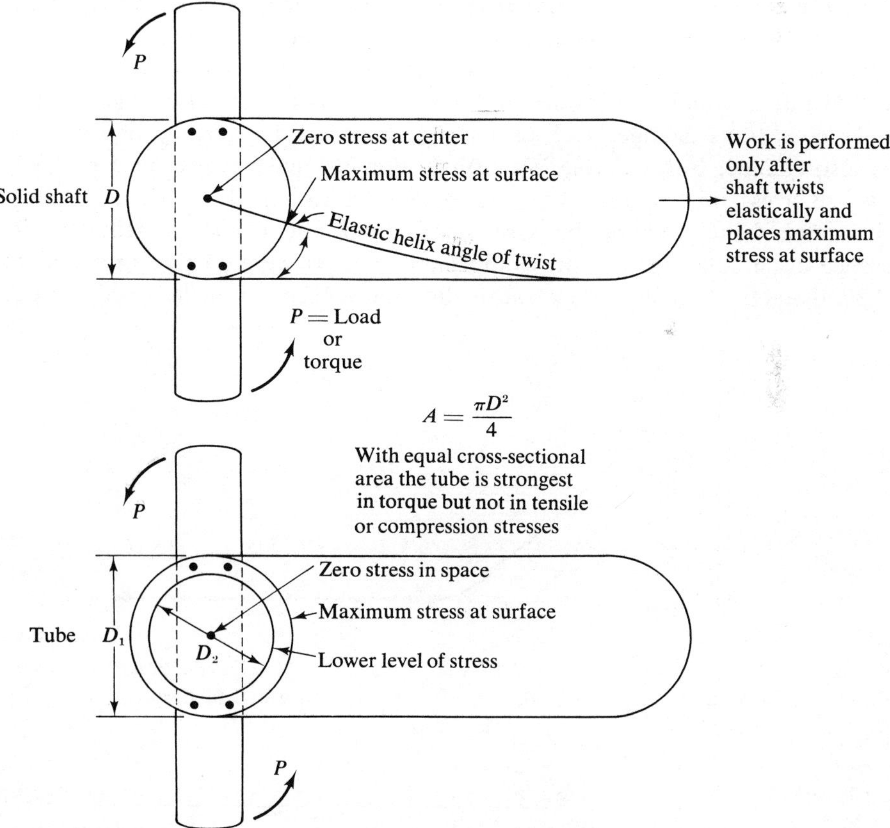

$$A = \frac{\pi D^2}{4}$$

With equal cross-sectional area the tube is strongest in torque but not in tensile or compression stresses

FIGURE 12–5 Torsional load and stress. As torque is applied to the shaft the metal elastically twists a given amount and then causes work to be done at the opposite end. With an equal area the tube is the strongest.

loading occur concurrently. Shear stresses also occur in a bolt which holds two plates of steel together in the presence of a bending load. With regard to the bolt, internal sliding of the metal occurs between planes of slip when loads cause deformation beyond the elastic limit of the metal.

Safe and Unsafe Stresses

The fatigue limit is the maximum stress that a steel will infinitely endure. Fatigue limit refers to endurance of the metal and in terms of time the endurance limit is the maximum cyclic stresses that continue to occur without causing metal failure. Fatigue limit and endurance limit refer to the same conditions. On the other hand, the fatigue or endurance strength refers to the particular stress which will cause failure after a specified number of cycles. As pointed out in Figure 12–6, the alloy steels were subjected to numerous fatigue tests at stress levels shown for the specified number of complete cycles from tension to compression and back to tension. The arrow on the curve shows the limiting stress that allows infinite loading without failure. As the stress is increased beyond approximately 75,000 psi fatigue life is shortened as indicated. The fatigue strength is then where the stress factor and cycle factor intersect on the curve resulting in fracture.

Data on the stress-cycle diagram are for the specified metal only. Stress-cycle and S-N (stress-number of cycles) are the same. As the metal changes and as the metal's conditions change, such as microstructure and geometry of the part, the data also change. For example, right angle dimensional changes in a part without fillets are stress raisers and lessen the metal's operating life. Surface notches also invite quick failure where the same loads could be carried indefinitely in the presence of smooth surfaces providing these stresses were carried in sound metal. As for dynamic loadings, they cause different effects in metals with respect to

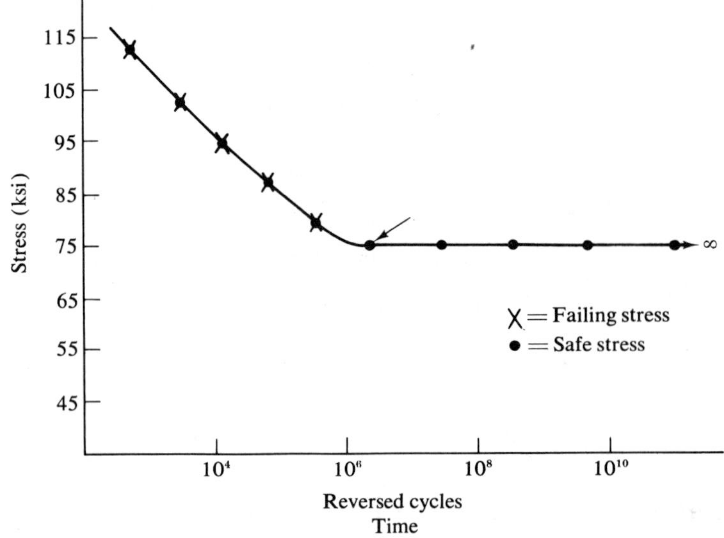

FIGURE 12–6 Stress-cycle diagram for an alloy steel which indicates that when the operating stress is reduced to 75 ksi an infinite operating life exists.

elastic and plastic deformation, the yield strength under static conditions not being a true guide. In static conditions a metal may safely hold thousands of pounds more than in dynamic situations over the same time period. It is then apparent from the data in Figure 12–6 that the specimen held more than 100,000 psi at the beginning of the test. Many steels have an endurance limit approximately equal to one-half of their tensile strength. If the tensile strength is 140,000 psi, then the endurance limit is about 72,000 psi. For nonferrous metals, a different type of endurance curve is produced; this group of metals seems to endure up to a given number of cycles. Often, the cyclic limit is stated as 500,000,000, which means that the particular metal will safely hold the load for the given number of cycles. This index shows the maximum stress that can be repeatedly applied for a given number of cycles without failure.

The Limiting Stress   According to data illustrated in Figure 12–6 the reduction in stress allows a longer life, and as stress is further reduced the endurance limit is reached whereby the time period moves toward infinity for that stress. Apparently the stress at the endurance limit is insufficient to start slip or fracture, and without slip or fracture safe conditions exist. Just what happens over the long periods of time while a ductile metal is exposed to cyclic stressing has not been definitely established. However, the forerunners of cracks may begin with surface imperfections in the part, existing cracks, or by slip, twinning, or cleavage of the metal. Accordingly, the data show strong indications that during cyclic stressing, striations or grooves appear in the slip planes of the metal and these often precipitate permanent atomic separations. These fatigue type of slip bands move to the metal's surface as wavy and unusually irregular bands as opposed to the smooth step type of band appearing from static stresses (Fig. 12–7). Ap-

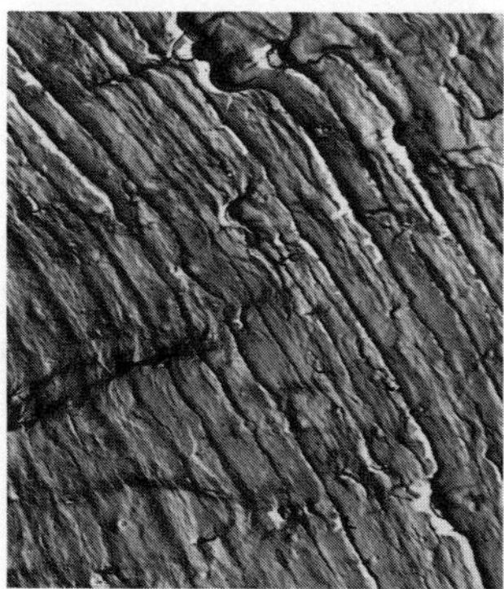

FIGURE 12–7   Striation pattern in a fatigue fracture of 2024 aluminum alloy. Magnification with electron microscope at 5000×.

parently these bands develop as the result of mass movements of dislocations along a favorable plane at a stress lower than the operating stress. Continued cyclic stressing accelerates the formation of these striations and increased sliding occurs on the atomic scale in the regions of the weakest bonds.

Striations Promote Metallic Separation   The striation usually starts early in fatigue if it is going to occur. As the striation's geometry changes and becomes more pronounced at maximum operating stress, a point in the lattice becomes over-stressed and subsequently metallic separation occurs. This fracture in the lattice is the starting point for ultimate failure of the metal under the same loading conditions. The discontinuity lengthens as the net result of a decreasing quantity of metal responding to a load which rhythmically overstretches the metal at the apex of the discontinuity, the metallic separation moving across the line of stress (Fig. 12–8). Because these unseen metallurgical conditions are not apparent to

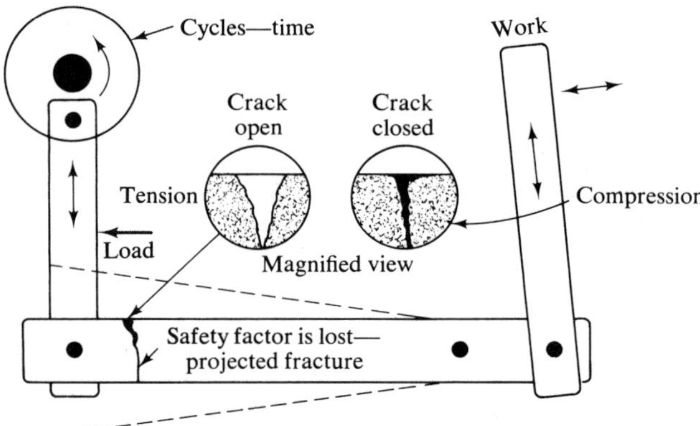

FIGURE 12–8   The enlargement of a crack in fatigue conditions within a heat-treated steel bar.

engineers and technicians, normal operations of the part may continue into days, weeks, months, or even years. But the separation grows toward its critical length and progresses in cyclic jumps, reducing the area of sound metal needed to sustain the steadily increasing stress. And suddenly, without warning, the critical length occurs and instant fracture and failure in the metal result, possibly after a year or more of satisfactory performance. Because the load is repetitive in cyclic intensity and because of the slow reduction of the safety factor to zero, the mathematical answer to the relationship among load, stress, and area to contain the stress is insufficient area. As the area of sound metal is slowly reduced the stress factor slowly increases. When the stress overcomes the strength of the remaining metal complete separation occurs. Obviously, this instant failure suddenly releases the load onto the connecting structure, if any, and this could cause a train of subsequent failures and possible catastrophe.

THE NEED FOR NONDESTRUCTIVE TESTING

Nondestructive testing engineers and technicians know of the probable existence of discontinuities in metals and are aware of the catastrophe which can occur in highly stressed parts which fail. Therefore, nondestructive testing is used to search for and detect these defects in stressed structures, and procedures are sometimes established to follow the defects under operating conditions. When a discontinuity is indicated it is analyzed and evaluated. Immediate action is then taken to reject the part if a dangerous condition exists or to accept it and follow the progress of the defect. Lengthening cracks in this example (Fig. 12–8) justify removal of metal in the affected area. In thick sections of cracked metal the region around the crack is removed and then filled with sound weld metal. On the other hand, the entire part or assembly may be replaced. It cannot be left to chance that the crack will not grow to critical length if it has tendencies to move across the stream of stress. Once the metal has started to separate under fatigue conditions it is only a matter of time, in many instances, until the crack reaches its critical size in the stress pattern, and this size is equal to instant structural failure. Obviously, without nondestructive testing methods and techniques available to find these discontinuities the state of the technology would be less advanced and the number of metallic failures would be dangerously large. Many failures would be catastrophic in material destruction, personal injury, and death.

PROGRESSIVE FAILURE

Fatigue failure is failure by cracking of the metal over an extended period of time. Evidence of ductility exists at the point of final fracture because of some plastic flow in all fractures except the cleavage types. Failure means that there is simply not enough metal available across the path of stress to hold the load as the crack reaches critical length. In the ductile failures some work hardness around the crack develops. An examination of the fracture presents a partial history of the cross section. Fatigue cracks develop in the intersecting regions of highest stress levels and weakest points in the slip planes. Also, many of these cracks begin at surfaces of grains where dislocations pile up, the granular surfaces being internal or on the surface of the metal part. Once the crack begins a series of ductile-brittle-ductile fractures occurs as the alternating stress opens and closes the void. The sequence ends in fracture and subsequently begins again with overstress and plastic flow. Plastic flow induces work hardness (if work is accomplished at a temperature below recrystallization) and, in turn, the deforming metal becomes so oriented that resistance and further slip are greatly reduced. Consequently, stress increases very rapidly to the fracture point, thereby relieving the stress momentarily. However, very ductile metals do not fracture at their highest stress or tensile, but flow plastically in large increments under rapidly decreasing stress levels to ductile fracture and zero stress intensity after an excessively increased elongation.

With regard to the fracture sequence, the harder metal which forms in front of the extending crack requires a higher stress to extend it further into the softer metal, which then becomes harder. Because the crack is opening and closing and is in the presence of complete reversible cycling including compression, the surfaces of the crack grind together and cause further deformation and increased hardness. At final fracture the natural crystalline structure of the metal is revealed on one side of the progressive crack. The contrasting appearance of the two fractures is pronounced, the initial crack having been very disturbed by both atmospheric conditions and possible compression loading. Crystallization of the metal, according to the view of the final fracture in Figure 12–9, did not occur and was not responsible for the break.

There is evidence that some microstructural conditions lead only to ductile failure, even though the overstress, plastic flow, work harden pattern prevails. In

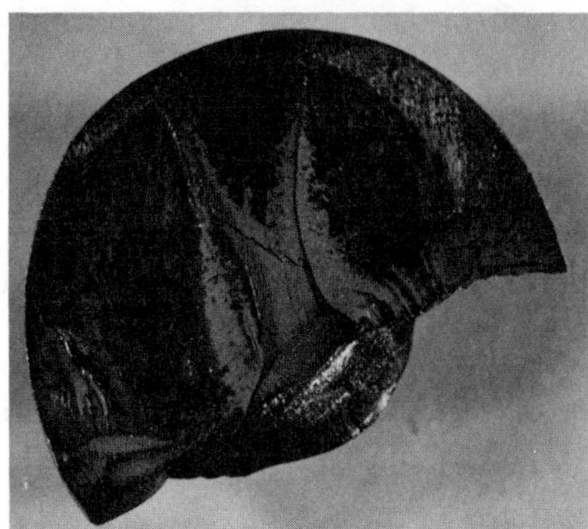

FIGURE 12–9   Cross section of fractured steel shaft showing progressive cracking under fatigue loading.

effect, some metals do not reach the brittle stage at the critical stress and fracture, but merely separate as if a total rupture of slip planes occurred whereby the critical reduction in area value was quickly reached.

METAL FAILURE

Metal failure is the separation of metal and is caused by sudden fracture as in impact loading, plastic flow as in over-loading, creep as in slow elongation to fracture, fatigue as in slow progressive cracking, corrosion as in rusting and metallic reduction, abrading as in severe wearing, cracking by heat treatment, and poor design. Temperature is often a governing factor relative to fracture, especially as the metal becomes hotter or colder than normal operating temperatures. Also, the surrounding atmosphere often has detrimental effects, especially when moisture and corrosive agents are present. Fracture is often the result of a combination of two conditions such as fatigue and corrosion.

Engineering designs take into account the operating conditions surrounding the metal such as temperature and atmosphere. To provide for unknown variables a reasonable safety factor is assigned to the metal's cross section by allowing it to have more metal than it really needs under known loading conditions. In service the metal is supposed to function at stresses far below its failing stress, and unforeseen loads are assumed to be absorbed in its safety factor of extra metal. But, as the records show, failures occur daily in all kinds of situations even though stresses remain below those designed for the part.

Impact and Overload

The collision of two automobiles demonstrates the sudden impact or fracture of metals. Rapidly accelerating compression and shear stresses overcome the metal's yield strength and failure occurs rapidly. The load on the metal may be applied at a slower rate, however. For example, a small bridge is overloaded when a slowly moving concrete mixer attempts to cross it. Fracture or severe deformation results after elastic conditions in the bridge's beams are exceeded and severe elongation begins, followed by rapid reduction in metal area and possible failure. The difference in these failures is the rate of loading; the latter example allows the metal time to respond. In impact loading plastic conditions may not have time to work harden and then fracture. The fast traveling stress is like a bullet and separates both ductile and hard metal. On many occasions, however, impact loading merely bends the metal without fracture in an overloading condition.

Creep Failure

Creep conditions exist as metal sustains heavy loads over extended periods of time. As the time passes, extremely small elongations sometimes occur and these accelerate as temperature increases. After initial elastic adjustments to the load are made in the metal's slip planes and grain boundaries, a very small but continuous strain often begins and lasts to destruction. This creep rate is the strain per time period and is influenced directly by temperature. As dislocations move along the slip planes they eventually reach grain boundaries and pile up under the needed stress. Such pile-ups at moderate temperatures induce resistance to further slippage, but as temperature increases further dislocation movements occur as the grain boundaries begin to rotate, causing accelerated rates of creep toward quick failure. Somewhere between the time that the grain boundary changes its resistance to dislocation movements to assistance in their movements is the compromising temperature, and this temperature varies as the chemical analysis of the metal varies. High melting temperature elements such as molybdenum increase this compromising temperature and reduce creep.

Fatigue Failure

Fatigue failure is the result of the time-load-cycle situation whereby small increments of slip in the metal occur due to movements of dislocations and plastic flow to metal separation as the tension and shear stresses increase to a critical magnitude. The generation of a crack by overstress creates stress at the crack's tip and, in turn, the metal work hardens and fractures or plastically separates. Progressive movement of the crack across the lines of stress leads to reduction in cross-sectional area of sound metal. As critical fracture approaches, critical stress moves into the last area of available metal and finds it insufficient to retain the overwhelming load. Reaction does not occur and failure is fast (Fig. 12–10).

Corrosion Failure

Corrosion is the loss of metal by direct chemical attack or electrochemical attack. When acids are exposed to a metal, for example, the metal usually responds by chemical reaction and loss of metal. However, when an anode and cathode are formed in the presence of an electrolyte, anodic corrosion occurs in the area of metal having the highest electrode potential with reference to the contacting dissimilar metal. Anodic corrosion reduces the quantity of available metal and causes a given stress to increase in magnitude. Such increases in stress levels ultimately lead to metal failure if corrosion continues. In this type of cathodic protection the larger the anode the less corrosion will occur at the anode, but in

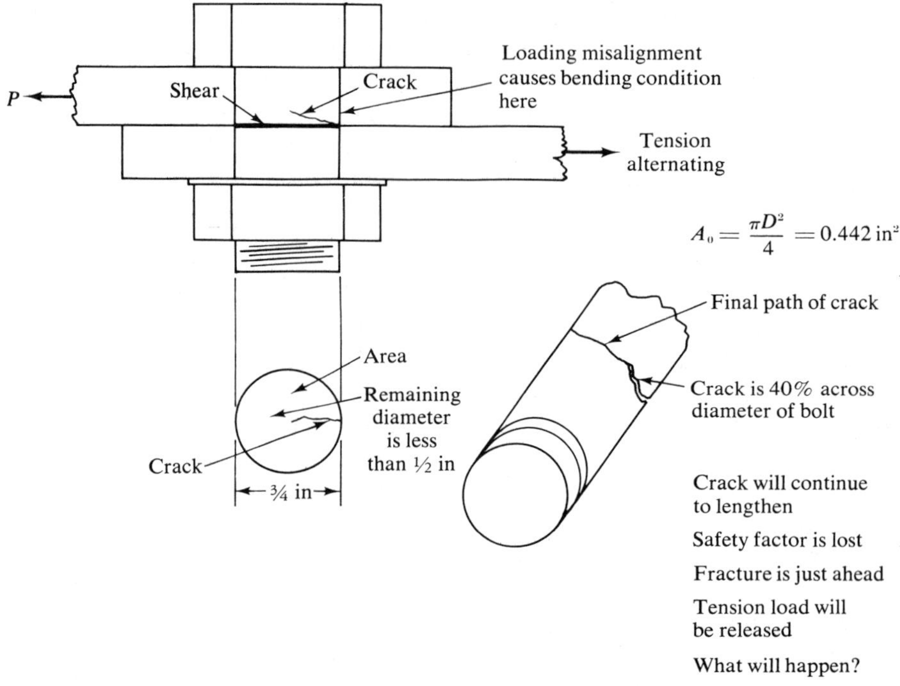

FIGURE 12–10   Crack growth during fatigue conditions to final fracture.

connections where the anode is small, fast corrosion results due to the rapid
dissolving of the anode. Deeply scratched tin plate, for example, will corrode at
the base of the scratch and further metal penetration results after a short period of
time due to cathodic protection for the tin.

Abrasion Failure

When two metals are in close contact and one rotates around another, metal
surfaces sometimes become dry and induce wearing of the contacting surfaces.
These wearing conditions ultimately lead to reduction in metal, and this leads
to metal separation and failure. Bearings which run out of lubrication begin to
freeze on the shaft while torque forces literally pull small pieces of metal from the
shaft or bearings. Such continuing action eventually reduces the bearing or shaft
to failure.

Failure by Heat Treatment

During heat treating operations some metals, especially the high carbon steels,
often crack after quenching because of a rapid increase of stress which becomes
greater than the strength of the metal. (Fig. 12–11). In this case and because
stress and rigidity are high in the presence of a lesser strength factor in the region
of the higher stress, metal separation occurs and oversatisfies the needs of stress-
strength balance in the zone of failure. Subsequent stressing of this cracked part
ultimately leads to complete failure. Overstressing often occurs in designs where

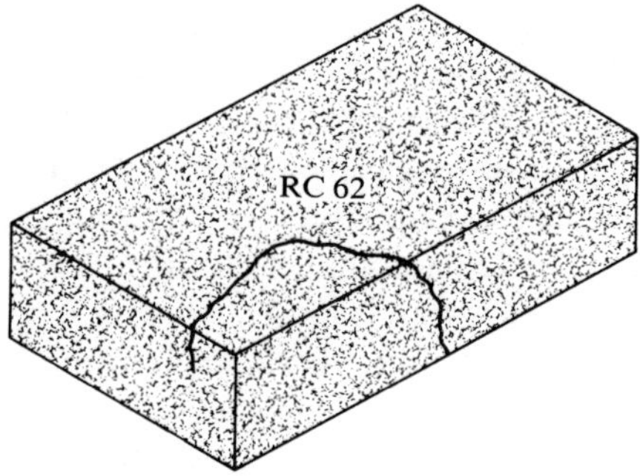

FIGURE 12–11 A heat-treating crack caused by failure to normalize prior to hardening this parallel made of high carbon steel.

holes are near an edge, where a geometric change occurs without a reasonable radius, where enough metal is not available to match the needs of stress, where stress collides with discontinuities, where uneven heating occurs as the metal's critical point is reached, where extremely uneven cooling occurs due to mass of metal differential, and where quenching procedures are too severe for certain alloys that do not plastically deform in the presence of excessive stress. Such a condition where exceptionally high stresses prevail leads to stress relief by bending or cracking of the metal.

Failure through Poor Design

Poor design, as previously pointed out, is believed to account for a large percentage of metal failures which occur because of excessive stress concentration and fatigue. A loaded part responds internally and its response creates patterns of stress which tend to follow surface contours and geometry of the part. When the paths of stress are caused to concentrate in small areas of the metal, failure frequently occurs because of internal overloading or an overcrowding of stress in a limited area of metal. For example, the notch effect frequently occurs where sharp change of surface exists. This internal condition is a reflection of the external surface condition such as the 90° root regions at the bottom of splines along a shaft. As stress reacts to the load the stress stream flows in concentrations at the right angle turns and builds a V notch pattern for potential failure.

The fracture which results from various types of design failures separates sections of metal by plastic flow in shear to fracture, or by brittle fracture in the absence of plastic flow and shear, or by a combination of plastic flow, which induces strain hardening, followed by brittle fracture. The shear type of fracture moves along the weakest path of resistance or series of slip planes from grain to grain across the grain boundaries to produce the final transcrystalline appearance. On the other hand, the intercrystalline crack occurs along the grain boundaries in a cleavage or brittle type of fracture. Such a fracture presents the natural crystalline structure of the metals whose grains were formed during their last appearance in austenite, if these are steels. Because of the physical differences between the grain boundary material and material in the grain, the face-centered cubic metals usually fracture in the transcrystalline manner. Apparently, the intra-granular material arrangements in face-centered cubic metals are normally weaker than their granular surfaces. Another example is illustrated by some of the stain-less steels and brittle fracture. Severe precipitation of compounds to the grain boundaries, such as carbide precipitation in austenitic steels during long time periods in temperature zones from about 1600 °F (871 °C) to 800 °F (427 °C) on cooling, can also allow cleavage fracture. In this respect, carbides fracture in a brittle pattern as opposed to the ductile shear deformation of austenite.

STRESS ANALYSIS

A study of stress patterns for assistance in stress analysis is easily accomplished with the aid of a photoelastic stress cycler (Fig. 12–12). This device focuses a

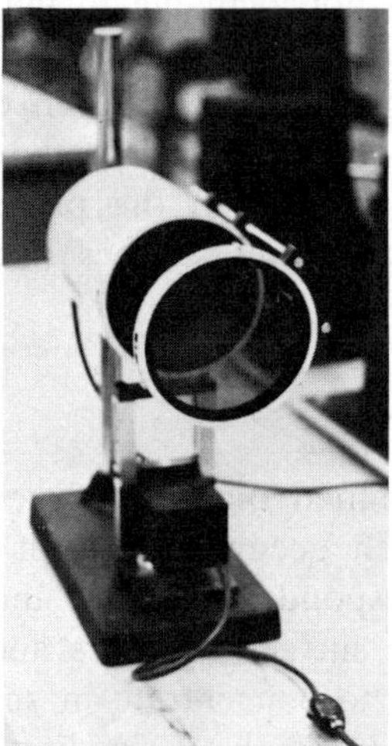

FIGURE 12–12   This photoelastic stress cycler enables the technician to observe stress flow in different designs.

polarized light beam on a part made from a special type of transparent plastic sheet stock about ¼ inch thick which allows the light waves to pass through. While the unstressed part is exposed to the light beam the plastic appears normal, that is, highly transparent. However, when a load is placed on the plastic, lines of stress appear and instantly respond to the load. As the load varies these stress patterns also vary. It is interesting to note, however, that stress will not jump across a hole, crack, or inclusion, but will detour around it, thus illustrating the elastic nature of stress. Because these observed lines of stress generally follow the surface geometry to the core of the part they must obviously meet somewhere between the several surfaces of the part. This meeting, or collision, of stress pat-terns then causes compromises. The stresses are elastic and they bend in semi-circular paths to avoid a hole or one another. Often, however, the load is so great that severe concentrations occur (Fig. 12–13).

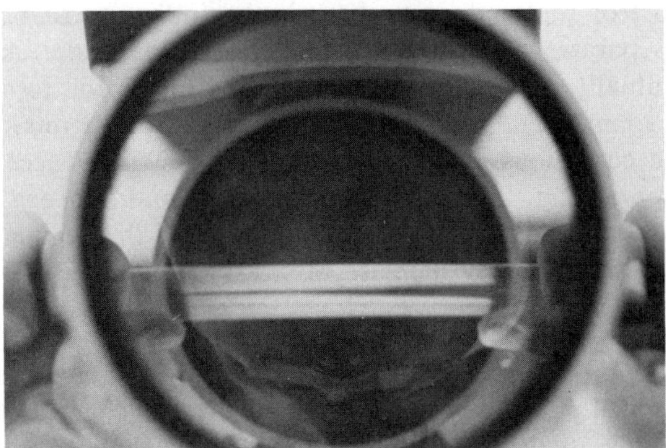

FIGURE 12–13   A rectangular bar of plastic is placed in a bending load. The resulting stress is observed as straight lines following the surface geometry.

Stress Must Remain Below Elastic Limit

When cyclic stressing occurs, as observed with the photo-elastic stress cycler, the amazing relationship of stress and load is revealed in all its patterns, compromises, collisions, and intensities. An analysis of these elastic flow patterns enables engineers and technicians to design their parts to avoid the undesirable stress concentrations which occur at sharp corners or when drilled holes are too close to an edge of the part. Strain gauges connected to cathode ray tubes and other instruments will bending stresses are too concentrated for a given section modulus or cross section of the part. Strain gauges connected to cathode ray tubes and other instruments will show these points of stress, but the paths and areas between these points must be recognized and evaluated because energy levels vary from point to point in a loaded material. At no time must the concentrated stress be equal to or greater than the elastic strength of the material in either static or dynamic conditions. Consequently, the designed load is always below the material's elastic limit.

Because a given stress exists in a material in response to an external load the stress pattern will only be smooth in exceptional designs where surface geometry is absent of stress raisers and where internal conditions are compatible with resulting stress compromises. The presence of properly oriented discontinuities or surface irregularities of the metal severely increases stress concentrations to well above the safe area stress and consequently increases the sizes of these irregularities by fracture, and this reduction in area frequently leads to complete metal failure.

CORROSION

Corrosion of metals occurs as a direct chemical attack, such as contact with acid, or by an electrochemical attack such as the contacting of two dissimilar metals in a galvanic cell system. In either situation corrosion causes a loss of metal along with the evolution of gases and solid by-products. The price of corrosion is high, amounting to many millions of dollars annually as the corroded parts are replaced.

Besides the cost of replaced parts there is the indirect cost of structural failures as the metal separates and further damage is done to the associated assembly. Corrosion eventually reduces the strength of the metal due to a gradual reduction in the area of sound metal. In the presence of load and stress, corrosion can lead to catastrophic conditions as the electrochemical actions accelerate and lead to failure without warning. In this situation a dual condition exists in the corrosion environment as the ionized metal deteriorates while the remaining metal is called upon to withstand the steady increase in stress intensity.

Direct Chemical Attack

The direct chemical attack on metals is simply an exchange of one material for another. For example, scaled steel is pickled in sulfuric acid to remove the scale. To accelerate the chemical reactions the acid solution is heated; rapid evolution of gas occurs along with the formation of a sludge as the base metal is exposed, clean and free from the previous scale. Continued exposure to the acid, however, results in further chemical attack as the acid removes small amounts of the metal and changes it to gases and solid residues. In other words, sulfuric acid corrodes steel. The scale on the steel in this example was initially produced by direct chemical attack of hot oxygen in the air as it combined with the metal's surface to form the iron oxide, possibly during a rolling operation. Indirectly and directly, metals become corroded by contact with corroding gases and liquids. The intensity of the attack depends on the relationship between the metal and the corroding medium such as strong concentrations of acids or alkali and susceptible metal. In effect, the metal is transformed into a compound such as a salt or oxide. Usually, these compounds are sulfides, carbonates, and oxides, but mainly oxides as the anode releases energy. Another example of direct corrosion is demonstrated when caustic soda contacts and destroys aluminum.

All Metals Corrode   No metal is exempt from corrosion, not even the noble metals, such as silver, gold, or platinum. It is true that these noble metals resist the attacks of oxygen and other active elements, but in some environments they too succumb to destruction. In recent times one of the biggest causes of corrosion has been the polluted atmosphere where aerosols form and become attached to a metal surface where corrosion begins. In this situation other forms of corrosion, such as electrochemical, may also occur. Basically, the direct chemical attack occurs wherever the environment promotes the association between an active chemical and a metal, the temperature being a critical factor. Normally, there is no significant current flow in direct chemical attacks on metals even though there are losses of electrons during the reactions.

Passive Film Protection   All direct chemical attacks are not totally destructive. The direct attack of oxygen on steel often leaves a protective layer of iron oxide on the surface which reduces or prohibits further destruction of the metal. So it is with many metals. Oxides form on copper and aluminum, for example, and as the depth of attack increases the intensity decreases to zero, leaving a coating of very fine oxide which retards attacks by other chemicals, including another oxygen attack. However, it must be remembered that corrosion is mainly concerned with the devastating attacks which bring eventual destruction to the metal. Direct chemical attack as a means of protection against corrosion is a secondary concern and is known as *passivation*.

Electrochemical Attack

The electrochemical attack on metals involves electrical current flow in the presence of chemicals. Such a type of corrosion includes two metals and an electrolyte. The two metals must be different and they must be in close contact with each other or relatively close through the electrolyte. With regard to their positions in the electromotive force series the more active metal becomes sacrificial to the less active. In other words, the metal having the highest single electrode potential becomes the anode and the other metal becomes the cathode.

THE ELECTROMOTIVE FORCE SERIES

In Table 12–1, the electromotive force series, each of the listed elements is arranged in a hierarchy from the highest electrode potential to the lowest. Any metal in this hierarchy which is lower than another will be protected by the other, to a degree, when in electrical contact. Such a situation brings eventual and total destruction to the metal which is highest in the series. This type of corrosion is common and is one of the biggest enemies of society.

To illustrate what happens when two dissimilar metals are placed in electrical contact, place a millivoltmeter in .contact with a specimen of magnesium and a specimen of steel so that an electrolyte wets the surface of each specimen (Fig. 12–14). Immediately, a voltage is shown which means that current is flowing from

TABLE 12–1 ELECTROMOTIVE FORCE SERIES

Metal	Symbol	Single Electrode Potential
Potassium	K	+2.92
Calcium	Ca	+2.77
Magnesium	Mg	+2.34
Beryllium	Be	+1.70
Aluminum	Al	+1.67
Manganese	Mn	+1.05
Zinc	Zn	+0.76
Chromium	Cr	+0.71
Iron	Fe	+0.44
Cadmium	Cd	+0.40
Cobalt	Co	+0.27
Nickel	Ni	+0.25
Tin	Sn	+0.13
Lead	Pb	+0.12
Hydrogen	H	0.00
Bismuth	Bi	−0.22
Copper	Cu	−0.34
Mercury	Hg	−0.79
Silver	Ag	−0.80
Platinum	Pt	−1.20
Gold	Au	−1.42

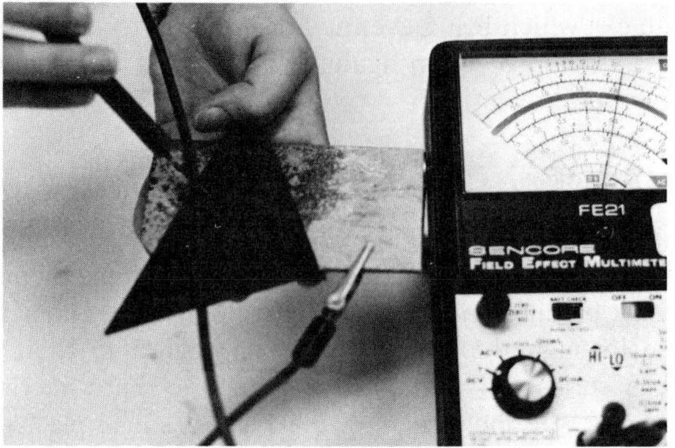

FIGURE 12–14 Two dissimilar metals are connected with saltwater. The clip is connected to magnesium and the electrode is placed against the steel. The instrument shows a voltage in the circuit.

the magnesium to the steel as long as the electrolyte connects the two specimens. Current will continue to flow for the duration of the connection. Loss of the electrolyte or the anode or passivation of the anode stops the current flow. The metals listed in the electromotive force series shift in hierarchy when their chemistry changes. In effect, a new hierarchy automatically forms when elements become alloys. This new governing list of metals is the galvanic series shown in Table 12–2. Notice that some metals are shifted upward or downward in relation to the electromotive force series. Such a condition is normal and is considered the governing criterion at the moment of dissimilar metal contact.

Anode-Cathode Process

An analysis of what is happening during electrochemical corrosion points out that electrons are moving through the connection from the magnesium to the steel

TABLE 12–2 A GALVANIC SERIES

Magnesium
Magnesium alloys
Zinc
Aluminum
Aluminum alloys
Cadmium
Steel or Iron
Stainless steel
Lead
Tin
Nickel
Iron-chromium-nickel
Brass
Copper
Bronze
Copper-nickel
Silver
Gold alloy
Platinum

(Fig. 12–14). This leaves the magnesium ions positive and the steel negative. A further investigation shows that hydrogen is liberated at the cathode and the steel's surface is clean. On the other hand, an observation of the magnesium anode shows pitting of the surface and loss of metal. The experiment further shows that when two dissimilar metals are connected with a salty solution or electrolyte, anode surfaces deteriorate and move into the solution. Current immediately seeks the cathode. Also, the anode becomes ionized (loss of electrons) and these ions flow to the cathode through the electrolyte as they are carrying positive charges. This situation occurs as the anode releases its valence electrons and the modified atom, or ion, becomes part of the electrolyte (Fig. 12–15). Obviously, this is an anode loss and the chemical change is the force or voltage which drives the electrons. Because anode and cathode actions occur concurrently the electrolyte is neutral.

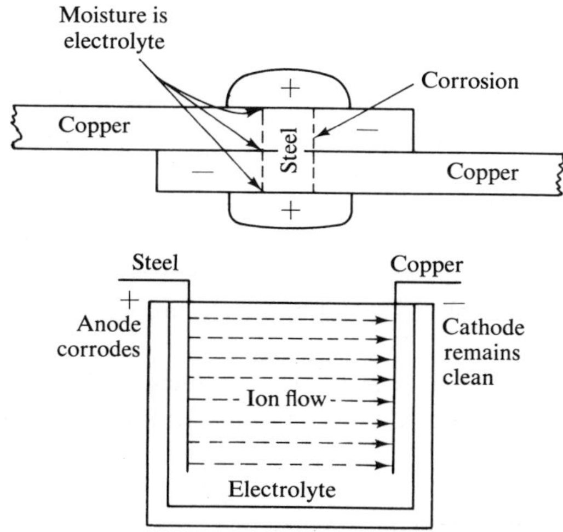

FIGURE 12–15   Anode-cathode process in corrosion at a riveted joint where a steel rivet connects copper plates.

OXIDATION-REDUCTION IN THE ELECTROLYTE

Electrochemical attacks on metal reduce the anode to electrical power, gas, and a sludge. When electrons move from the anode to the cathode, cations (or positive ions), flow through the electrolyte to the cathode. At the same time anions, or negative ions, flow to the anode as the result of gaining electrons. Hydrogen is liberated. Accordingly, the metal losing the ion is slowly reduced to zero as it releases itself to the lower electrode potential. Corrosion of the anode results because it is oxidized by a valence increase and loss of electrons while reduction occurs elsewhere in the system by gain of electrons due to a reduction of valence. In other words, the anode is oxidized by slow destruction while the cathode is reduced by constant protection. Normally, this is the concept of anodic corrosion and cathodic protection, but if the active metal becomes passive, the cathode is subject to corrosion. In this respect, the term *passive* refers to an oxidation situation in which the effects of anodic corrosion are so modified that further oxidation ceases, and this automatically stops the reduction process at the cathode. But if oxygen at the anode is able to break though this passive and protective film for some reason, the whole process of destruction again functions.

ANODE-CATHODE RATIOS

The phenomenon of electrochemical attack on metals comes about when the anode is smaller than the cathode area. As the ratio between anode and cathode increases, the attack on the anode also increases. Current density is measured in amperes per square foot of cathode area which is immediately reflected upon the anode's area. Basically, when the cathode calls for positive ions the intensity at the anode attempts to match the needs at the cathode. A demanding cathode, as it controls the anode, can quickly deplete all the ions of the anode. The smaller the anode with respect to the cathode the greater the surface loss at the anode. This loss of surface is relatively small when the anode is large but rapid loss occurs when the surface is small. Also, the greater the potential difference between the anode and cathode the greater the intensity at the anode. In other words, the farther apart the two electrodes are on the electromotive force scale the greater the flow of ions.

If dissimilar metals are to be connected, the area of the anode must be equal to or, hopefully, greater than the area of the cathode, for it is the anode that must be sacrificed to protect the cathode. Consequently, structural fabrication must account for this galvanic problem by designing the anodic materials in areas greater than those of the cathodic materials. One solution is to make the smaller connectors, such as bolts or rivets, of cathodic materials. Such a design provides for a small cathode and a large anode, but the design must still allow for adequate strength of the connector.

CATHODIC PROTECTION

Much has been recently discovered regarding the anode-cathode system of corrosion. For example, it is now known that metallic designs can be cathodically protected over a long time period by strategically located anodes in the electrochemical system. Such a discovery provides for long-term protection of the cathode.

Foundations for buildings and bridges can be protected for years by carefully placed anodes which can be replaced when necessary. Steel is still king in the construction industries with regard to buildings and bridges. In order to preserve these basic steel structures and maintain them safely from destruction by corrosion, strategically located anodes are placed in an electrical system. To illustrate, magnesium anodes are buried in the soil and electrically connected at specified distances along the steel members of the foundation. The ion flow from magnesium to steel is continuous but slow because of soil dampness. The rate of ion flow is governed by the surface areas of the steel members within the protective limits of the anodes.

Instead of using replaceable anodes where it may be impractical, a direct current generator is sometimes connected to a parallel system of buried anodes so that the positive electrode is connected to all anodes. The generator's negative electrode is connected to the steel structure. In operation a continuous positive ion protection functions and the surface of the steel within the ion field is cathodically protected.

Sacrificial Anodes

One of the largest corrosion prevention designs ever devised is the pipeline system of protection. As steel pipe is welded and laid in the soil anodes are also placed in the soil a few feet away from the pipe and connected to the pipe with an electrical cable. Anodes are placed at strategic intervals, depending on the soil condition and size of the pipe. The distance between anodes varies with the system's needs; for example, some anodes are laid at intervals of several hundred feet or even miles for adequate protection. The same concept of corrosion prevention is also used for the protection of steel hulls of ships. Zinc slabs are attached to the steel plates below the water line at required intervals and the seawater is the electrolyte (Fig. 12–16). Periodically, this system of protection requires replacement of the anodes.

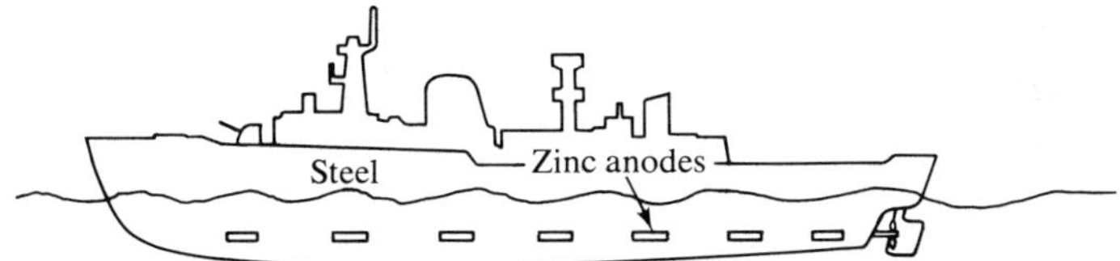

FIGURE 12–16   Zinc anodes attached to the steel hull of a ship prevent corrosion of the hull.

Because metal plates or electrical cables allow electrons to flow from anode to cathode, ionization movement through the water or the soil occurs when zinc, aluminum, or magnesium positive ions seek the negative surfaces of the building, ship, bridge, pipeline, or whatever is to be protected.

CONCENTRATION CELLS

Two types of concentration cells are the metal-ion and oxygen concentration cells. Because either of these two types of corrosion can occur at a joint an understanding of what happens can often alleviate or remove this costly situation. Again, as in galvanic corrosion, two electrodes are formed, the anode and cathode. Because a metallic joint must include at least two metals, usually three, consideration should be given to the design with regard to the gaseous or watery surroundings of the joint. If two different metals are in contact at the joint, galvanic corrosion occurs. But if the joint is constructed of the same metals, concentration cell corrosion can occur instead, if the electrolyte is present, and of such chemistry that ion flow can occur. The design of the joint and the condition of the environment within an electrolyte then determine if an attack by concentration cells will occur.

Metal-Ion Concentration Cell

A bolted or riveted joint made of the same metals and submerged or enclosed in an electrolyte may be receptive to corrosion if the electrolyte is stagnant or restricted in flow within the joint. For example, a riveted joint allows the two outer areas of the metal to be farther apart than the inner areas. Ions in the electrolyte will then be more concentrated in the inner regions of the joint than at the periphery because a movement of the electrolyte occurs at the outer edges. Such an environment promotes anodic corrosion at the outer edges of the joint (Fig. 12–17)

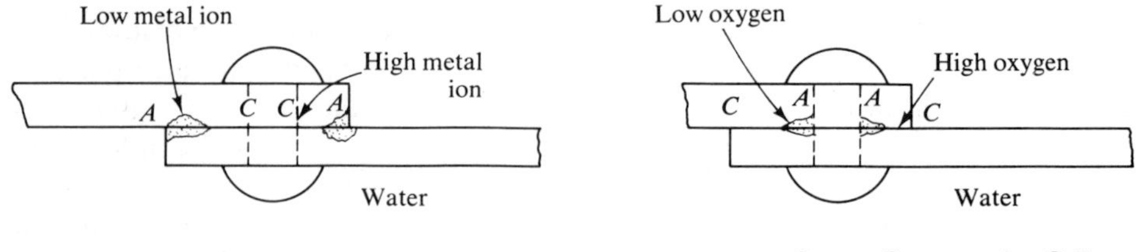

FIGURE 12–17 Two concentration cells showing the anodic-cathodic process in action whereby the anode is depleted to save the cathode.

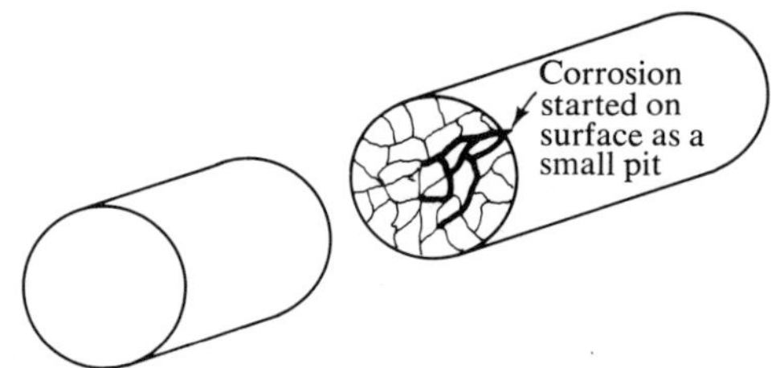

FIGURE 12–18 Intergranular corrosion in a bisected bar of steel whereby strength is lost in the cross section.

because the cathode forms in those regions where ion build-up exists. As long as a differential in electrolyte concentration exists about a metal, the anode and cathode will form. Immediately, electrons move through the metal to the cathode while positive ions move through the electrolyte to the cathode. Disintegration of the outer edges of the joint occurs while protected areas remain in the inner regions. If good circulation is established, the rate of corrosion attack will be reduced.

A variation of the joint type of corrosion occurs on metallic surfaces which contain dents or in mechanical arrangements which contain water. Even a drop of water occasionally picks up different quantities of soils and oxygen. Being on the surface and fairly stable in shady areas, a one-drop concentration cell often starts corrosion as its center becomes lower in ions than its perimeter. The central anode lessens the oxygen content and a double-acting cell quickly generates a pit on the steel's surface. Eventual evaporation of the cell occurs, but the chemical differences between the metal's surface in the pit and the metal just below the pit are sufficient to start intergranular corrosion by the anode-cathode system (Fig. 12–18). Deposits of rust on steel reflect the corrosion residue. This type of corrosion in aluminum, however, is evidenced by a white powder or a pimple. A probe inserted into the pimple verifies the accuracy of the suspicion. Severe cases of this type of corrosion cause laminations to form in the metal, reducing the strength level to dangerous conditions. The path of this type of corrosion follows the grain boundaries deep into the metal and if not discovered in time, a dangerous reduction in the metal's strength occurs.

Oxygen Concentration Cell

Another example of corrosion is the oxygen concentration cell whereby the reverse situation occurs. Again, the joint is made of the same metals and surrounded with an electrolyte. Because oxygen is present in the air and in water a chance exists that oxygen may be more concentrated at the outer regions of the joint than in the inner regions. Stagnation of the electrolyte promotes low oxygen concentration when compared to a higher concentration at the edges of the joint. Consequently, the anode-cathode system begins as soon as the concentration differential occurs (Fig. 12–17). Lower concentration regions provide the anode and higher concentrations establish the cathode. The flow of current is variable, as it is in the metal-ion cell, but the corrosion effects in both types of cells at the anodes continue, even with variable flows of current which respond to either ion concentration or oxygen concentration differentials.

With regard to liquids the oxygen concentration cell difference between the surface of water, for example, and at a given depth below the surface is often enough to start the functioning of a cell in the presence of metal. Again, as in the metal-ion concentration cell, better circulation techniques can greatly reduce the tendencies for metals to corrode in the presence of potential corrosion regions.

STRESS CORROSION

The combination of corrosion with tensile stress is often fatal to metals in short periods of time. Stress may be residual, such as that existing in hardened and tempered steels or heat-treated magnesium alloys. Again, stress may be externally produced, as in any tensile loading operation. Corrosion may start due to the environment by direct chemical attack or by galvanic action, and it must be at the surface to combine with the stress. Pressure vessels such as boilers often emit solid chemical residues which enhance the corrosion process when pressure and high temperatures are present. On the other hand, corrosion may be precipitated directly by tensile stress in any structural design such as that occurring in heat-treated magnesium alloy structural members which have not been properly surface protected. When both stress and corrosion are present in a metal it is only a matter of time before sudden fracture occurs due to metallic separation across the main line of stress.

Several means are available for corrosion to move deep into the metal under the pressure of stress. For example, the anode-cathode principle may be instigated when the granular boundaries and precipitates become anodic to the cathodic interiors. As penetration continues concentration cell attack may start and supplement the disintegration of metal along the granular boundaries or in a shortcut across the grains. If stress in an appreciable magnitude is not present, the intergranular corrosion process will follow the grain boundaries in any direction. In the presence of tensile stress the attack will move to travel at a near right angle to the flow of stress.

Stress Accelerates Corrosion

Because stress corrosion does not allow passive film protection the deteriorating process proceeds to satisfy the electrochemical needs of the cells as well as metallic shifting in an attempt to block the killing stress. With both deterioration and metallic shifting working together the countdown to fracture continues. Cold-working operations leave metals in strained conditions, and areas where stress differentials occur are potential regions for anodic action to begin. In fact, any chemical differences in regions of the metal are possible regions for corrosion to begin; but normally, other factors, such as load on the metal, must also be present. For example, a common failure occurs in cold-drawn brass and other copper alloys when chemicals and loads appear together. Pressure vessels containing liquids have been plagued with this type of failure (Fig. 12–19).

FIGURE 12–19 This pressure vessel finally split because of stress corrosion.

Preventive Measures

Possible preventive measures against stress corrosion include stress relief treatments of the metal after fabrication to reduce the magnitude of the residual stress. Another method of prevention is to make the part or assembly from those metals which have good resistance to the several corrosions. Also, certain kinds of surface treatments are available to retard or eliminate the initial nucleus of corrosion. A very effective surface treatment is shot peening whereby the surface is cold worked into compression stress. Before tensile stress can begin the load must exceed the residual compressive stress.

CORROSION FATIGUE

Another dual type of corrosion occurs when cyclic loading induces stresses along the surface areas of the metal where potential corrosion can begin. If the corrosion exists, the elastic stretching of the metal breaks any passive film formations and allows nuclei regions for corrosion to start. Any part which is exposed to any type of corrosion is especially receptive to stress resistance and, of course, this is the beginning of the end of the part. Numerous cases of this multiple type of corrosion occur because it automatically includes stress corrosion, both static and cyclic.

A typical failure occurs in hot water pipes and steam pipes which are exposed to corrosive films and soils. When the pipes are restricted in movement and cannot expand and contract with fluctuating pressures stress concentration regions begin, and in the presence of corrosive conditions the expanding metal invites the corrosion. As time passes the remaining sound metal can no longer effectively resist the rising stress and instant fracture occurs across the line of stress. An examination shows a final transcrystalline fracture which matches the metal's final attempt to stabilize the load.

OTHER TYPES OF CORROSION

Often, it is not possible to determine the type of corrosion which dooms the part to failure because many failures involve multiple attacks. An erosion type of failure frequently occurs where high velocity fluids pound on the metal's surfaces, such as at bends in pipes and tubing and in the cavities of impeller blades. Even though the quantity of metal removed initially is not measurable, the continuous bombardment eventually removes substantial quantities. If any type of corrosion starts in the presence of erosion, the reduction in available sound metal will accelerate, partly because the attempted formation of passive films is blasted away. The formation of cavities on metallic surfaces along with corrosive pitting eventually brings on failure.

Pitting

Pitting of metallic surfaces is the result of multiple attacks by any of the corrosive media. Passive film breakdowns in random surface areas permit localized corrosion to begin. The surrounding passive film may not allow spreading of the corrosion on the surface, but it has no control when the action moves inward. Lightly stressed and static structures are often fatally corroded by the intergranular corrosion which begins in a pit and follows the winding boundaries of the grains. Because the anodic boundaries disintegrate the remaining metal must sustain the load. Obviously, when corrosion consumes the metal's safety factor, fracture occurs instantly.

Dezincification

A form of brass sometimes is attacked by galvanic action between regions of solid solution differences, such as one volume of metal which is richer in zinc than an adjacent region. Because of the chemical differences numerous anodes and cathodes are established whereby the zinc atoms are ionized from the copper-zinc solution, leaving a spongelike appearance. When stressed, failure occurs because insufficient metal remains to hold the load. A very common form of corrosion is elevated temperature scaling brought on by a long-term exposure of the metal to high temperatures. Gradual surface chemistry changes occur when red-hot temperatures allow hot oxygen to combine with a metal's constituent such as the case when steel is attacked and iron oxide forms. A less common type of corrosion occurs when stray electrical currents flow to buried metal structures and then proceed back to the direct current source. Corrosion of the metal occurs at the point of exit from the buried pipe or other part.

CORROSION PREVENTION

Numerous means are available to alleviate or eliminate corrosion of metals and they vary from painting to cathodic processes. Somewhere in the corrosion prevention system is climate control. Air conditioning removes moisture from the atmosphere and therefore reduces the tendency for an electrolyte to form, but all places cannot be air-conditioned. Often corrosion occurs as a multiple attack and the effects are more damaging. Because metals must operate in some kind of environment and under some kind of load (even the weight of the metal) it becomes a matter of careful selection of the metals to help prevent corrosion. Stress is always present to some degree and increases directly with the load that the metal is to carry. The environment then is the big variable which must relate to some kind and magnitude of stress and operating condition.

Because the environment often provides the potential for several types of corrosion the engineer and technician must be on the alert to prevent its occurrence. Galvanic corrosion is very common partly because of the lack of knowledge regarding the problem of design vs. corrosion. The galvanic series clearly points out the corrosion potentials between any two metals. Of course, the best solution is to construct the joint from the same metals or, if different metals are to be used, to select those close together in the series or design the connector or smaller areas with cathodic metal. In effect, maintain a large anode and a much smaller cathode. For example, galvanized iron (zinc-coated steel) provides a core of steel for strength and a case of zinc for the steel's protection. Should the zinc surface be penetrated to the steel, zinc ions quickly move to protect the steel. As long as a sufficient quantity of zinc remains the iron base alloy will be protected. On the other hand, should the tin surface of a food can be penetrated to the steel, the iron ions quickly move to protect the larger area of tin. Such a situation accelerates the corrosion of the steel can whereas the zinc protects the steel. The anode should be large enough to protect the cathode.

Elimination of the Electrolyte

In the situation where concentration cells may form, the metals should be constructed to eliminate places where moisture can collect to prevent the forming of an electrolyte. When liquids are used in a design, such as for a pump system or other metal part which is submerged in water the circulation should attempt to reach all regions of the liquid and not allow stagnant places where either the ion or oxygen cell can form. Naturally, in the regions of the connection circulation will be poor; therefore metals in this type of design should be those of the stainless types which have inherited corrosion resistant capabilities. Some of these include the austenitic steels, copper-nickel alloys, copper-zinc alloys, copper-tin alloys, and many of the nickels. One of the more effective discoveries for preventing corrosion is the use of a nonmetallic liner such as plastic in the tank or connection.

Replacement Anodes

Other methods of corrosion prevention use the outside anode system or sacrificial anodes whereby long-term protection is given to the cathode. Any steel structures which are buried, such as pipelines, bridge columns, or building foundations, for example, can be made the cathode by electrically connecting them to buried anodes in strategic locations. Or the direct current generator system can be used whereby the generator's negative terminal is connected to the structure to be protected (Fig. 12–20). The connection provides the means for electrons to flow to the

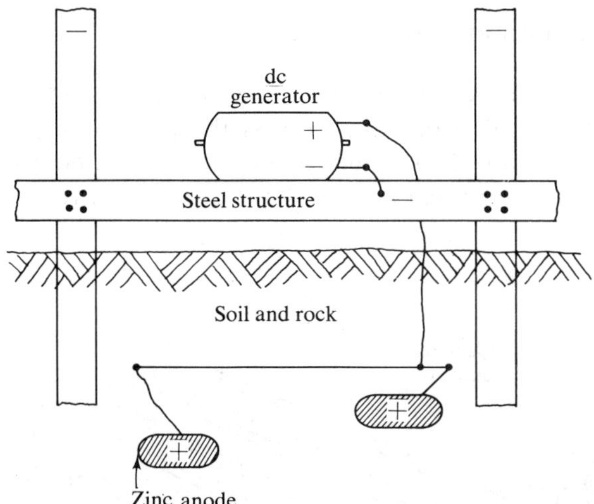

FIGURE 12–20   Direct current generator connected to the steel structure of a building to provide protection against corrosion.

cathode. The positive terminal is connected electrically to the numerous anodes over long distances. Recently it has been observed that moving steel assemblies and parts which are submerged in water, such as ships, can be protected by attaching slabs of zinc, aluminum, or magnesium to the steel hull. When the painted steel plates are scraped by foreign objects, ions from the attached anodes flow to the unprotected cathodes and provide long-term protection for the hull and propellers. The same principle is used in closed circulation systems such as water heaters and liquid circulation systems.

Surface Protection Measures

Possibly the most common corrosion prevention methods involve the use of paint, cladding, and plating. Numerous types of paints, such as the red leads and zinc, aluminum, and copper types, in addition to related products are used to protect metallic parts. Steel bridges carry large quantities of red lead or aluminum powder paint. Also, resin base paints are very popular as well as the endless varieties of varnishes, shellacs, plastic dips, asphalts, and tars. These materials are basically organic, but some types include inorganic additives such as the powdered metals.

Ceramic-coated metal parts have proven to be very successful in resisting corrosion, especially at elevated temperatures where corrosive gases bombard the part's surface. A typical example is the ceramic-coated exhaust member in expensive engine exhaust systems. The ceramic is a type of cladding, or coating. Another example is the ceramic coating of bathtubs and related items.

Some of the most effective types of cladding for large sections of metals are the coatings of aluminum, zinc, and tin. Thin sheets of aluminum can be rolled onto steel and bonded. Many metals can be sprayed onto another while molten. Also, dipping a metal part into a liquid metal results in a coating which acts as a protection of the base metal.

Electroplating and Anodizing

Two of the more sophisticated processes used in corrosion prevention are plating and anodizing. In the electroplating process the part to be plated and protected, such as a steel bolt, is attached to the cathode while both electrodes are submerged in an electrolyte in the presence of a flow of direct current (Fig. 12–21). The

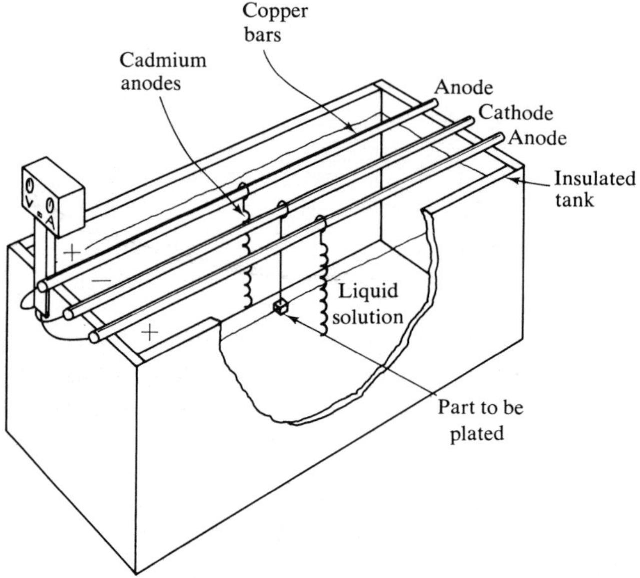

FIGURE 12–21 Schematic drawing for an electroplating process. The part to be plated is electrically attached to the cathode while the positive anodes furnish metal ions to the liquid solution.

anode is any metal which is to become the plating. The main difference between electroplating and galvanic corrosion is the action at the cathode. In plating anode ions move to the cathode and become permanently attached, and in galvanic corrosion the cathode remains clean. When anodizing, the part to be protected, such as an aluminum bracket, is attached to the anode while both electrodes are submerged in the electrolyte. As direct current flows corrosion occurs on the bracket's surface, but it is a very fine type of oxide film and acts in a passive manner. When the correct depth of corrosion is obtained the part is removed and washed. In use the passive film is corrosion resistant. Further details of electroplating and anodizing are given in another text, *Manufacturing Processes*.

Questions

1. Define fatigue and fatigue failure, listing the three main factors which cause failure.

2. Why must tensile stress be present during the fatigue failure or progressive failure process?

3. How does design influence fatigue and progressive failure?

4. Differentiate between endurance limit and endurance strength.

5. How does a striation form? What is its significance to progressive failure?

6. What is a slip plane? Explain its place in the permanent deformation of metals.

7. Define critical fracture.

8. Discuss the several types of metallic failures.

9. Explain the creep of metals.

10. How does good design reduce the tendency of progressive failure?

11. Define corrosion.

12. Explain the significance of knowing the hierarchy of metals in the galvanic series.

13. Describe the process of galvanic corrosion.

14. Describe the processes of metal-ion and oxygen concentration cell types of corrosion.

15. What is the main difference between a direct chemical attack and an electro-chemical attack?

16. Why is intergranular corrosion one of the most dangerous types of corrosion?

17. Differentiate between stress and fatigue corrosion.

18. List several methods for preventing corrosion.

19. Differentiate among the mechanisms of corrosion, electroplating, and anodizing.

20. How do alloys influence the elastic limit of metals in comparison to the plastic deformation capabilities of the metals?

13

Nonmetals in Engineering

Many common applications in engineering designs require the use of nonmetals such as plastic, graphite, wood, adhesive, ceramics, concrete, glass, and rubber. Some of these materials are used as an independent component; other applications require various combinations of several materials. The composite of epoxy and aluminum strips bonded together as a laminate is a typical example of nonmetals and metals being used together so the design benefits from the mechanical and physical properties of both materials. This discussion will include the metals whenever they are used in direct conjunction with the nonmetallic materials.

A nonmetal is often essential for a particular job, such as a plastic barrier or windshield or a concrete road. Rubber is used as an insulating material in the electrical industries, and automobile tires are used by the millions. Glass-covered cases of instruments provide dust-free operating conditions for mechanisms, and clay products and ceramics provide the many bricks and pottery goods. The hardest commercially used cutting tools are made from pressed and sintered aluminum oxides. In order to benefit from the several materials in a composite, a laminate is made whereby fibers of graphite and epoxy are laminated with thin sheets of heat-treated titanium alloy to produce high strength structural parts. Various combinations of metals and nonmetals are used in all kinds of unusual situations. One of the fastest growing applications is the use of composites, and the number of possible uses is nearly endless.

COMPOSITES

A composite is a mixture of different materials. At least two materials are included in the mixture and each of these materials maintains its identity or purpose in the finished product. However, chemical changes may occur in some composites during processing, such as water disappearance in the hydration of concrete or compounds formed due to heat in the sintering of metallic powders. A typical example of a composite is concrete, the mixture consisting of gravel, sand, cement, and water. The cement, sand, and water combine to make a bonding material around the individual rocks which are much harder than the other constituents. After proper mixing, hydration, and curing, the new material is strong enough for use as structural shapes such as columns and beams as well as for its use in paved roads (Fig. 13–1). In general, the typical composite is a finished shape which consists of different materials arranged as an aggregate, a laminate, or a fibrous pattern.

FIGURE 13–1   A highway bridge system of concrete beams and columns.

Purpose of Composites

The purpose of a composite is to gain the most benefit from the particular arrangement of constituents. For example, concrete is an aggregate of four materials which is unlike any of its constituents. Because of hydration and the resulting bonding of the particles of the aggregate into a single mass, a structural member is made available that incorporates a set of mechanical properties completely unlike those of the constituents. Both the mechanical property potentials and the physical properties of the individual constituents are available to the composite. Another example of a useful composite is a grinding wheel. The particles of hard silicon carbide or aluminum oxide are bonded together with a resin into a particular shape so that the fast turning wheel will cut away small pieces of metal on contact and eventually cut the part to size. When ground tungsten carbide is mixed with nickel powder and subsequently compacted into the shape of a lathe cutting tool a valuable tool is available as soon as sintering of the briquette is finished. Sintering causes fusion of the nickel powders to the carbide and results in a cemented carbide which has limited competition in the metals cutting industries. The main constituent is the carbide and the other constituent, nickel, literally diffuses into the carbide particles and rigidly holds them in place for use as a cutting tool. Basically, then, composites are produced for nearly an unlimited number of reasons, including corrosion prevention as in plating, wear resistance as in hard surfacing (Fig. 13–2), cutting capability as in a grinding wheel, ease of shaping as in powder metallurgy (Fig. 13–3), lightweight structural shapes as in laminates, heat shielding as in silicon carbide, veneer on plywood for furniture, rubber and nylon cords for automobile tires, and hundreds of other material combinations.

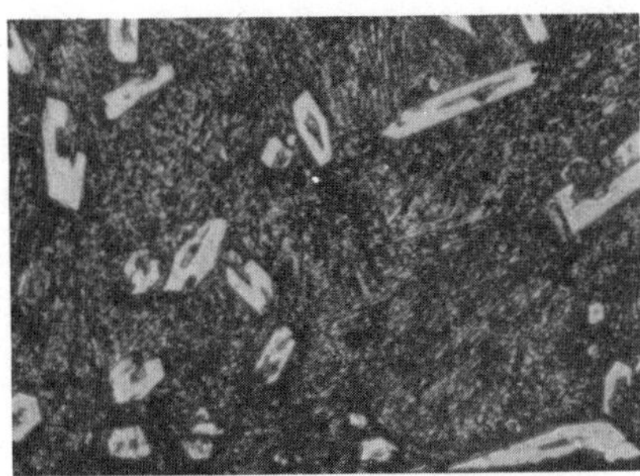

FIGURE 13–2 Microstructure of hard surface bearing (Stellite no. 1) applied by oxyacetylene procedure to produce surface hardness of RC 54. Electroetch in phosphorice acid, 150×. (Courtesy of Dresser Industries)

FIGURE 13–3 The ram has lifted from the die, leaving the sprocket briquette ready for sintering. The sprocket is a product of powder metallurgy. (Courtesy of Martin Sprocket and Gear)

Agglomerates

The most common of the composites are the agglomerates, or aggregates. In these mixtures the constituents are more often randomly arranged such as the mixtures of concrete or powdered metals. Random arrangements of the constituents usually result from intensive mixing prior to final manufacture. Each constituent contributes its part. The total physical arrangement is then unlike the laminated or fibrous types.

Clay Products and Ceramics

A mixture of alumina, silica, and water forms the basic mixture for bricks and similar products of clay. Because of the availability of numerous kinds of clays care is used to make the best selection for the product to be made. Fine china and porcelain are produced from carefully controlled amounts of silica and alumina or the kaolin type of clay which is a decomposition of feldspar. Common building bricks and tiles are made from the calcareous clays which contain lime and those clays containing iron and other earth materials. The special types of fire clays contain silica, alumina, iron oxide, and other earth alkali. The basic clay and any additionally required constituents such as water are subsequently mixed and processed into desired shapes and then fired for hardness and strength.

Shaping is accomplished by routine casting and specially designed methods. Bricks, for example, are produced by slicing the soft and rectangular shaped extrusion of clay into lengths as it emerges from the die. Subsequent drying and firing at temperatures up to 2000 °F (1093 °C) and sometimes 3000 °F (1649 °C) change the mass of clay into a hard and strong building material. Because of water evaporation some porosity in the product occurs as the strength increases. Depending on chemistry of the product and manufacturing conditions, compression strengths range from around 3000 psi to more than 8000 psi. Bricks are hard, stiff, and brittle, and when failure occurs it is either by crumbling for the softer bricks or shattering for the harder and stronger bricks.

Refractory Bricks   Special types of bricks are waterproofed or glazed while other clay products are coated with porcelain enamel. Furnace bricks are produced to withstand the temperatures of molten steel and are either basic refractories of magnesium and oxygen or silica refractories made from a silica-bearing material such as ganister. A different type of refractory brick is the product of fused coke and silica or silicon carbide. This particular ceramic is extremely hard, fairly strong in compression strength, and resists temperatures created at earth entry speeds of space vehicles. Therefore, heat shielding of these vehicles is frequently formed from silicon carbide as well as aluminum oxide.

Diversity of Products Made from Ceramics   The ceramic industry uses all types of compounds in the production of high alumina parts such as sanitary ware, insulating objects, pottery refractories, special bricks, and pump, valve and cylinder parts such as bodies and pistons of special mechanisms. Also, compression and tension springs which operate at high temperatures, grinding mill pebbles and deburring components, furnace parts, grinding wheel grits, electronic components, cutting tools, and missile nosecones are produced. The low percentage aluminas include the several clays, talc, feldspars, and flints along with the needed fluxes such as magnesium carbonate. Various mixtures are made of these raw materials, even the high alumina content clays are used. After mixing, the pastelike material is formed by one of several methods which include slip casting, pressing, extruding, centrifugal casting, and throwing.

Slip casting is merely pouring the raw material into a prepared mold, and wet pressing involves the shaping of the part by pressure. Dry pressing of alumina powders is performed in metal dies using the powder metallurgy techniques, followed by a sintering operation. The pressed briquette is heated to a temperature which diffuses the particles into a solid mass. When metal powders, for example, are mixed with the powdered intermetallics and pressed and sintered, a cermet is produced such as a cutting tool. Extrusion is a common method of forming the pastelike material, and centrifugal casting uses the force of a spinning mold to shape the part. Throwing is a method of shaping by use of the potter's wheel. After forming, the ceramics are dried at specified temperatures up to as high as 350 °F (177 °C) and are subsequently fired at white temperatures to cause vitrification and the formation of a glass bond on the surface of the ceramics.

Concrete

Concrete is a mixture of Portland cement, sand, gravel, and water. The cement is a manufactured product of calcium carbonate and aluminum silicate. When a pulverized siliceous material such as clay, limestone, and special constituents is roasted at temperatures up to 2650 °F (1454 °C) consolidated masses of hard clinkers result and these are subsequently ground into the fine cement powder. The powder is again mixed with a material such as gypsum and then ground to its final size. A careful blend of 1 cubic foot of cement, 2 cubic feet of sand, 3 cubic feet of an aggregate such as gravel and about 6 gallons of water makes an excellent concrete for general construction use. Too much water severely decreases the strength of the concrete, but too little is unsatisfactory. After a thorough mixing the concrete is often given the slump test to determine its fitness for use. Such a test involves the filling and rodding of a 12-inch conical shaped mold which is open at the top. The cone is filled first at the one-third level, then at the two-thirds level, and finally completely. Rodding is the packing of the pasty material with a round rod having a conical point. When the conical container is lifted upward, the pasty concrete slumps because there are no retaining walls. The number of inches of slump indicates workability; a 3-inch slump, using a standard test, is often acceptable for general paving, but the percentage of slump varies with the design.

The size and shape of the particles of the composites of concrete justify the study of the relationship among the aggregate particles. For example, if all the larger aggregates are nearly round, it is obvious that only about three-fourths of a given space will be filled, the remaining one-fourth being air spaces or, subsequently, sand and cement. In this case aggregates with smaller particles should also be included so that a nonuniform line of aggregate surface associations will result; this random surface relationship will reduce the tendencies for the mixture to crack. A mixture of several sizes and shapes will help provide a stronger bond throughout the composite.

Proper Curing Is Essential   After concrete is poured it should be allowed to cure for several days before loading; often 30 days is required. During the curing period it should be kept damp if possible. As hydration proceeds a chemical change between water and cement occurs, temperature increases, and as the mixture sets it becomes permanently bonded to all particles of the sand and gravel. With respect to large masses of concrete, internal cooling is required to help prevent overheating and water evaporation which may cause cracking. Coolant pipes are laid in the concrete forms during their construction. Obviously, they remain. Concrete is nearly always poured in a form which contains structural steel rods, shapes, or wire mesh. Because concrete holds its loads best in compression stress the rods are placed mainly where tensile stresses exist. As an example, a loaded beam resting on two columns will deflect downwards near its center. The bottom layers of concrete will be in tension, but the nonslipping steel rods will take the stress and support the concrete. On the beam's top side compression stresses exist and, in this respect, concrete is capable of holding the load in the presence of the stable rods. In effect, when steel rods or other structural forms are tightly gripped with the surrounding concrete along the bottom areas of the beam, safe conditions exist, both in tension and shear. Equal compression stresses, opposite to the tensile, exist along the beam's top layers of concrete in symmetrically shaped beams. As for columns, steel rods contain stresses along the axis of the column and help prevent buckling.

Powder Metallurgy

One of the most important composites is the part made from powdered metals. Basically, several different metallic or nonmetallic powders are blended by proper mixing and are subsequently compressed into the proper shape. After shaping the compacted briquette is sintered so that diffusion of the individual particles of materials occurs. Figure 13–4 is a photomicrograph of a mixture of tungsten carbide

FIGURE 13–4   Microstructure of tungsten carbide used in oil well drilling bits. Etchant was potassium berricyanide, 1500×. (Courtesy of Dresser Industries)

and cobalt. The density of this compaction is 13.9 g/cc. According to the view, 84% of the material is tungsten carbide and 16% is cobalt. The transverse rupture strength is 420,000 psi, the ultimate compressive strength is 550,000 psi, and the impact resistance is 25 in-lb Charpy. In effect, a single piece of material results. For the metals mechanical properties are comparable to cast or wrought parts, even though these properties are sometimes lower than those resulting from other process forming. For example, a composite of powdered bronze and graphite is compacted by compression forces into a sleeve type self-lubricating bearing. After compaction into shape the bearing is sintered at a temperature below the melting point of the material whereby a close relationship of the two constituents occurs. In use the turning shaft draws on the microscopic size particles of graphite as a lubricant.

Powder metallurgy provides the means to make parts from dissimilar materials so that the benefits of each material are available. (Details of powder metallurgy are discussed in another text, *Manufacturing Processes*.) Because aluminum oxide and tungsten carbide are extremely hard and heat resistant materials and because they cannot be shaped by conventional methods, the powder metallurgy process offers a quick and effective means of shaping. The ground material of given size and shape is placed in a die having the required shape of the needed cutting tool. Thousands of pounds of pressure are directed onto the powders, resulting in at least a compaction ratio of 3:1. The tightly packed briquette is subsequently placed in the sintering furnace at a temperature which diffuses the individual particles into a solid mass. After sintering the extremely hard shape is available as a cutting tool with cutting capabilities unequaled by nearly all other materials. The ease and accuracy of this process is illustrated by the spur gear (Fig. 13–5). The cavity in

FIGURE 13–5 This spur gear is a product of powder metallurgy. Simplicity of production is demonstrated by a single movement of the ram system into the die where the powder is placed.

the die has this shape. The objective of the composite is to allow the constituents the opportunity to provide their qualities; therefore, during the manufacturing process no melting of the major constituent occurs. In the case of cemented carbides, however, the metal powders melt and cement the hard carbide particles together.

Miscellaneous Composites

Several other important composites are available and include the familiar grinding wheels, a mixture of abrasive particles and a resin or glass. Elevated temperature heating provides a strong and effective cutting tool. Silicon carbide and aluminum oxide constitute most of the heavy-duty grinding wheels; however, other materials such as diamonds are available. A completely different and very common composite is the asphalt used in paving highways. An aggregate of sand, small gravel, and petroleum, asphalt forms a kind of semistable composite known as asphalt concrete.

Laminates

When two or more layers of materials are bonded together in some manner a laminate results. These simple shapes range from the effects of metal spraying to bonded structural members. When tungsten is sprayed onto a hot steel shaft for purposes of increasing surface hardness, a kind of composite results because two separate materials exist even though the added surface clings to the steel core.

Electroplating is a composite operation in that an independent metal is deposited onto another metal as a layer. Several different layers of plating can be accomplished. Dipping and cladding are composite operations because the new surface produced by each is a completely different material, and the identities of the base material and surface material are very clear. A modification of the composite concept is the carburizing and nitriding of steels. In these two heat-treating operations only a chemically modified surface results, not the addition of a completely new material. Iron carbide, chromium carbide, or iron nitride is developed in the steel's surface due to chemical actions between carbon and iron, carbon and chromium. or nitrogen and iron at elevated temperatures. Figure 13–6 is a photomicrograph of the carburized case of a stainless steel oil well friction bearing

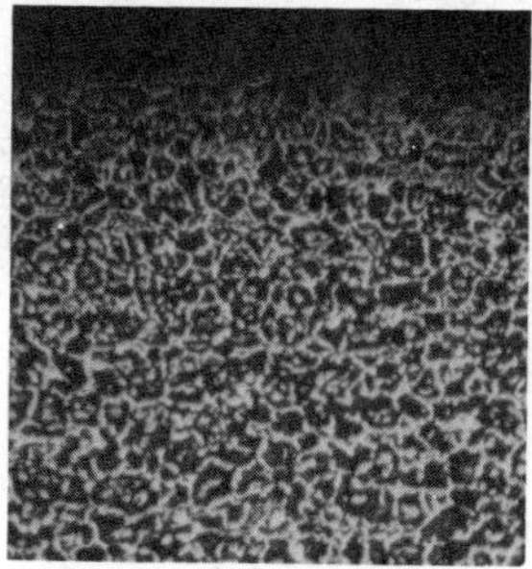

FIGURE 13–6   Carburized and heat-treated case of AISI 431 stainless steel. This friction bearing bushing is used for oil well drilling bits. Etchant was cupric sulfate, 100×. (Courtesy of Dresser Industries)

bushing for a drilling bit. The case is RC 65 and the core is RC 44. In effect, these microscopic size particles of the chemical compounds are dispersed throughout the original case of the steel, the case matrix being present all the time. The painting of the steel with red lead paint is also a type of lamination.

Experimentation in earth and space environments has shown that some composites are more effective in resisting creep at very high temperatures than the typical solid solutions. For example, an alloy of nickel and aluminum oxide which remains a mixture at 2100 °F (1149 °C) has shown satisfactory capability in sustaining needed loads. As a comparison, the tough nickel-cobalt solution shows inferior creep resistance at this temperature. Another composite containing magnesium oxide has successfully withstood temperatures higher than 5000 °F (2760 °C). Because of the increased capabilities in manufacturing processes and the availability of structural and new types of refractory materials along with their strength and cryogenic properties, interplanetary travel is feasible. It stands to reason that a normal living environment can be sustained in the spacecraft for an extended period of time and be insulated from the outside adverse environment of radiation, heat, or cold.

Honeycomb   The laminates are rapidly increasing in use because of the increased benefits in mechanical properties. Layers of metals and nonmetals are being fabricated into all kinds of structural shapes, even the load-bearing structures of aerospace vehicles. A typical laminate, for example, is the helicopter

FIGURE 13–7   A helicopter rotary blade being removed from the bonding mold. (Courtesy of Bell Helicopter Company)

rotary wing which is composed of layers of different metals, epoxy, and honeycomb (Fig. 13–7). Such a configuration provides for essential strengths and lightness in weight. In effect, the entire vehicle and its payload depends on this intricately shaped airfoil in flight. Honeycomb is one of the most effective weight reduction laminates. The resulting hexagonally shaped cells originate from the numerous alternate layers of very thin aluminum sheets and narrow strips of high strength bonded epoxy. But, as expanded and standing alone, honeycomb cannot sustain its own shape (Fig. 13–8). When the section of honeycomb is epoxy

FIGURE 13–8   Expanded honeycomb being prepared for a bonding operation. (Courtesy of General Dynamics, Fort Worth Division)

bonded in an autoclave under heat and pressure between two sheets of aluminum alloy, however, a strong and stiff structural part is produced that is many times lighter in weight than a solid metal of the same dimensions. Honeycombs are made from aluminum and its alloys, titaniums, stainless steels, fiberglass, and even paper. A more recent modification of honeycomb use includes graphite fiber outside panels which sustain the shape of the honeycomb. The graphite honeycomb assembly is much lighter in weight than the all-metal types. Figure 13–9 illustrates an advanced stage of honeycomb preparations for bonding.

FĬGURE 13–9   Great care is used in the shaping process because of the nature of the honeycomb cells. (Courtesy of General Dynamics, Fort Worth Division)

Ablating Materials   A special type of epoxy-bonded honeycomb made from organic and inorganic materials has been used as a heat shield ablator for vehicles returning from space. This silicone type of elastomer absorbs the very high temperatures generated from contact with the earth's atmosphere and starts a carbon-silicon chemical action whereby the heat causes charring conditions and the increasing temperatures become lost in the slowly ablating material from the shield. Protection of the vehicle behind the shield is measured in time during ablative conditions. Time, then, is measured in inches of thickness of the shield and this thickness is equal to slightly more than the calculated time during penetration of the atmosphere.

Structural Members and Special Shapes

Much of this advanced technology is a result of the aerospace industries involvement in research and development. Two main reasons for the development of laminated structural members are the need to reduce weight and the need to terminate cracks once they have started. The structural laminate does both of these in addition to supplying the needed strengths with reasonable safety factors. For example, two of the original and well known laminates are the phenolic and the fiberglass. Both materials have similarities in construction and both consist of layers of clothlike material and epoxy. After curing both laminates have adequate strengths for their intended purposes and both are light in weight for given sizes. But neither of these laminates is strong enough to safely sustain very heavy loads imposed on structural members. Consequently, stronger laminates have been sought such as designs of laminated metals and nonmetals. These new configurations are increasingly being used as structures in some of the fastest moving aircraft.

Laminated Structural Members   The combination of minimum strengths, crack control, light weight, and cost has culminated in laminated beams, plates, and other shapes of different materials. Laminations of thin sheets of heat-treated aluminum alloy and epoxy which have been pressed into the required shape not only have the needed strengths, but offer the designs at lower prices than the larger single sections, along with weight reduction and better crack control (Fig. 13–10). In this regard, parts which are used in cyclic loading are subject to fatigue cracks which often begin at the surface of the part. If a crack begins at the surface of a single section of the laminate, it has a good chance of ending at the other side of the same material and not progressing across the whole section. However, laminates do no guarantee that a crack will not move into the adjacent layer of material.

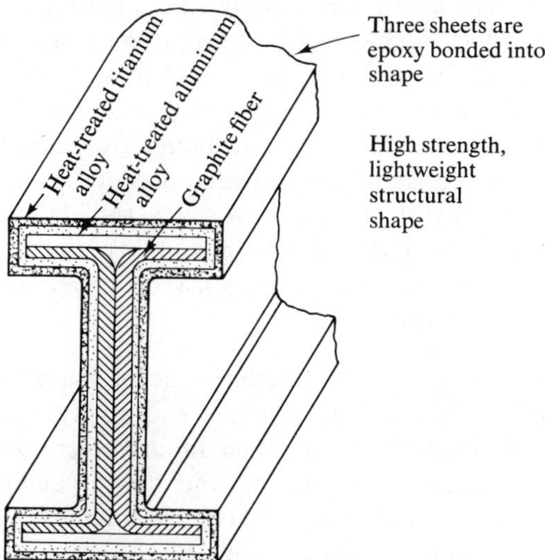

FIGURE 13–10   A laminated structural I-beam fabricated from sheets of titanium alloy, aluminum alloy, and graphite fibers.

Graphite Fibers   Another material to move into the aerospace industry is cloth made of graphite fibers. When these fibers are laid in one direction, woven (will be explained), and coated with epoxy a strong and flexible cloth results which is as light in weight as any common cotton cloth (Fig. 13–11). When layers of graphite fiber cloth are bonded with layers of epoxy to form a laminated structural member an entirely new material is produced which is exceptionally light in weight and is a substitute for designated metal members because of the

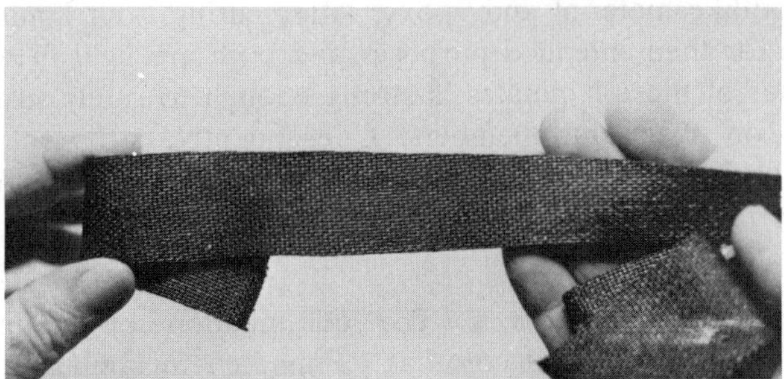

FIGURE 13–11   Graphite cloth, a raw material used in the fabrication of laminated structural members.

acceptable strength and stiffness properties. This graphite laminate is being used for main aircraft structural members as well as numerous other components. Such a process provides a whole new concept for high strength-lightweight designs.

Still another combination is the bonded sandwich built of alternate layers of heat-treated aluminum alloy, epoxy, and graphite cloth. Arrangement of the laminations into specified geometric shapes increases stiffness in the outside layers along with lightness in weight and flexibility without brittleness. In fact, this type of graphite fiber is not brittle but is more like a cotton fabric. A modification of the structural sandwich type of high strength member uses layers of heat-treated titanium alloy, epoxy, and graphite. Heat-treated aluminum alloy is also included in the multiple layer sandwich, the higher strength alloys being placed at the outside surface of the shape in order to better resist the bending loads. Such lightweight and deflection resistant shapes have acceptable mechanical properties and are replacing some alloy steels. For aerospace vehicles, this type of material allows more weight for the payloads. Machinability and formability of these laminates are excellent and they safely maintain their mechanical properties within seasonal temperature fluctuations.

Boron-Coated Tungsten Rods   A different type of laminate is the sheet made of boron-coated tungsten rods laid side-by-side and epoxy bonded. Some of these rods are only a few thousandths of an inch in diameter. Succeeding layers are arranged to change the direction of the rod and this procedure produces a thick section of exceptionally stiff and fairly lightweight material. The stiffness factor is related greatly to the 50 million psi modulus of elasticity of tungsten in tension. Because of the rotation in the cross section of the material the high stiffness value is in all directions. Sheets of this laminate have shown success in aerospace use.

Fibrous Patterns   Graphite cloth, as has been pointed out, consists of woven fibers of graphite bonded into a thin sheet with high strength epoxy. This cloth is available in various widths, its weave pattern being typical to regular cloth. When arranged as a bonded laminate into channels and other shapes acceptable mechanical properties are displayed for use as stress carrying main members. Limitations of this laminate and all epoxy-bonded laminates include a temperature environment below the destructive temperature for the particular epoxy of 300–400 °F (149–204 °C). However, these epoxies successfully withstand subzero temperatures.

Carbon   A more recent material to enter the construction field is carbon plate and sheet having the texture appearance of phenolic but the color of carbon. When a phenolic is burned under controlled temperature and pressure conditions the remaining residue is carbon, but carbon in the same physical pattern as the original phenolic. Having already burned, the material can never burn again; therefore, it is fireproof. The resulting shape is a duplication of the shape of the raw material as it enters the pressure vessel or autoclave. Apparently, thickness of the new material and area dimensions depend on the size and facility of the material processing machine. Original weave of the cloth is clearly observable in the new material, even the several layers of the cloth which constituted the original phenolic shape.

Tests on the carbon shapes show little brittleness. In fact, low impact loading shows little shattering, only the ability to bend somewhat before crumbling into powder when overloaded. Compression loading, developed from high speed and high velocity wind, allows satisfactory use of the material on leading edges of fast moving aircraft and incoming space vehicles. With regard to the generated heat, tests on this material with the oxyacetylene torch show successful ability to maintain shape in the presence of the inner cone of the flame (Fig. 13–12).

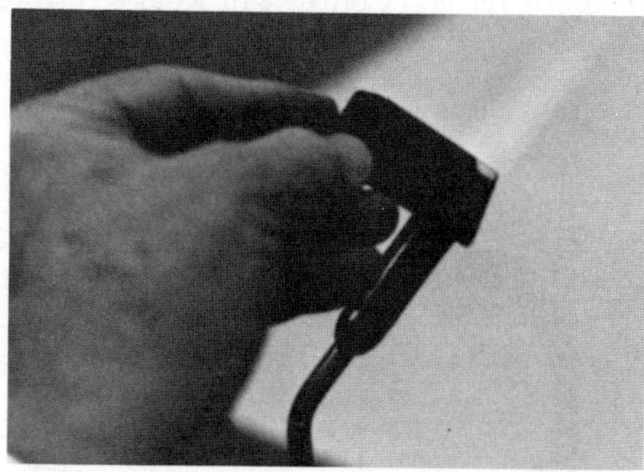

FIGURE 13–12   A carbon compact used in space vehicles, The torch has produced a white heat at the corner, but the ability of this material to transfer heat is very poor.

For aerospace use the carbon compacts are shaped to all heat affected structures such as leading edges of air foils and the nose of the vehicle (Fig. 13–13). These areas include the forward edges of wings, stabilizers, control surfaces, and nose regions of the vehicle's body. Then, to assure safe flight in the atmosphere where temperatures exceed 3500 °F (1704 °C) silicon carbide is diffused around the outside areas of the carbon compacts. This one-piece material (Fig. 13–14) is not brittle and it resists thermal shock to a satisfactory degree as it effectively withstands temperatures in excess of 3500 °F (1704 °C). Materials behind the heat shield carry loading stresses; therefore, the purpose of heat shielding is to save the vehicle from destruction by high temperatures during atmospheric penetration.

Space vehicles in current operation use the best materials available and certainly those in the future will improve upon those materials available today. Due

FIGURE 13–13 The space shuttle *Orbiter's* nose and underfuselage glow with heat on entering the atmosphere following a mission into space in this artist's concept. The *Orbiter* will maneuver in space like a spacecraft and enter the atmosphere to make a landing similar to other aircraft. Heat shielding preserves the vehicle for reuse. (Courtesy of Rockwell International)

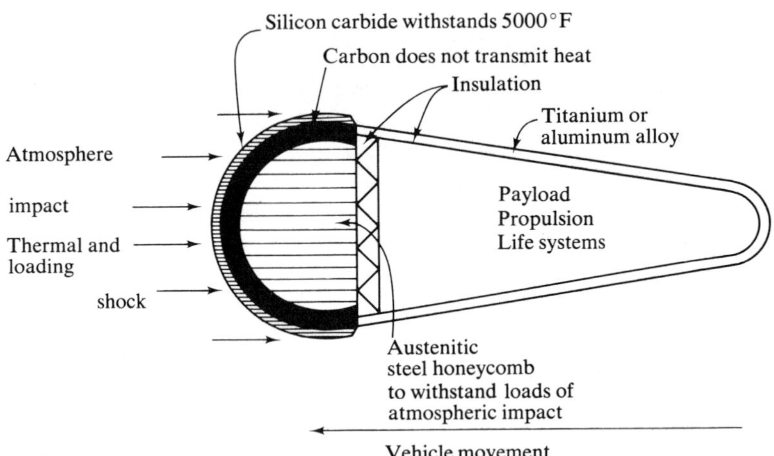

FIGURE 13–14 Silicon carbide heat shielding diffused on carbon protects the payload as the spacecraft enters the atmosphere.

to the nature of materials and because all materials are different, each having its particular advantage, a vehicle moving into deep space will continue to use a combination of metals and nonmetals. Metals provide for the several strengths of the load-carrying structures while some of the nonmetals assure safety in flight by insulating the load-carrying metals from extremely high temperatures encountered at the beginning and ending of flights. The face-centered metals perform very well in the cryogenic temperatures of space. With regard to radiation, unless exposure is long-term, only personnel-inhabited areas need shielding.

PLASTICS

The plastic family of synthetic materials is very large and includes the macro-molecule and chainlike ingredients which allow plastic deformation into solid objects. Elements such as chlorene, ethylene, nitrogen, oxygen, hydrogen, carbon, and others are united to form the nearly endless number of compounds for subsequent processing into the specific plastic. Due to the large molecule type of structure, the manufacturer is able to control the several chemical reactions which occur after mixing certain formulas and process the resulting gas and fluid into a solid material. The final material is either a thermoplastic or thermosetting plastic and is available in several physical forms.

Resins Are Basic Ingredients

The basic material in all plastics is the resin. This fluid is of an organic origin and is divided into two main kinds, the thermoplastics and thermosets. The basic differences between these resins are their chainlike makeup, processing procedures, and final bonding arrangement. The *thermoplastics* are large masses of giant molecules linked together in single chains from molecule to molecule. Such a structural arrangement allows physical deformation of the mass in the presence of temperatures slightly higher than the boiling point of water. The process of converting the elements to the plastics is called *polymerization* and begins with complicated laboratory apparatus which ultimately transforms gases and liquids to the solid plastic. When the linear chains of the thermoplastic group of molecules become cross-linked, an entirely different plastic results and this is the *thermoset*. Once set, this plastic is not reshapable.

Molecular Bonding

The resulting molecular structure forms through the processes of additions and condensations among the linear or cross-linked materials. Atomic forces hold the molecules together and include the primary, or chemical, bonds and the attractive, or secondary, bonds. Indentification of the plastic originates from its primary bonds because these bonds cannot be broken without destruction of the plastic's identity. In other words, the primary bonds, being chemical, are extremely strong and cannot be re-established if broken. On the other hand, the secondary bonds are much weaker in their attractive forces and allow shifting of the linear linked molecules, but not the cross-linked. Such is the result when themoplastics are heated somewhere between 200 °F (93 °C) and 350 °F (177 °C) whereby plastic deformation occurs when forming pressures are greater than the secondary bonding strength of the material. If the same temperature and pressure is applied to the cross-linked plastic after its initial cure, shattering may result. Consequently, thermopalstics are reshapable indefinitely by heat, but the thermosets can be shaped only once.

Granular vs. Amorphous Materials

In the study of metals it has been pointed out that atomic bonding holds the mass of metal together in a form of lattice network throughout a single grain. In most plastics a different atomic arrangement exists in that the large grouping of atoms forms molecules and these exist throughout a nongranular structure or amorphous material. In the metals the atomic organization is broken at each grain boundary throughout the entire mass of metal. In the amorphous materials, much as in a liquid, the molecular arrangement exists throughout the mass of material or, actually, one grain or one piece of plastic. An exception to the general statement that plastics are amorphous in structure occurs in some kinds of plastics during manufacturing processes. Atomic structural arrangement is reflected directly into the material's physical and mechanical properties.

From Raw Materials to Finished Product

Even though the plastics are easily controlled during their production by chemists and engineers, a long and complicated procedure is necessary before the final basic raw material is produced. This means that plastic manufacturers produce their products in convenient raw shapes so that further manufacture can be completed in turning out finished parts. As previously pointed out, gases, liquids, and solids constitute the original raw materials. These are systematically processed according to their formulas into the specific plastics. Because the resin is the fundamental ingredient, it is produced both as the thermoplastic and the thermoset. Heat, pressure, chemical changes, and additives eventually result in conversion of the raw materials to the desired resin.

Some of the thermoplastics include acrylics, acetals, cellulosics, polyamides, styrenes, vinyls, polyolefins, and fluorocarbons. Some of the thermosets include the silicones, phenolics, polyesters, urethanes, epoxies, and aminos. Each of these basic plastics is manufactured into numerous trade names, such as Lucite, an acrylic produced by DuPont, and Plexiglas, an acrylic produced by Rohm and Haas. These two plastics are of the thermoplastic type. Two common thermosets include Bakelite, a phenolic produced by Union Carbide, and Laminac, a polyester produced by American Cyanamid.

Resin Additives   Many of the plastics use fillers such as fibers, powders, and various chemicals to produce a reinforced resin for higher strengths or for heat or light resistance. Some fillers provide electrical conductivity and some provide translucent qualities. When a stabilized condition is essential in the plastic a stabilizer is added, such as a chemical, to retard attack by atmospheric conditions. To appeal to aesthetic values, many plastics are modified with a particular dye to add color. When plasticity in shaping operations is required, a plasticizer is added. A solvent is used when welding of sections of plastic is necessary. During the mixing of resins and additives, manufacturers and other personnel are cautioned to follow safety procedures because of the possibility of rapid chemical reactions and even explosions. When the proper formula is completed the modified material is subsequently processed by one of many means into its commercial shape.

Commercial Shapes   Both thermoplastics and thermosets are produced in conventional shapes; that is, in the same shapes as metals and wood. Because of the available manufacturing processes in the plastics industries, shapes such as round bars and rods, square and rectangular bars, round and rectangular tubes, sheets and plates, and other geometric shapes are available. Further, the foamed type of plastics are also available in the same shapes. Because local manufacture often requires fusion of parts, the several solvents are also available. A large plastic warehouse resembles a metal storage area because of the similarity of shapes of the materials.

Properties of Plastics

Chemistry of plastics varies extensively. Such properties as color, weight, dimensional stability, damping characteristics, viscosity, and electrical, optical, and thermal characteristics are pertinent to the particular plastic and are labeled as physical properties. The mechanical properties such as strength are directly related to a given set of physical and chemical properties.

Physical Properties There are numerous colors of plastics and these are produced in relationship to the several available dyes. Most plastics are nearly twice as heavy as water (1.0 specific gravity), yet a few such as the polyethelene and foam types will float on water.

With regard to dimensional stability such as warpage and shrinkage, most plastics have a high degree of stability. However, some plastics, possibly due to design, are subject to warpage by stress factors built into the design. The stiffness factor of many plastics is only one-half million pounds per square inch and this extremely low property is subject to attack by design configurations. The shrinkage factor also exists is some thermosets as cooling occurs in and out of the mold. A reduction in the plasticizer in some thermoplastics causes a slight shrinkage in the finished part. In designs where vibrations are present, plastics act as good dampers and absorb much of the vibration.

A large variable regarding plastic resins is their rate of flow, or viscosity. If water is considered as 1.0 centipoise at 70 °F (21 °C), then among the liquid resins the plastics are many times stiffer in centipoise values than water. For example, penetrating oils may be 15–50 times less mobil than water while many liquid resins have the consistency of syrup. It is evident that the resins of both types of plastics are fairly stiff when compared to water.

As far as electrical properties are concerned, most plastics act as insulators from electrical current. However, insulation is only relative in value and depends entirely on dielectric values. Plastics with high dielectric strengths are acceptable electrical insulators. Most plastics do not absorb water and are therefore not subject to lost of dielectric strength. Many electrical insulators are consequently made from certain plastics. On the other hand, some plastics can be produced to transmit electrical current.

Some plastics have excellent light transparency properties because of their natural colorless condition. For example, the acrylics are clearer than most of the glasses, being capable of transmitting more than 90 per cent of the light in contact with them. These clear types of plastics also transmit other light waves such as infrared and ultraviolet. With regard to thermal properties, most plastics transmit little heat. Plastics are one of the poorest heat conductors and find ready use in the refrigeration and cooking industries. Hot or cold conditions tend to remain with the source in contact with the plastic.

Mechanical Properties   Properties related to hardness and strength are considered to be mechanical properties. These properties are directly related to the plastic's chemistry and physical condition. Compared to metals, all plastics are relatively soft. In comparison to the copper alloys, however, some of the hardest plastics are comparable to brasses and some bronzes. The modulus of elasticity in the hundreds of plastics is fixed with a specific plastic and all are relative low in comparison to the metals. This means that a steel, for example, has a stiffness value of 30 million psi while a typical plastic has an E value of 450,000 psi. These types of plastics cannot replace the steels with regard to loading capability. It is evident that some manufacturers attempt to do this, however, and only place inferior plastic parts on the consumer market which quickly shatter when used a few times. With reference to tensile and compression strengths, most plastics have only one-fourth the strength of the weakest steels. Shear strengths are also low. It must be pointed out, however, that plastics have their place in industry for use in nonstressed parts. There are some exceptions such as fiberglass and phenolic configurations which have tensile strengths that reach the strengths of some steels. Some aerospace applications with laminates and epoxies have demonstrated very high tensile strengths equivalent to structural steels.

Mechanical properties are related to the particular plastic at a certain temperature. As the temperature increases from room temperature, or 70 °F (21 °C), mechanical properties decrease. Most plastics lose a significant amount of strength beyond 300 °F (149 °C), yet a very few will function at 500 °F (260 °C). With respect to subzero temperatures (−60 °F, or −51 °C), most plastics are acceptable and safely carry their loads. Impact strength of many plastics is fairly good, even in the presence of cold temperatures. When lightly loaded, most plastics begin to creep and show dimensional changes over a period of time. And in reference to fatigue, the same general rules of time and fluctuating tensile load apply to plastics as to metals. Vibrations bring a quick end to some plastics which are poorly designed.

Forming Processes

Once the basic raw shapes of plastics have been produced, along with needed solvents and adhesives, the secondary industries take these products and process them into thousands of consumer items. Some of the typical processes used in making the various parts will be discussed. Plastics are formed into their final shape by means of casting, thermoforming, reinforcing, foaming, and molding. Regardless of the method used, the plastic must first be made soft enough to flow plastically, be able to move into a closed container which sets its shape, and then be able to maintain the shape after setting. In order to bring about a final shape, there must be a preferred temperature for causing fluidity or plasticity in both basic types. For thermosets, there must be a means to cause the chemistry to react and cause rigidity while the plastic is under heat and pressure. In this respect, there needs to be a machine to force the plastic material into the desired shape, such as a mold, roll, or mechanism. After shaping in the mold the thermoplastic is usually water cooled; the cooling rate of the thermosets makes no difference. A means must then exist to eject the finished part. Much of this process is automated and requires less skilled help.

Casting   One of the most common production processes is casting. In many ways this process is similar to the pouring of liquid metal into a mold. The casting process involves the pouring of liquid plastic into a suitable mold and allowing the full mold of plastic to cool to a lower temperature, often room temperature. Melting temperatures range from about 250 °F (121 °C) to 350 °F (177 °C). Molds are made in two pieces of various materials and include the cavity for shaping the material. The melted liquid is poured into a hole which enters the mold's cavity until the mold is full. The split mold allows the solid plastic to be quickly removed. Sheets and plates are formed by pouring the liquid plastic into cavities between mold plates, and rods and other shapes are poured into split tubes. A split-tube type mold allows the casting to be easily removed. Numerous kinds of molds are available.

Cast plastic products are sometimes superior to pressure molded products in strength, but most cast metals are weaker than the wrought. Polyesters, epoxies, and acrylics are commonly cast. A type of casting, for instance, includes the covering of electrical components with a cast epoxy. The plastic gives the required strength along with good insulation properties. Another example of liquid pouring (casting) is the embedding of objects in a clear plastic, such as items of jewelry. Many geometric shapes are also produced by casting. A recent process combines parts with the plastic such as parts in electronic systems.

Plastisol   Even though the application of liquid plastic to the surface of a metal part is a different casting operation, there is a relationship that necessitates its description at this time. A plastisol is a ground mixture of fine particles of plastics such as polyvinyl chloride and plasticizers which fuse into a homogeneous liquid when heated to approximately 350 °F (177 °C). Three main methods of applying the liquid plastic are dipping, slush casting, and rotational casting.

Dip casting involves the dipping of a part into the liquid plastisol and then removing it. Freezing of the plastic occurs on withdrawal. In this respect, some parts are oven heated at 350 °F (177 °C) after withdrawal from the dip and then cooled, often by a jet of water. Dipping is a routine process. Many hand tools are dipped in plastisol for purposes of both grip assistance and for electrical insulation. Also, wire mesh screens and kitchen dish drainers are dipped in plastisol, and hollow plastic parts are made by dipping a pattern, which is often the part, in the plastisol and then stripping the flexible part from the pattern. Glueing the edges follows.

Slush casting involves the pouring of the plastisol into a hot, hollow mold, followed by quickly pouring the remaining liquid from the mold. Thickness of the part's wall is determined by length of exposure time; usually three minutes gives a useful wall thickness. The entire mold is placed in a 350 °F (177 °C) oven for a few minutes to cause complete fusion and then water cooled. Slush casting is a common casting process. Many hollow toys and consumer products are produced by this method.

When an enclosed object is to be produced without seams, rotational casting is used. A measured quantity of plastisol is poured into a heated split mold. Immediately, the mold is two-dimensionally rotated so that the liquid moves completely around the inside surface of the heated mold. When the mold is separated, the part is removed, fully enclosed, seamless, and hollow. Hollow balls and numerous kinds of toys are produced by this process. Thickness of the part's walls is dependent on the quantity of plastisol.

Molding   Molding is the most widely used production process. The molding processes are varied and include injection, extrusion, transfer, compression, laminating, cold, and calendering.

Injection molding is used primarily with the thermoplastics. In its simplest form, the pellets of plastic are forced into a heating chamber which causes the material to become liquid. Under high pressure the liquid is injected into a cold mold having the shape of the part. When cooled, the part is removed. Many molds are water cooled. Two common plastics used in this process are the acrylics and vinyls. Typical products include consumer articles and household wares along with toys and appliance covers.

Extrusion is similar to metal extrusion in that a die having the exact shape of the extrusion's cross section is used. Powders or pellets of thermoplastics are moved into the heating chamber at approximately 400 °F (204 °C) and then pushed through the die in the plastic state. A conveyor system moves the length of extruded plastic to another processing station for cutting into the desired dimension. Cooling occurs along the line, and stiffness occurs in the part when cooled to room temperature. Standard types of shapes, as for metals, are producible by this method.

Compression molding is a simple process of pouring a thermoset powder into a heating chamber at approximately 300 °F (149 °C) and then transferring it to a closed hot mold. As the plastic condition turns to liquid, a pressure is placed on the liquid and curing occurs. Pressures vary according to the size of the mold. A typical one-square-inch mold requires approximately 4000 pounds of pressure while at a temperature of 300 °F (149 °C) to form it. The plastic hardens by curing and is ejected hot. If a set of dies or molds can be made, the part can be produced. This is a fast and economical process because the hot mold is used over and over. Both small and large parts are produced.

Transfer molding is a modification of compression molding. The thermosetting plastic material is poured into the top portion of a transfer mold where it forms a liquid when heated to its liquifying temperature. This liquid is then pushed through the sprue into the mold's cavity, taking the shape of the part. The mold's heat sets the thermosetting plastic as it hardens, and the part is immediately ejected. Typical parts produced by this process include many of those previously discussed, such as electrical insulators in power transmission lines and in automotive electrical accessories. Sophisticated designs are also produced by this process.

Laminating is a common method used to produce higher strength plastics such as the phenolics. Basically, a stack of laminating material such as paper or cloth is arranged so that alternate layers consist of a thermosetting resin. When the stack is placed in a press and heated while being squeezed, bonding and curing occur because of the combination of heat and pressure. Many kinds of higher strength parts are produced by laminating. Structural parts to support electrical transmission equipment, for example, are made from laminated phenolic. Often, the lamination is a linen cloth and resin. The printed circuit is a typical example of the laminated board.

Cold molding is a simple process which includes the pouring of the thick mixture of plastic into a mold and placing the mass of cold plastic under pressure. No heat is used during the shaping process. After removal from the mold the part is cured at approximately 425 °F (218 °C) for several hours or even for several days to cause fusion. Products include those which have no special requirements other than the shape. Cold molding produces a wide dimensional tolerance, but a rough surface finish. A typical product is the common doorknob.

Calendering is a sheet-producing process that uses several rolls or calenders to form the plastic. Fundamentally, the soft mixture of thermoplastic is poured into a mechanism which feeds the plastic material into heated rolls. As they turn, the distance between the rolls is adjusted to determine the thickness of the thin plastic sheet. A cooling roll subsequently prepares the long length of sheet for the final roll which twists the material into large coils. The vinyls constitute a large percentage of materials for calendering. Products of this process are nearly unlimited and range from flooring to trash bags, and from curtains to equipment covers.

Foaming   The foaming of plastics has resulted in many types of new items on the market, especially those in the decorative (construction) and toy categories. Both thermoplastics and thermosets are foamed, including the urethanes, styrenes, cellulosics, silicones, and phenolics. Any practical shape can be produced that is subject to shaping by pressing or casting. Both physical and chemical foaming processes are used to form the beads, or foam. During the manufacture of the particular plastic a gas-forming material or gas is combined with the plastic's chemistry. Subsequently, during forming operations and while being heated, gas is liberated and expands, producing the beads, or foam. When the pre-expanded beads of polystyrene, for example, are blown into a mold's cavity and heated to nearly 300 °F (149 °C), fusion occurs among the beads along with further expansion, filling the mold properly. Water cooling of the mold follows. The result is the shaped part. Some shapes are also produced outside a mold.

Chemical foaming results when two materials are mixed which causes a reaction in the form of a foam. A resin expands and forms beads when another constituent causes liberation of gas within the resin. As an example, foamed polyurethane is produced by chemical foaming and is shaped by pouring the gas-forming constituent and resin into a mold, allowing the foam to be made by gas expansion, curing the filled mold by time or with water cooling if the mold was heated, and then removing the finished product. Many shapes of foam are produced by merely pouring the proper mixture of materials into a container having the desired shape; the mixture produces beads until the reaction is complete. Some dimensional finishing is often required.

Reinforcing   Many plastic objects, such as boat or automobile bodies, are not strong enough when the typical resin is used. Therefore, an additive is used. Reinforcing is the use of some additional material such as a filler or laminate to produce a desired shape. Boat hulls, for example, are often fiberglass structures. Reinforcing can be accomplished by several methods, the simplest being applying a layer of fiberglass or some other material onto a coating of epoxy and then hand rolling the plastic material to form the bond. A mold or pattern is used which is coated with a release material to prevent sticking of the resin. When the combined resin and fiber material are applied the plastic begins to cure at room temperature. After curing the part is ready for further processing according to its use, such as sawing or drilling operations.

Hand rolling is just one way to apply the resin. Spraying is another method used to apply the mixture of plastic materials. Several coats are sprayed until the proper thickness is obtained. Besides hand brushing and spraying, there is the method of pouring a premixture of fibers and resin into a mold, followed by an application of pressure and heat. Subsequent curing results as in other similar processes. Other methods of applying resin include vacuum bag and pressure bag molding. Either vacuum or pressure causes a bag to force the resin mix against the contours of a mold. Hardening follows as in other applications. Another reinforcing is a squeezing procedure. When a section of the resin mix is placed between two parts of a mold and squeezed, the match molding results in the desired shape of the part or other configuration.

Fundamentally, reinforcing the resin provides greater strength properties than the resin alone. Glass cloth or glass fibers mixed with the basic resin provide higher strength materials for molding numerous articles and shapes. Automobile body parts, boats, household hardware and appliances, toys, and sporting goods are some of the products produced by the reinforcing method. The object is to provide adequate strength in the design.

Thermoforming   Large shapes are often produced by thermoforming. The thermoforming process involves heating a section of thermoplastic sheet stock to approximately 300 °F (149 °C) whereby it collapses to the shape of the mold. The mechanical shaping process allows the heated sheet to rest on the mold as it plastically deforms to the contour of the mold. When cooled the newly shaped part is ready for use. Any reasonable shape can be produced by this rapid and efficient method.

A faster thermoforming method uses a different principle which relies on a vacuum to form the part. Vacuum forming uses atmospheric pressure to shape the part. To perform the operation a heated sheet of thermoplastic stock is placed over the mold and sealed around the periphery. Air is then withdrawn from the space between the mold and the sheet; the sheet falls onto the mold and adheres to the fine details of the mold.

Thermoforming lends itself to several procedures based on the principle of pressure. A further modification includes the use of pressure without vacuum. When a part is made by a pressure which is higher than atmospheric, the heated sheet is quickly blown onto the mold and takes on the contours and details of the mold. This method is known as blow forming. Exhaust holes in the mold allow escape of the trapped air. Pressures vary up to nearly 300 psi in shaping parts by this process. Because the thermoplastic sheet will deform into a shape as if the plastic were nearly liquid, many different shapes are possible. Air cooling after forming normally enables stiffness of the part to quickly occur, but in some situations water cooled molds are required. Design of the part is reflected into the design of the overall system.

Machining Operations   Both the thermoplastic and thermoset types of plastics lend themselves favorably to machining operations due to the nature of the cured resins. In general, the plastics are machined by the same processes used for metals or woods. However, a main difference exists in that any cutting operations which generates heat will cause some of the plastics, especially the thermoplastic type, to soften and become a sticky solid. This condition immediately reduces cutting efficiency and often ruins the part. Consequently, cutting tools of high speed steel must be sharp and have a special rake angle, depending on the operation. A negative rake angle is often necessary so the plastic chip is scraped off rather than being neatly sheared. The speeds and feeds vary somewhat from those used in cutting metals.

Some of the machining operations include sawing, turning, drilling, countersinking, boring, milling, punching, routing, filing, sanding, buffing, reaming, threading, grinding, blanking, shearing, and other special operations. Of course, all of these operations are not applicable to all plastics because of the brittle nature of some thermosets. Manufacturer's specifications must be followed, therefore, when machining a specific plastic. Such operations as punching and shearing tend to crack some plastics, while others are sheared with ease.

Other Forming Operations   In addition to machining operations, some plastics such as the thermoplastic Delrin (Cadillac Plastic and Chemical Company) are weldable by several processes. The ultrasonic welding method using about 20,000 Hz provides a hot film at the two surfaces to be joined and solidification follows. Hot plate and spin welding are also applicable in this joining operation. Another joining operation for this particular plastic is press fitting while cold. Hot heading at 315 °F (157 °C) also provides satisfactory joints. Delrin, like other plastics, is subject to residual stresses after forming operations. These stresses can be removed by annealing at 315 °F (157 °C).

Finishing operations on many plastics are common. Using Delrin again for our example, the surface may be painted, but a specially prepared surface is essential. Chrome plating also can be performed on Delrin surfaces and the resulting brillance is competitive with those parts having a metallic base.

The advantages as well as some disadvantages of plastics have been discussed. It is again pointed out that the organic nature of plastics places severe limitations on their use when compared to metals and ceramics. Possibly the greatest limitation exists in the area of high temperature effects. At elevated temperatures, beginning around 300 °F (149 °C), most plastics begin to disintegrate and some convert to poisonous gases. Most plastics are not as strong as metals, and the stiffness factor in plastics is very low. On the other hand, some of the polyimide polymers have been used in the high temperature regions of rocket engines, and a phenolic resin has been utilized in the heat shield of some spacecraft as an ablative material.

ADHESIVES

The purpose of an adhesive is to cause one material to adhere to another by intimate contact between molecules of the adhesive and of the material to be bonded. The materials in a high strength joint which are to be bonded must be thoroughly cleaned. Not even a finger print is to be left on the material prior to application of the adhesive. In order that the materials be properly bonded in these high strength joints the adhesive must wet all areas of the contacting surfaces and then be exposed to a controlled heat and pressure treatment. On the other hand, in low strength joints less precautions are taken in cleaning and wetting because the strength of the joint to be produced is usually not critical. Adhesives range from the common glues to the epoxies and involve cohesive forces within the bonding material or the strong attraction between the molecules of the materials to be bonded. Adhesive forces are subsequently established between the parts to be bonded and the adhesive. With regard to those resins which cause softening and mixing of plastic surfaces the process is a fusion or welded type of joint whereby the molecules of the parts and the resin mix. This is not adhesion among three or more parts.

Thermoplastic

Several types of adhesives are available and include the two types pertaining to engineering designs, thermoplastic and thermosetting. Some of the thermoplastic types include the several polyvinyls, acrylics, and those associated with the large group of thermoplastics. Many adhesives for labels and tapes are produced from polyvinyl acetate and are often called cements. Often a solvent is used with these plastics to form the bond between the parts to be attached. Cohesive forces prevail among the materials of these joints. Neoprene and polyvinyl butyral are common cements for general use purposes, one being a bonding agent with glass. All of the thermoplastic types of joints are low stressed joints and are influenced by surrounding conditions, especially temperatures from 200 °F (93 °C) to 300 °F (149 °C).

Thermosetting

The thermosetting adhesives include the several epoxies, phenolics, urea formaldehydes, and compounds of organic materials and acids. By far the most common and strongest of the adhesives is the epoxy, and many kinds are available. Some are produced as thin tapes or wide sheets protected with a peel-off cloth. These forms of adhesives are cut with scissors to the desired sizes and placed on the prepared surfaces which are to be bonded. Epoxy is common to the bonded honeycomb and numerous laminates of metals and nonmetals. With respect to the aircraft adhesive, it is normally maintained at 0 °F (−18 °C) until ready for use.

When sheets of aluminum alloy are placed between thin ribbons of epoxy strips and subsequently bonded in an autoclave, a thick core or stack of metal sheets is produced. When this core is properly expanded, the hexagonal honeycomb results (Fig. 13–15). Only those bonded surfaces retain the adhesive stresses; therefore, the metal between the bonds is plastically deformed to the new permanent shape during processing. Then, when a sheet of metal is bonded to each surface of the honeycomb a strong structural member exists. This type of adhesive requires bonding in the presence of low pressure and a temperature of 300 °F (149 °C) to 500 °F (260 °C) for curing. Besides honeycomb bonding, the many new types of structural laminates use the epoxy high strength adhesive.

Another type of epoxy is produced as two separate resins, the resin and the hardener. When parts of each are mixed hardening results, permanently bonding the parts together. Some mixtures require the addition of catalysts and accelerators, each being mixed with the basic resin separately to avoid a possible explosion.

FIGURE 13-15 Expanded honeycomb is drawn over a pattern for subsequent forming operations. (Courtesy of General Dynamics, Fort Worth Division)

These air-curing epoxies are easy to mix and apply and are used for many types of adhesive bonding. For example, hard spots in honeycomb paneling may be required for attaching a bolt or merely strengthening the particular volume of honeycomb metal. The liquid resin is poured into the hole and it sets around the bolt, providing a solid fastener.

Uses of Adhesive Bonding

The use of bonded parts reaches into all areas of construction and manufacturing. Building materials such as panels of insulations; honeycombs of treated paper, cloth, or metal; foamed plastic panels; laminated structural members; and numerous kinds of sandwich shapes are extensively used. Adhesives join metal to glass, wood to metal, metal to metal, and metal to plastic. In fact, all solid materials can be adhesive bonded after proper shaping and surface preparation. Some bonds result by air curing and others result by the application of heat and pressure. The list of uses is nearly endless. Appliances, automotive parts, sporting goods, hardware assemblies, bicycles, helicopter rotor blades, and honeycomb panels for all types of aircraft include adhesive bonding in their makeup.

Mechanical Properties

Mechanical properties for stressed joints vary with the type of adhesive, its curing process, and the design of the joint. The basic joints are mainly limited to the butt, scarf, half lap, double lap, plate, joggle lap, tension lap, and double plate. Butt joints are commonly used in low stressed assemblies. The shear type of assembly provides high strength in the bond. The shear joint is a lap type, and shear strengths of 5000 psi are typical with the high strength adhesives such as the phenolics and epoxies. Epoxy-bonded joints are resistant to subzero temperatures and to those elevated temperatures up to 300 °F (149 °C). A very important point related to the joint is that the workmanship and fit of the parts must assure that the adhesive will contact all required surfaces.

GLASSES

Most glasses are hard, brittle, solid materials having some properties of a liquid. Their structure is amorphous, or noncrystalline, but special glasses can be produced in crystalline form. Most glass is transparent, is a nonconductor of electricity, is weak in tensile strength but stronger in compression, is responsive to thermal shock, and is resistant to most fluids. In the presence of hydrofluoric acid glass is rapidly attacked and quickly destroyed, as often evidenced by a white smoke.

Types of Glass

Different kinds of glasses are available, each having its particular qualities. Only a slight change in the chemistry produces a different glass; therefore, there are thousands of different specific kinds. Most glasses are commercial, however, and these can be divided into six main types. Other glasses vary from the hard and brittle to the soft and very flexible.

The most common type of glass is a product of limestone, soda ash, and silica sand, and is known as soda-lime glass. Variations in its chemical proportions result in several different hardnesses. Products such as common tableware and related receptacles, decorations, enclosures, plates, and shapes such as window panes and lightbulbs are produced from this basic glass. Shaping is accomplished by methods similar to shaping methods for metals, that is, by casting, rolling, pressing, extruding, and drawing. Lightbulbs and similar hollow shapes are blown with special machines. When temperatures around 1500 °F (816 °C) exist the glassy mass is very plastic and easily deforms by pressure to whatever shape is needed. Most other glasses are formed in the same way, with the exception of the high temperature types.

Another kind of glass is fused silica and is a product of silica sand. Because of its high forming temperature, about 3000 °F (1649 °C), it is reserved for special applications requiring glass with a purity of 99%. The thermal expansion of fused silica is low, but its thermal shock resistance and optical qualities are good.

Borosilicate is used to make optics for telescopes. It is also used for products which will be subjected to heat, such as Pyrex cookware. Much of the glass goods in laboratories is made from the borosilicates. This glass also has a very good resistance to thermal shock along with a low thermal expansion. Another exceptional quality is its ability to quickly radiate its heat.

The 96% silica glass is a product of silica and boric acid which gives it excellent thermal and chemical resistant properties. Because of its ability to retain thermal shock in conjunction with high temperature strength it finds extensive use in the aerospace industries for observation windows and special types of optical lenses.

The aluminosilicate type of glass includes silica, lime, aluminum oxide, boric acid, and magnesium oxide. This glass also has excellent thermal shock resistance. It is ideal for cookware, heating apparatus, and thermometers.

\The lead alkali silicates produce good light control characteristics and offer excellent capabilities for design configurations where both geometry and sparkling brillance are reflected from its chemistry. Because of the lead content, a radiation shielding capability exists as well as electrical conductivity.

Manufacture

The manufacture of glass is similar to metal production in that the several constituents are placed in a furnace and heated to the melting temperature. Special handling mechanisms remove the plastic glass from the furnace and process it into desired shapes. Immediately, the shaped glasses are cooled slowly to prevent stress formation. Reheating is required sometimes to assure stress removal. Following annealing some glasses are tempered by reheating to a specified temperature below the melting point and cooled in an air blast or liquid coolant. Such a rapid cool causes the surfaces of the glass part to become stressed in compression. The input of the retained compressive stress decreases the bending factor when a tensile stress moves into a compression zone. Tension stresses are fatal to glass, but compression loading is satisfactory within limits, as is the case for other brittle materials. Normally the typical glass is brittle when exposed to sudden shock. There are numerous other glasses, however, which are flexible and elastic and are not subject to load shock. The term *glass* means little more than a combination of constituents which have solidified in an amorphous condition, simulating the initial liquid. Remember, however, that some glasses are crystalline.

WOODS

Wood is an organic material consisting of long bundles of tubular materials of a high cellulose nature which are bound and held in a rounded form by a cement-like material called lignin. The general growth of a tree provides increased length and width of the wood over a period of several years. Width is accompanied by the growth of new annular rings while the grain pattern moves upwards along the length (Fig. 13–16). This natural arrangement of the growing process provides three major axes from which cuts of wood are produced. The width, or radial axis, is across the tree's diameter and provides wide boards containing short segments of the rings. The tangential axis runs at a right angle to the radial axis and at a tangent to the rings whereby only parts of a few rings are included in the cut. The longitudinal axis runs the length of the board, but usually includes parts of several grain patterns as the cut moves from grain to grain when a straight-line cut is made.

Types of Wood

The two types of wood are soft and hard. The general statement that softwoods come from trees with small and needle type leaves and that hardwoods come from trees with flat and often larger leaves is not always true. Some of the softwoods include pine, spruce, cedar, fir, cypress, soft maple, soft mahogany, and redwood.

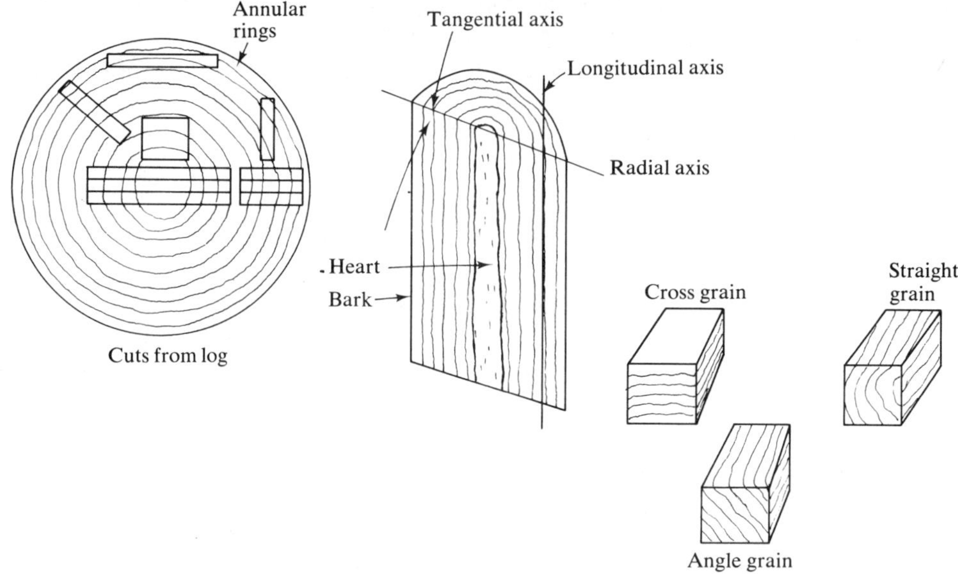

FIGURE 13–16 Characteristics of wood grain and wood cuts.

Hardwoods include oak, hard maple, hickory, walnut, and mahogany. However, a line of demarcation cannot be drawn between hardness and softness or rigidity of the fibrous material in the wood. Together with the cut of the wood, the grain compactness influences the strengths and uses of the boards. Straight-grained boards are cut from the log to allow the grains to run parallel, or nearly so, to the board's surface. An angle cut causes the crosswise grains to rest at angles to the surface due to orientation of the annular rings. And the cross-grained board provides grains at 90° to the face of the board. The larger the tree the more radial cut boards are available in large widths. The relation of the grain to flow of stress when a board is placed in service reflects its strength.

Drying after Sawing

After logs are sawed into the several shapes such as planks, studs, and joists, the green, or wet, wood is dried by controlled processes in a kiln so that the moisture content does not exceed about 25%. For indoor use moisture contents are somewhat reduced compared to the outdoor use types. Air drying is a common practice for seasoning wood to an approximate moisture content. The moisture content is determined by dividing the water weight in wood by the oven dry weight of the wood and multiplying by 100. Moisture contents greatly influence the mechanical properties of the wood.

Mechanical Properties

Mechanical properties of wood are related to other solid materials; however, because wood is a biological product it is used mostly in engineering designs as columns and beams involving heavy structural loading. Because beams deflect under load, compression stresses on one side of the beam offset tensile stresses on the opposite side. When tensile and compression stresses are present shear appears also. Wood structural members frequently fail in horizontal shear when overloaded. This failure is observed as a long split along the midportion of the timber. Vertical shear sometimes occurs across the cross section. When compression loading occurs against the ends of the tubular fibers, as in a column, the tubes must maintain rigidity through the elastic limit. A flexed beam includes both tension and compression stresses, however, one stress balancing the other, in addition to vertical and horizontal shear. With respect to strength, white oak fails in compression loads of approximately 9200 psi. Specimens of wood were sawed to 3-inch lengths with cross-sectional dimensions of 1 inch by 1 inch to test their strength. The white oak specimen failed in a type of shear, typical of hard material under compression loading (Fig. 13–17). Other woods tested in compression include those shown in Table 13–1. The modulus of elasticity of wood varies from 0.8 million to around 2 million psi. As the moisture content of wood increases the strength decreases, and this relationship refers directly to the specific gravity of the wood. Also, the modulus of elasticity decreases with an increase in moisture. Grain structure of wood varies as illustrated in Figure 13–18.

FIGURE 13–17 Wood specimens, 1 inch by 1 inch by 3 inches, tested in compression in both longitudinal and lateral loading. Results are shown on specimens.

TABLE 13–1 COMPRESSION FAILURES OF SEVERAL WOODS

Wood	Compression Failure (psi)
Hard maple	9500
Soft maple	5900
Magnolia	7900
Yellow pine	5400
Fir	5000
Soft mahogany	5100

FIGURE 13–18 There are hundreds of different woods, and each has its particular grain pattern.

Uses of Wood

Wood is used universally in applications in many segments of society. Its uses range from columns in a bridge to apple crates and firewood. One of the greatest uses of wood is in house construction where all common shapes of wood are used, including the laminated plywoods and veneers and the several types of hardboards. Very long beams and rafters of a laminated design allow large spans of space, such as are found in churches and large stores. Plywoods consist of wide sections of thinly cut wood with alternate layers of a resin or glue bonded together into one firm sheet. More expensive laminations include a thin veneer bonded to a plywood. The hardboards are pressed mixtures of wood fiber and resin and are used in similar applications as plywood. The several cuts of wood include studs for compression loading, joists, rafters, and beams for flexural loading, planks for flooring and roofing, molding for trimming, and plywood for covering large spaces including cabinet construction. Table 13–2 lists various types of wood and some of their applications.

Wood is often purchased by the board foot, which is 12 inches square and 1 inch thick. The common formula for determining board feet is

$$\frac{\text{thickness} \times \text{width} \times \text{length}}{12}$$

The length is expressed in feet while thickness and width are in inches.

The quality of wood determines its use. Common wood, or yard lumber, is a general purpose wood. Timber is a much better grade and is used for structural purposes. Smaller sections and highest quality types of wood are used for mill work and furniture. The quality of wood is determined by a combination of factors such as knots, insect holes, porosity, and fractures present. Also, the size of lumber affects its classification.

For grading purposes lumber is divided into hardwood and softwood classifications and each has its standards for grading. Hardwoods are classified as firsts or

TABLE 13–2   KINDS OF WOOD

Name	Hard	Soft	Use
Ash	X		Wall paneling
Balsam		X	Life preservers
Beech	X		Mill products
Birch	X		Fine mill work
Cedar		X	Chests, posts
Cherry	X		Fine furniture
Chestnut	X		Rustic furniture
Cottonwood		X	Cheap cabinets
Cypress		X	Silos, construction
Elm		X	Cheap furniture
Fir		X	Boats, construction
Gum	X		Fine cabinets
Hackberry		X	Cheap furniture
Hickory	X		Tool handles
Locust	X		Cabinet work
Magnolia	X		Furniture
Mahogany		X	Fine furniture
Maple		X	Good furniture
Maple	X		Fine flooring
Oak, red	X		Railroad ties, construction
Oak, white	X		Fine flooring, construction
Pine, white		X	Mill work
Pine, yellow		X	House construction
Poplar		X	Plywood
Redwood		X	Fencing
Spruce		X	House construction
Sycamore	X		Millwork
Teak	X		Shipbuilding
Walnut	X		Fine furniture
Willow		X	Cheap furniture

seconds and these are graded from the poor side. Firsts must include a minimum size or equal a minimum of square feet without defects. Seconds must provide a slightly smaller square footage of surface without defects and remain within a minimum size. Selects require a smaller width and smaller length with a minimum of defects. Next in the classification is the number 1 common, followed by number 2 common. Most hardwoods are destined to become furniture.

The softwood classification has the categories of yard, structural, and shop lumber. Yard lumber includes select and common grades. Select grades are further divided into *A* grade for the best and subsequently into *B, C,* and *D* for poorer grades. Common grades begin with 1 as the best, followed by lower grades of 2, 3, and 4. Structural lumber is destined for stressed members and must be acceptable for these purposes. Shop grade lumber is a general purpose lumber.

Because wood is organic it should be protected by a surface coating. Varnish, paint, or a treatment such as creosoting, which is given to electric line poles and columns, can be used to protect the wood.

RUBBER

Rubber is an elastomer which is unlike most other materials due to its peculiar molecular arrangement along with its chemistry. Natural rubber is a type of tree sap which is subsequently treated to thicken it so that solid shapes of unrefined rubber can be made. Further processing produces a tough and flexible material which is manufactured into numerous objects requiring a natural rubber. Natural rubbers have limitations with regard to attack by liquids such as oils and gasolines and they must be used at temperatures well below the boiling point of water. Consequently, the artificial elastomers are being produced which, in many instances, have superior properties such as use at higher temperatures, resistance to petroleum products, and ability to be manufactured into longer lasting products in the electrical, chemical, and mechanical industries.

Variable Mechanical Properties

Rubber, both natural and synthetic, is usable at subzero temperatures. Mechanical properties of rubber vary considerably because of the chainlike molecular structure which lengthens when stressed in a manner unlike other solid materials. Synthetic elastomer molecules may become cross-linked under deformation and remain rigid, much like the thermosetting plastics. A vulcanized joint, for example, between two sections of rubber is made with the help of sulfur and certain accelerators. The thermoplastic type of material then becomes cross-linked and, therefore, stronger.

The manufacturers of the synthetic rubbers use the additions of fibrous materials to the thick liquid along with the required chemicals to produce the specific type of rubber. After thorough mixing the mass of rubber is fed into a rubber processing machine which produces the desired shape. Rolling, extruding, pressing, and shaping by other methods provide the numerous parts and articles of rubber. Rubber tires, for example, are produced by a combination of processes including the bonding of sheets of rubber to a strong fabric containing nylon or other cords. Wire cords are also molded into the rubber for strength purposes.

Shaping

Rubber is processed into shapes in similar ways as the plastics; keep in mind that the addition of accelerators and sulfur in the presence of heat causes cross-linking. Just as an epoxy resin is bonded to a sheet of aluminum alloy, rubber is also bonded to different metals. Molds which are heated to around 240 °F (115 °C) to 300 °F (149 °C) while the part and adhesive are under a pressure of 600 psi cause rubber to bond to metal. Basically, in producing the synthetics some particular property is attained, such as polychloroprene's good electrical resistance, butyl's resistance to solvents, and the silicone's all-around temperature use. A natural rubber tire has excellent wear resistance and is not severely affected by temperatures generated in use, but petroleums are detrimental to the rubber.

Questions

1. What is a ceramic?
2. Define a composite.
3. What is alumina?
4. Describe the manufacture of a typical clay product.

5. What is the difference between a building brick and a refractory brick?
6. What is concrete?
7. How is the slump test performed?
8. Describe the curing process related to concrete.
9. What is powder metallurgy?
10. Describe a laminate. Why is this type of design important?
11. Name two important purposes of producing the several structural type laminates.
12. How is honeycomb produced? Describe its importance in structural fabrication.
13. Differentiate between the thermoplastic and thermosetting plastics.
14. Differentiate among the several kinds of adhesives.
15. Describe the process of adhesive bonding.
16. What is glass? Describe the six main types.
17. Why are some types of wood considered to be engineering materials?
18. Define rubber.
19. Name some reasons why laminates are used in the aerospace industries.
20. List the advantages of using a composite.

Processing Effects on Metals

Because metals are unlike all other materials in nature they require special handling, treating, and shaping in order to gain the many benefits from their physical and mechanical properties. Metals are actually unique among other materials in that most can be shaped into desired configurations by heat or pressure. The new shape then withstands the many attacks from its environment, until in some cases it is overwhelmed and destroyed by the forces man brings upon it. As has been pointed out, most metals are nonferrous; together with the ferrous, they constitute most of the elements. When two or more of the elements are combined some kind of chemical compound or alloy is produced, and when these compounds and alloys are refined into desired chemistries many thousands of metals are available (Fig. 14–1). Details of how the earth's ores are processed into metals and then shaped are described in another text, *Manufacturing Processes*.

MAIN METHODS OF SHAPING METALS

It can be assumed that metals begin their useful life at some time after being refined in the liquid state and cooled to the solid state to become a casting. The new shape is then used as it is or it is reheated and forced into a new shape by some kind of pressure, such as hammering (Fig. 14–2), bending, rolling, or extruding. Another shaping method is the process of powder metallurgy whereby metal powders are pressed into desired shapes and sintered for strength. Machining may be involved in all forming processes. Whether the shape results in a casting, a wrought product, or a pressed and sintered part, each shape is affected by temperature, time, and cooling speeds from the hot-forming temperatures. Therefore, the technician must be aware of the metal's chemistry, its shape, its intended use, and its environment in order that a part may be successfully designed to operate throughout its lifetime. For example, if the part is to be used in static conditions, then its mechanical properties will be unlike those necessary for dynamic conditions. If corrosion con-

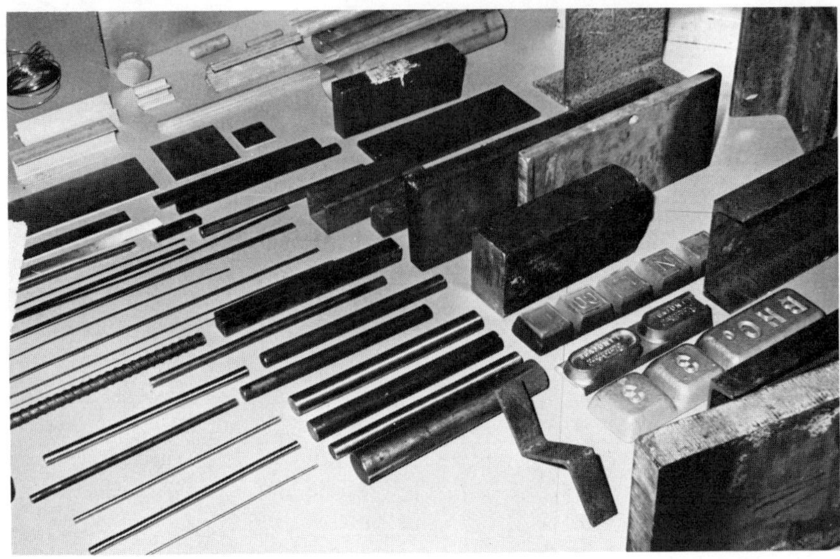

FIGURE 14–1   An array of metals used as raw stock for subsequent manufacturing operations.

FIGURE 14–2   Different stages of a forging. The finished part is on the right.

ditions are present, close attention must be given to physical properties. If overloading conditions are possible, special emphasis on design is mandatory. Much of what is applicable to the operating conditions of metals is also applicable to the nonmetallics, especially in regard to dynamic operating conditions.

EFFECTS OF SHAPING ON MECHANICAL PROPERTIES

Cooling rates and forming pressures have varying and sometimes drastic effects on mechanical properties of metals. The chemistry of the metal influences its fixed physical properties more than anything else; mechanical treatments and varying temperatures influence the mechanical properties of the metal. Because metals are granular and because these grains occur in different sizes caution must be used to control their size and shape and to provide a desired microstructure within the boundaries of these grains. Basically, mechanical properties are direct reflections of the metal's chemistry, the chemical arrangement in the many grains, and the size and shape of these grains. Consequently, environmental conditions must be controlled during the processing stage in order that exact properties result which are in accordance with predetermined requirements.

Recrystallization

Recrystallization of metals occurs on heating and varies with the metal and its metallurgical condition. At recrystallization new and small grains are established. When cooled in any manner these small and desirable grains remain. But if temperatures increase beyond the point of recrystallization, grains grow in size and, again, when cooled remain at room temperature at a size commensurate with the size at elevated temperature. Large grains are less resistant to impact loading than fine grains. Another factor which affects impact loading is the shape of the grains. For example, cast grains are often large and include either the dendritic or equiaxed shapes. Both conditions are inferior to deformed shapes such as those resulting from pressure forming. In deformed shapes elongation and grain boundary overlap occurs along with more homogeneous conditions inside the granular microstructure. Because of granular changes on heating, procedures must be established to assure that lowest temperatures are used during the processing stage.

Hot and Cold Working

Some metals are formable at either cold or hot temperatures. Cold processing temperature is temperature below the recrystallization temperature and hot working temperature is temperature above the recrystallization temperature. Working conditions such as hot rolling and forging cause granular deformation into a part's desired shape, but are accompanied by grain boundary collapse, constant recrystallization, and possible grain growth. No increase in hardness or strength occurs at hot processing temperatures, but because of plastic conditions of the hot metal lower forming pressures are used and danger of tearing is reduced. On the other hand, forming temperatures below the recrystallization temperature cause work hardening and increased strength in the microstructure. In this situation plastic deformation is permanent rather than temporary. Higher forming pressures are required to cause plastic flow, and a danger of tearing the metal exists. Cold forming is often carried out at room temperature if the metal conforms to this condition because scale does not occur and higher strengths are produced. For example, the recrystallization temperature in carbon steels is approximately 1333 °F (723 °C), but the addition of alloying elements changes this temperature. Lower melting metals recrystallize at lower temperatures and consequently also mark their points between hot- and cold-working operations.

Cold-working processes increase mechanical properties such as hardness and strength and hot-working processes destroy previously established cold-worked or heat-treated microstructures and result in homogeneous conditions within and among the many grains of metal. But hot-worked metals are not usually used in the "as produced" condition because of their soft and weak microstructures. As a result, machining and some forming operations such as bending are conducted in the metal's "as produced" condition since its hardness is in the Rockwell B scale or softer. Machining and bending or cupping operations require soft metal in order that cutting or plastic flow will occur. After cutting or forming the part into its required shape, heat treating is often performed to increase the hardness or strength to the required value.

MANUFACTURING PROCESSES

The field of manufacturing processes is so large that only a short summary will be given. Because manufacturing processes begin with raw materials and form these materials into desired shapes it can be seen that the overall process is vast, complex, and diverse. Manufacturing materials have been discussed in this text and it is these materials which are shaped by some manufacturing process. With regard to metals, processing is accomplished either hot or cold and mechanical properties are directly affected.

Casting

The most widely used manufacturing process is casting because it is the casting which is subsequently used in most processes. Casting is merely the pouring of liquid metal into a container called a mold, and the resulting solid metal is called a casting. However, when this casting is to be reshaped it is called an ingot. Many types of casting processes are available and include sand and permanent mold types such as automotive engine blocks and heads, die cast parts such as carburetors, centrifugal cast parts such as cast iron or steel pipe and train wheels, shell molded parts having many shapes in the hardware areas, investment cast parts such as jewelry and false teeth, and variations of these processes. The objective of casting processes is to provide a solid shape by allowing liquid metal to flow into a pre-formed cavity. Because casting is only one of the main shaping processes some means must exist to provide for wrought operations. Again, it is the casting, or ingot, which subsequently becomes the raw material for pressure forming.

Wrought Processes

The purpose of wrought processes is to provide shapes other than castings and, in many situations, to produce those shapes not practically formed by casting processes. Basically, the ingot is run between rolls which plastically reduce it to other shapes for further shaping operations. Or the hot ingot is placed on an anvil so that a forging hammer can force it into the desired shape. Another wrought process pulls or draws the plastic metal through a die which has the desired shape. On the other hand, the extrusion process forces or pushes the plastic metal through a die which has the exact shape of the required product. The sheet metal industry uses numerous types of bending and stamping operations which shape the part to its required dimensions. In summary, rolling operations produce sheets, plates, bars, and I-beams. Forging operations produce tools such as the handles of adjustable wrenches and crankshafts. Drawing and extrusion operations produce exact shapes such as round rods or H shapes. Bending operations produce hundreds of parts such as household appliances, and stamping operations produce household hardware and cookware.

Powder Metallurgy

When metal is powdered by one of several methods it is available for use in pressing operations whereby it is shaped into desired configurations by high pressures. Powders are placed in a mold or die and a ram compresses the powder into raw briquettes. These brittle shapes are subsequently sintered into strong and usable shapes by heating to a temperature which is normally below the melting point of any constituent in the briquette. However, cemented carbides such as tungsten carbide are heated to the melting point of the nickel powder, which is then cemented to the many particles of carbide. Excellent cutting tools are produced by powder processes.

Cutting Operations

The most diverse of the forming operations is cutting and this operation is most often classified in the machine shop category. Machining includes the use of more different types of machines and operations than other shaping processes. No other process can replace the cutting operations because casting, wrought, and powder processes are unable to produce finished parts in all kinds of shapes. Cutting operations remove metal from previously produced shapes. Machines such as saws, drills, lathes, mills, shapers, grinders, broaches, slotters, borers, planers, and the numerous automatics quickly reduce the workpiece to its required dimensions by removing the exact amount of metal. Special types of machining operations include ultrasonic, electron beam, laser, electrical discharge, electrochemical, and chemical. Conventional cutting operations use tools and cutters especially designed for the specific cutting operation. Other cutting operations use high velocity impact particles, electron impact, high energy concentration, anode-cathode sparking, anode-cathode ion loss, and direct chemical attack to develop exact dimensions. Obviously, in all of the cutting operations there is a loss of metal which becomes scrap.

Joining Operations

Metals are joined by mechanical or fused processes. Some of the mechanical processes include bolting and riveting of two or more parts together. The strength of the joint depends on the design of the joint and the shear value of the bolts or rivets. Other mechanical type joints include soldering and some types of brazed joints in which the liquid metal adheres to the surfaces of the parts to be joined. The epoxy joint is a mechanical joint; the epoxy strongly attaches itself to the cleaned surfaces of the materials being bonded. The strength of the joint is the shear strength of the bonding material. The strongest of the joints is the fused, or welded, joint in which one of the metals flows into the other, forming a solid metallic union between the two or more parts. Several types of welding operations are available and include oxy-acetylene, ac-dc electric, gas tungsten arc (TIG), gas metal arc (MIG), carbon arc, thermit, submerged arc, resistance, electron beam, laser, plasma arc, and more. Designs of welded joints vary and include lap, butt, tee, tubular, and modifications of these. Because the fused metal is cast consideration must be given to the resulting joint's strength when wrought metals are joined by welding. Further consideration must be given to the effects of heat on wrought-formed shapes or heat-treated shapes. As previously pointed out, heat reduces the strength of many metals which have been increased in strength by cold forming or by heat treatment. However, heat can also increase the strength of many metals.

Surface Operations

Many metals are prepared to resist corrosion by painting, oxidation, dipping, spraying, cladding, or by electroplating. Such operations depend on the metal being prepared for treatment and the specific treatment. Surface protection treatments are usually accomplished to resist some kind of corrosion; however, other types of surface treatments such as case hardening or shot peening drastically influence mechanical properties. Some metals are naturally resistant to corrosion, such as brass, gold, and austenitic steel. Other metals, such as carbon steel and low alloy steels, require surface protection from various corroding conditions. Aluminum is highly resistant to corrosion but its alloys may corrode somewhat. When a layer of aluminum is rolled onto and bonded to an aluminum alloy the combination of corrosion resistance and high strength are provided. Another example shows that the anodized surface of aluminum alloy is more resistant to corrosion than the normal surface because the resulting aluminum oxide acts as a passive barrier to further oxidation. Galvanized iron is merely zinc-coated steel sheet in which the zinc acts as a protector of the steel's interior. Case-hardened parts have had their surfaces chemically altered so that high hardness, high strength, and higher corrosion resistance exist. The surface hardness is high and the tough core carries the load. When the metal's surface is bombarded with steel shot (shot peening) a compressive stress is built into the surface along with hardness and, in turn, fatigue life of the part is increased.

PROCESSING AND STRESS RELATIONSHIPS

As metals are processed into their final shapes by one or more of the manufacturing processes variable relationships exist among shape change, temperature, and stress. Hot working induces no residual stresses in most metals and leaves the newly shaped metal free of stress at room temperature, but may cause surface alterations such as scale. Scale is easily removed by pickling; if cold working is to follow, it must be removed prior to the forming operation. The purpose of hot forming is to change the shape of the metal with reduced loads while plastic flow is at its maximum. Some internal flaws such as blow holes and gas pockets are often welded shut during these operations. Stress becomes a constant value from the yield on the stress-strain diagram, assuming the metal's core is at the same temperature as its surface. And as temperature increases the yield strength decreases while plasticity increases. Sometimes, hot working is the only method of reshaping a mass of metal by wrought processes.

When higher hardnesses and strengths are required in the new shape and when clean and smooth surfaces are desired the forming process is performed while the metal is cold, or at some temperature below recrystallization. During cold rolling or cold forming higher loads on the metal are incurred because of the rise of the yield strength as compared to a lower yield during hot forming. Whether forming operations are accomplished on hot or cold metal the deforming load must be beyond the elastic limit of the mass of metal in order that permanent deformation occurs. It is obvious that deforming loads lower than the yield are only elastic, and when the load is removed springback occurs and no dimensional change results.

Cold Forming Increases Stress

Mass of metal, load, shape, time, and temperature are involved in the cold forming of metal. If the rising stress increases to the tensile strength in a given volume of metal during deformation by plastic flow, tearing or instant fracture occurs in that area to relieve the stress. This metallic separation can occur on the surface as well as internally. When internal bursts occur in a bar they may never be recognized and consequently will pose a serious threat to subsequently highly stressed parts machined from the bar. The burst occurs when stress differentials are incompatible with ductility factors and time. For example, during a cold-rolling operation, a given load satisfactorily shapes the rod according to surface inspections. But an ultrasonic inspection shows a large internal discontinuity which can be confirmed by X ray and ultimately by parting the rod. Apparently, with regard to the burst, the given time limit is insufficient for plastic flow to react to the controlling stress in the failure area and the only reaction to the stress is conformance by either metallic flow or separation. Because the area or a volume of metal is subjected to rapidly rising stress values as the center of the workpiece is approached from a given surface load, consideration must be given to the area factor which must house and accommodate the stress. Fundamentally, from a given load in pounds, as the area decreases the stress in that area increases.

Compression Stresses

When compressive loads are placed on a metal instant stress reaction occurs to equalize the external and internal loads so equilibrium is reached where the sum of the forces is zero. As loads increase beyond the elastic limit of the metal, plastic flow or fracture occurs. Malleable metals flow plastically even though work hardening occurs. It is in this range of biaxial compression that plastic flow formulas must show a high percentage in order that fracture does not occur. A point to keep in mind is that the increase in plastic flow below the ultimate and above the yield results in an increase in hardness, which is strength, pointed toward ultimate fracture.

Tension Stresses

Tension loads on the metal also induce instant reacting stresses pulling in the opposite direction to save the metal from destruction. As the loads increase stress increases and, again, a point is usually reached at which the sum of the forces is zero and equilibrium is attained. Stresses beyond the yield in very ductile metals cause permanent plastic flow and work hardening or increases in strength directed at the ultimate strength and fracture. Percentage of reduction in area formulas must show reasonable ductility factors to prevent tearing and ultimate fracture.

Shear Stresses

Shear loads are cutting loads and always occur in designs where near collision forces slide past each other. A bending beam includes shear loads at every reaction point. Bolts and rivets in a lap joint are in shear stress when members move into compression or tension stress. Every bending material develops tension and compression stresses concurrently, and these stresses vary with the magnitude of the deforming loads.

In summary, manufacturing processes must be compatible with the mechanical properties of the material. Deforming loads which establish stresses greater than the strength of the material cause failure in the overstressed area. Consequently, the engineer, technician, mill and forge operators, metallurgists, and designers must be cognizant of the projected stresses in a metal when deforming loads are applied. Many metals require heat treating after shaping for one or more reasons.

HARDNESS AND STRENGTH PROCESSING

At some stage in the manufacturing process some parts will take their final shape and be made ready for their strength treatments. Many objects which have been cast, rolled, forged, bent, pressed, or cut, however, are used in any of these "as fabricated" conditions in which the hardness and strength of the objects are usually at the minimum. Parts and assemblies which fall into this category are never subjected to heavy loading or highly stressed conditions. Most products in the home and many objects in the commercial hardware areas are released to consumers in this "as fabricated" condition.

On the other hand, many materials are further processed by some kind of metallurgical treatment involving heat treatment. Any time that hardness and strength properties are required beyond the natural properties of the "as fabricated" types of consumer items heat treatment is required. These parts must be produced from the proper materials when subsequent heat treatment is required. Consequently, manufacturing personnel are cognizant of the necessity for great care in the choice of material and process. Heat treatments of ferrous and nonferrous metals have been previously discussed in generalities, but specific treatments which must be used to obtain exact mechanical properties will be investigated (Fig. 14–3).

Heat Treatment of Specific Metals

After shaping processes are completed or when heat treatment must be performed the parts are dispatched to the metallurgical rooms for processing. Because a tremendous quantity of steel is shipped from the mills to the consumers in the soft condition only the annealing or normalizing treatments are given to the metal so that subsequent shaping operations can be accomplished. Steels which have been furnace cooled from austenite result in a Rockwell B-scale hardness value. Many of these raw stocks include sheets, plates, bars, rods, tubes, pipes, extrusions, and wire. When completed parts are produced by manufacturing from this array of raw stock further heat treatments are performed.

Heat treatments of ferrous and nonferrous metals in furnaces similar to those shown in Figure 14–4 are accomplished by many of the major and basic industries, including aerospace, shipbuilding, railroad, foundry, mill, armament, automotive, earthmoving, tool, machine, household, space, toy, sporting goods, and numerous others. Many heat-treated items are castings of different types and many forgings

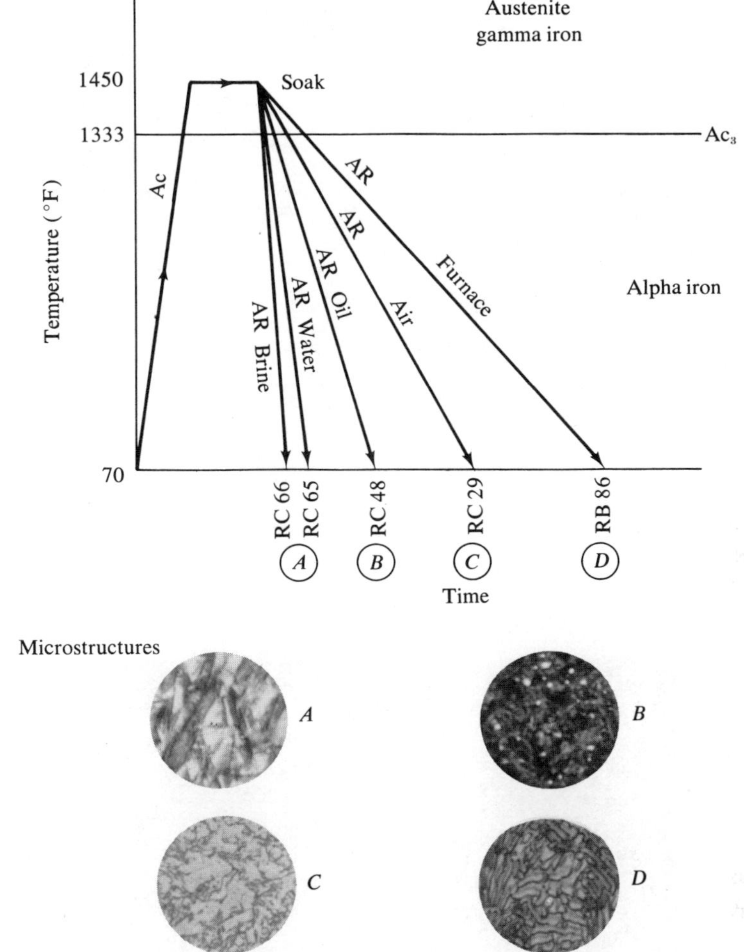

FIGURE 14–3   Varying cooling rates produce different hardnesses in an AISI 1095 steel.

are hardened and tempered. By far, the machine shops produce the greatest numbers of items for heat treatment even though the foundries and forge plants also produce a huge tonnage of cast and forged parts. Rolled structural shapes are produced in both the normalized and hardened and tempered conditions. Many of these shapes include railroad rails, columns, beams, pipe, and plate. The sequence of operations needed to produce completed steel parts is as follows: annealing, shaping, normalizing, hardening, and tempering. If electroplating is to be performed then plating and baking follow. With regard to the nonferrous metals, annealing, solution treatments, and precipitation treatments are required for many.

Annealing, Normalizing, and Hardening   Steel parts are processed through the normalizing treatments to remove processing stresses and are then hardened by heating and quenching operations (Fig. 14–5). Refer to the iron-iron carbide diagram for steel (Fig. 8–28) in order to equate the heating and slow cooling

FIGURE 14–4   Gas type heat-treating furnaces. (Courtesy of Lindberg Hevi Duty)

FIGURE 14–5   A box type heat-treating furnace. (Courtesy of Lindberg Hevi Duty)

processes with resulting microstructures. Data shown in Table 14–1 also point out specific austenitic temperatures needed in heat treatments for several typical steels. Further information on other metals is available in the *Metals Handbook* published

TABLE 14–1 ANNEALING, NORMALIZING, AND HARDENING
TEMPERATURES FOR SELECTED AISI STEELS

AISI Steel	Annealing Temperature (°F)	Normalizing Temperature (°F)	Hardening Temperature (°F)
1035	1550	1625	1550
1045	1525	1600	1525
1055	1525	1600	1525
1065	1500	1600	1500
1080	1450	1525	1450
1095	1450	1525	1450
2330	1500	1625	1500
3130	1500	1625	1500
4130	1525	1625	1525
4140	1525	1600	1525
4340	1550	1600	1550
5130	1525	1600	1525
6150	1550	1600	1550
8650	1525	1600	1525
9850	1525	1600	1525

by the American Society for Metals. The furnace shown in Figure 14–6 is a typical
small furnace for many heat-treating operations. Special types of heat-treating
processes which require a vacuum atmosphere are conducted in furnaces similar to
the one shown in Figure 14–7. After stress removal by normalizing, hardening is
performed by quenching the red-hot steel in water, oil, or, in some cases, air.
Austenitic temperatures shown in Table 14–1 are average and should suffice for
proper hardening operations. After hardening, the tempering operations are
accomplished.

FIGURE 14–6 Electric type heat-treating furnace at 1450 °F (788 °C).
The thermocouple can be seen at top center and heating elements are embedded in the ceramic sides.

FIGURE 14–7   This vacuum furnace provides controlled surface conditions of the part during heat treatment. (Courtesy of Lindberg Hevi Duty)

Tempering   Possibly the most critical of the heat treating operations is tempering because any deviations in tempering result in a change in mechanical properties. As previously pointed out, tempering effects begin around 200 °F (93 °C). Tempering can be conducted at temperatures as high as 1250 °F (677 °C), but the temperature never exceeds the critical points of the steel. A typical tempering furnace is shown in Figure 14–8. Because tempering is accomplished at temperatures below the scaling temperatures an oxidizing atmosphere maintains a clean surface on the part. The data in Tables 14–2 through 14–12 point out how specific hardness and tensile strengths are obtained. These carbon and low alloy steels covered in the tables are typical among those used in engineering designs. Table 14–12 provides the needed data for common tool steels.

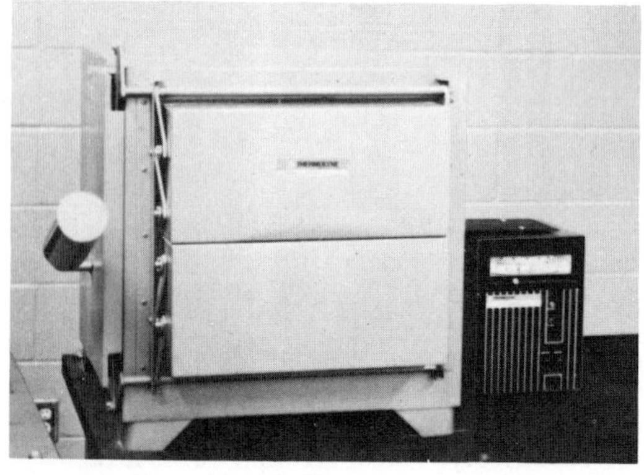

FIGURE 14–8   This box type electric furnace and pyrometer are used in heat treating small parts in an oxidizing atmosphere.

TABLE 14–2 TEMPERING TEMPERATURE VS. HARDNESS FOR AN
AISI 1035 Steel

Tempering Temperature (°F)	Rockwell Hardness (C)	Tensile Strength (psi)
1200	B 92	93,000
1100	20	108,000
1000	23	115,000
900	27	127,000
800	30	136,000
700	38	171,000
600	42	194,000
500	44	207,000
400	49	246,000

TABLE 14–3 TEMPERING TEMPERATURE VS. HARDNESS
FOR AN AISI 1045 STEEL

Tempering Temperature (°F)	Rockwell Hardness (C)	Tensile Strength (psi)
1200	B 93	96,000
1100	20	108,000
1000	26	123,000
900	28	129,000
800	35	157,000
700	41	188,000
600	47	229,000
500	49	246,000
400	52	273,000

TABLE 14–4 TEMPERING TEMPERATURE VS. HARDNESS
FOR AN AISI 1055 STEEL

Tempering Temperature (°F)	Rockwell Hardness (C)	Tensile Strength (psi)
1200	24	117,000
1100	33	148,000
1000	34	153,000
900	36	162,000
800	37	166,000
700	40	182,000
600	49	246,000
500	54	291,000
400	55	301,000

When steel parts are heat treated they are normally hardened and tempered after normalizing. A simple means for helping to determine the tempering temperature is illustrated in Figure 14–9. Because tempering temperatures are so closely associated with microstructure very little deviation in temperature is acceptable. According to Figure 14–9*a* a triangle is constructed and labled as indicated. The questions are asked in sequence from *A* to *D*. Answers to these questions are in part *b*. If two answers are known, then the third is supplied automatically from tables. The

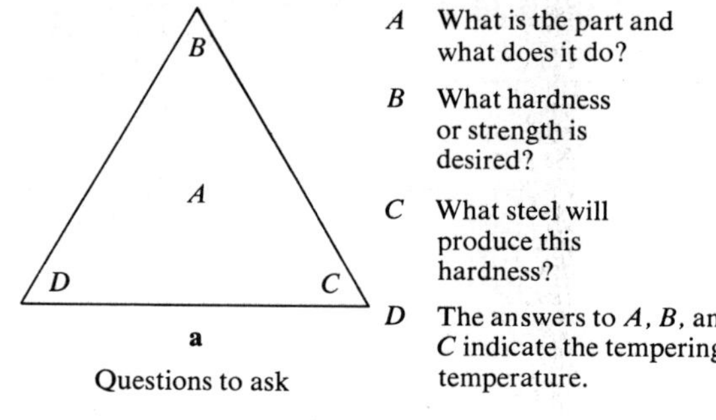

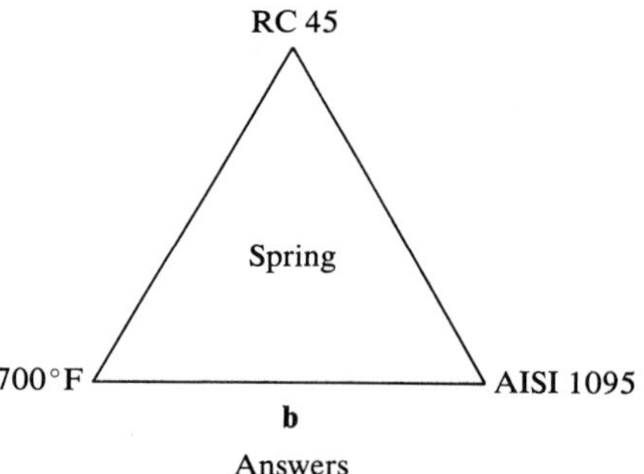

FIGURE 14–9   How to choose a steel for a given use. The triangle represents three things to do.

TABLE 14–5   TEMPERING TEMPERATURE VS. HARDNESS
FOR AN AISI 1065 STEEL

Tempering Temperature (°F)	Rockwell Hardness (C)	Tensile Strength (psi)
1200	25	120,000
1100	32	144,000
1000	36	162,000
900	38	171,000
800	39	175,000
700	43	200,000
600	52	273,000
500	56	280,000
400	57	301,000

tempering temperature for a given hardness or strength changes as the analysis of the steel changes. Therefore, an assortment of heat-treating temperatures is required to fulfill the needs of the thousands of manufactured parts. An example of the need for varying heat treatments is given in Figure 14–10.

Special Heat Treatments   Operations such as case hardening include a charcoal type of treatment whereby the low carbon steel part is soaked at 1700 °F (927 °C) in the muffle (Fig. 14–11). When the box is sealed carbon monoxide gas combines with iron and forms iron carbide in the surfaces to a depth depending on time at

FIGURE 14–10 This small aircraft engine with accessories consists of many parts heat treated to exact specifications.

FIGURE 14–11 A muffle which is used to pack charcoal around steel parts to be carburized and subsequently case hardened.

TABLE 14–6 TEMPERING TEMPERATURE VS. HARDNESS FOR AN AISI 1080 STEEL

Tempering Temperature (°F)	Rockwell Hardness (C)	Tensile Strength (psi)
1200	30	136,000
1100	36	162,000
1000	37	166,000
900	39	175,000
800	40	182,000
700	44	207,000
600	52	273,000
500	56	280,000
400	59	*

*Not predictable for this steel.

TABLE 14–7 TEMPERING TEMPERATURE VS. HARDNESS
FOR AN AISI 1095 STEEL

Tempering Temperature (°F)	Rockwell Hardness (C)	Tensile Strength (psi)
1200	29	132,000
1100	33	148,000
1000	42	194,000
900	43	200,000
800	44	207,000
700	45	214,000
600	51	264,000
500	57	301,000
400	60	*

*Not predictable for this steel.

TABLE 14–8 TEMPERING TEMPERATURE VS. HARDNESS
FOR AN AISI 2330 STEEL

Tempering Temperature (°F)	Rockwell Hardness (C)	Tensile Strength (psi)
1200	B 90	89,000
1100	23	115,000
1000	27	127,000
900	31	140,000
800	34	153,000
700	36	162,000
600	43	200,000
500	44	207,000
400	46	221,000

TABLE 14–9 TEMPERING TEMPERATURE VS. HARDNESS
FOR AN AISI 4140 STEEL

Tempering Temperature (°F)	Rockwell Hardness (C)	Tensile Strength (psi)
1200	26	123,000
1100	32	144,000
1000	34	153,000
900	40	182,000
800	44	207,000
700	46	221,000
600	49	246,000
500	52	273,000
400	56	280,000

1700 °F (927 °C). Approximately 0.062 inch of high carbon steel is produced in the surface after 6 hours. After cooling the carburized part is grain refined at 1625 °F (885 °C) by air cooling. Subsequent soaking after preheating at about 800 °F (427 °C), at 1450 °F (788 °C) and water quenching produces an RC 65 on the surface and about an RC 23 in the core. Another case hardening process is flame hardening, shown in Figure 14–12. As the sprockets turn, the gas torch heats

TABLE 14–10 TEMPERING TEMPERATURE VS. HARDNESS
FOR AN AISI 4340 STEEL

Tempering Temperature (°F)	Rockwell Hardness (C)	Tensile Strength (psi)
1200	30	136,000
1100	33	148,000
1000	40	182,000
900	43	200,000
800	47	229,000
700	49	246,000
600	51	264,000
500	53	282,000
400	57	301,000

TABLE 14–11 TEMPERING TEMPERATURE VS. HARDNESS
FOR SEVERAL AISI STEELS

AISI Steel	Tempering Temperature (°F)	Rockwell Hardness (C)
3130	1200	B 90
3130	800	33
3130	400	45
4130	1200	20
4130	800	37
4130	400	46
5130	1200	20
5130	800	38
5130	400	47
6150	1200	32
6150	800	47
6150	400	59
8650	1200	27
8650	800	44
8650	400	54
9850	1200	28
9850	800	43
9850	400	52

FIGURE 14–12 These sprockets are being flame hardened. (Courtesy of Martin Sprocket and Gear)

TABLE 14–12 TEMPERING TEMPERATURE VS. HARDNESS
FOR SEVERAL AISI TOOL STEELS

AISI Steel	Tempering Temperature (°F)	Rockwell Hardness (C)
W1	400	61
W1	550	57
D3	400	61
D3	550	60
T1	1000	64
O1	400	61
A2	400	62
D2	400	61
M2	1000	65
S1	300	61
O2	400	61
S5	300	61
F2	400	63
H11	1000	55
403	400	40
420	400	52
440	400	57

the surfaces of the teeth. Water quenching follows. This type of heat treatment which heats only the surface to austenite requires a high carbon steel. Other case hardening operations have been discussed and include liquid and gas carburizing, nitriding, and induction hardening. A different type of operation, known as spheroidizing, causes any steel's carbides to form in spheres so that severe bending operations can be performed. Micrographs in Figure 14–13 show how pearlite and the spheroidized structure appear. Plates of carbide will fracture if the metal is bent too far whereas the spheres move about in the ferrite.

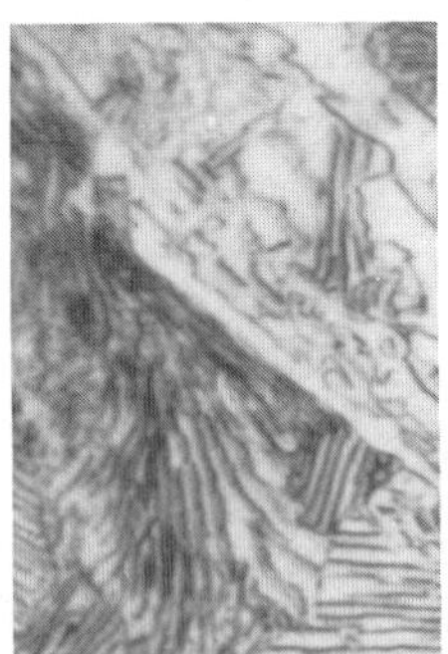 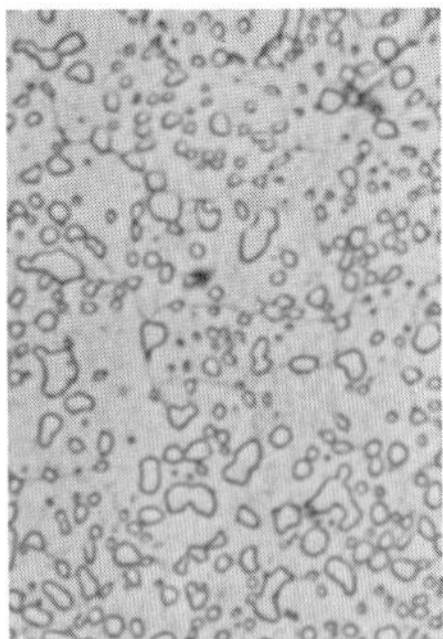

FIGURE 14–13 Soft pearlitic steel is spheroidized to allow bending and cupping operations in a high carbon steel which will subsequently be hardened and tempered.

Solution and Precipitation Treatments   There are many nonferrous metals which can be heat treated for increased mechanical properties. Most can be annealed for forming purposes. The aluminum alloys and several copper alloys, for example, are annealed by soaking at 800 °F (427 °C) and furnace cooled. Some can be softened by air cooling. Solution and precipitation temperatures for hardening purposes for

TABLE 14–13 HEAT TREATING TEMPERATURES FOR
SEVERAL NONFERROUS METALS

Metal*	Solution Temperature (°F)	Precipitation Temperature (°F)	Time (hr)	Temper
AA 2014	935	340	10	T6
AA 2014	935	70	48	T4
AA 2024	920	70	48	T4
AA 6061	970	350	8	T6
AA 7075	910	250	24	T6
MA AZ63A	725	425	10	T6
MA AZ91C	775	335	10	T6
MA ZK61A	930	265	10	T6
BC (Be 0.35% + Ni 1.5%)	1700	840	2	
Ti-6A1-4V	1700	950	3	
Ti-4A1-3Mo-1V	1650	1000	8	

Code
*AA   Aluminum alloy
 MA   Magnesium alloy
 BC   Beryllium copper
 Ti   Titanium alloy

several alloys are listed in Table 14–13. Alloys of aluminum, magnesium, copper, nickel, and titanium constitute most of the heat treatable nonferrous metals. The forged aluminum alloy shown in Figure 14–14 is a typical example of the results of heat treatment that produces strengths in the metal equal to structural steel with only one-third the weight.

FIGURE 14–14 The main rotor grip holds the rotary blade of a large military helicopter. The metal is forged and heat-treated aluminum alloy.

MANUFACTURING MATERIALS—FOR HOW LONG?

There must be raw materials in order for manufacturing processes to produce the millions of consumer parts. The only present source of these materials is the earth and the supplies are being rapidly reduced. Mines, the sea, and the atmosphere provide increasingly more raw materials. This means that the total available materials are slowly being reduced. As the population increases and as more items are produced it is only a matter of time before many of the primary raw materials are exhausted. And what then? An examination of the table of elements, primarily the metals, points out the required constant reduction of many of these metals from the earth so that manufacturing can continue and help sustain the good life. But why does man discard many valuable wastes often called trash and garbage? A worn-out table fork is not trash and should not be placed in the garbage container.

Reclamation Is a Requirement

Even though some progress is being made in the area of reclamation, it is not fast enough. Too much reliance is placed on individual ingenuity to save the precious waste materials instead of primary governmental actions. Laws are needed to stop this mass destruction of waste material. A tremendous quantity of waste material is discarded yearly from every residence and from hospitals, schools, parks, manufacturing plants, service stations, and ships. Waste usually is buried permanently or burned so that man can never again use it in any form. Why? Possibly, this procedure is the easiest and quickest way to separate waste from that which is "good." But waste is not useless just because it is different. The millions of tons of waste materials are nothing more than mixtures of all those elements pointed out in the table of elements.

It is not impractical to reclaim this material, but it does require someone to do it. The number of reclamation enterprises is nearly insignificant, but the metals reclamation is improving. The main problem apparently exists with organic materials. The reclamation of organic material provides the potential for a significant replacement of energy consumption. Proper processing of organic materials in each city could greatly reduce the oil and gas consumption because the organic materials can be used as fuel and the resulting by-products can be made into building materials such as bricks and boards. The result would be a reduction in the drain of energy which is so greatly needed for manufacturing purposes.

New Sources of Energy Must Be Found

Along with accelerating efforts in the reclamation of materials there must be a significant effort in the area of energy discoveries and inventions. Unlimited sources of energy exist in the power of the winds, the waves, the sun, and the atom. The wind blows most of the time in certain regions of the earth, and in these regions thousands of windmills could be used to harness the wind and convert this tremendous energy into electricity. Once the initial cost is covered, realistic operation would be worthwhile. Tidal waves along seacoasts in numerous regions of the earth reach very high points relative to their low points. It is this numerical differential constantly in operation that contains the potential for unlimited kilowatts of electrical energy. And then there are the deserts and cloud-free regions on the earth where the sun shines all day. A thousand acres of solar cells could significantly reduce the gas and oil consumption rate in these areas. Finally, fusion and fission of the atom must be pursued as an increased energy source as rapidly as measures of safety permit. The amount of electrical power from atomic power plants is fantastic. In all, the total output from these four sources of new energy plus effective reclamation could greatly reduce the consumption of oils and gases, making the oils and gases more available for automotive use.

Questions

1. Describe some of the processing effects on the mechanical properties of metal.
2. Describe some of the manufacturing processes which make the materials available to society.
3. Name some new methods of shaping metals.
4. How does the chemistry of a metal relate to its physical properties?
5. How does the chemistry of a metal relate to its mechanical properties?
6. What is recrystallization in metals?
7. Differentiate between hot- and cold-formed operations in metals.
8. What is work hardening? How can it be removed for further forming?
9. List some wrought-forming procedures.
10. What is the difference between a casting and an ingot?
11. List several casting operations.
12. Describe powder metallurgy.
13. Why are cutting operations necessary in some of the major shaping operations?
14. List several joining operations.
15. Describe several surface treatment operations.
16. Why is heat treatment of metals often essential?
17. Describe the main heat-treating operations.
18. Why is tempering a very critical heat-treating operation?
19. Discuss the need to reclaim waste materials.
20. Describe several means for obtaining new forms of energy.

Glossary

Ablate To melt away by vaporizing.

Accm The temperature at which the excess cementite goes into an austenitic solution in hypereutectoid steels.

Age hardening The changing of a mechanical property in a metal by temperature and time whereby a chemical or intermetallic compound moves from solution to a mixture condition.

AISI American Iron and Steel Institute.

Allotropic The reversible atomic cell lattice in a grain of metal which allows the properties of the metal to be changed. The alpha-gamma-alpha transformation is an allotropic change.

Alloy With reference to metals, the combination of two or more elements, such as steel, which is formed from iron and carbon. It may be a combination of a metal and a nonmetal.

Amorphous A noncrystalline material such as glass.

Angstrom unit Equal to one ten–millionth of a millimeter; used in measuring the wave length of light.

Anisotropy Directionality of crystallographic planes within a grain of metal.

Anneal To soften a metal by controlled heating and cooling processes.

Anode The positive electrode where an electric current flows.

Anodic The sacrificial or positive electrode where galvanic corrosion occurs in a corrosive environment.

Anodizing An electrolytic surface treatment for metal that prevents further corrosion such as the anodizing of aluminum.

Area   The quantity of material available to receive and sustain stress, calculated as follows:

$$\text{Triangle:} \quad A = \frac{1}{2}\,b \times h$$
$$\text{Square:} \quad A = b \times h$$
$$\text{Circle:} \quad A = \frac{\pi d^2}{4}$$

As cast   The microstructural condition of a metal resulting from casting; no other treatment is applied.

ASM   American Society for Metals.

ASNT   American Society for Nondestructive Testing.

As rolled   The metal's hot or cold, rolled microstructural condition resulting from the rolling operation.

ASTM   American Society for Testing and Materials.

Atom   The smallest particle of material that is identifiable as an element.

Austenite   The solid solution of carbide in gamma iron.

Austenitic steel   Nonmagnetic stainless steel of the chromium-nickel type which forms solid solutions.

Bainite   A microstructure in steel with a very close platelike or feathery appearance resulting from an intermediate quench such as oil from austenite.

Body-centered cubic   The lattice structure of a metal in which the central body region of an iron cell contains one central atom in addiion to the corner atoms. When heated to approximately 1333 °F to 1420 °F, the central atom leaves the body and joins the face-centered atoms.

Brittleness   The property of a material that allows the material to shatter or fly apart when exposed to impact, bending, or tensile or compression loads.

Carburize   To add carbon or carbide to the surface of low carbon steels for the purpose of subsequent case hardening.

Case hardening   The heating and cooling operation in the presence of a designated chemical for the purpose of inducing a hard case over a tough core.

Catalyst   A material causing a change in other materials without being directly affected itself.

Cathode   The negative electrode where an electric current flows.

Cathodic   The protective electrode or negative area.

Cell   The smallest volume of metal that is representative of other volumes in the grain.

Cementite   An extremely hard chemical compound consisting of free iron carbide; is independent of iron carbide in pearlite.

Centroid   The center of gravity of a given volume of material.

Chemical properties   The quantity of all elements constituting a material.

Cleavage   The brittle fracture of a material.

Coefficient of expansion   Linear expansion (or contraction) per inch of length per degree change in temperature. Mild carbon steels have a value of 0.0000067 inch per inch per Fahrenheit degree change in temperature.

Cold working   The forming of metals below their recrystallization temperatures.

Compound   The chemical combination of two or more elements resulting in a new material; for example, iron and carbon to produce iron carbide.

Copper aluminide   The hard chemical compound consisting of one part copper and two parts aluminum.

Creep   The dimensional change of a material while under load at an elevated temperature.

Critical fracture   The separation of a material under load as the rising stress becomes greater than the material's strength at the point of fracture.

Critical point   The temperature in a steel that causes a major change to occur (excluding tempering) such as recrystallization or the change from a mixture of constituents to a solid solution. There are three critical points designated as Ac_1, Ac_2, Ac_3.

Cryogenic   Temperature environments where temperatures are colder than $-200\ °F$.

Crystal   The very small grain of a metal that is visible to the eye; has significant effects on a metal's mechanical properties.

Crystalline   A granular material such as metal.

Cyanide   A very poisonous salt of hydrocyanic acid often used in heat treating and electroplating operations.

Dendrite   The coarse grain or pine tree shape of crystal formed in castings as solid nuclei are formed from the liquid; it has a central axis with differing masses of materials forming at $90°$ to the central axis.

Density   The amount of material in a given volume. Water is the standard for liquids and solids and hydrogen is the standard for gases.

Deoxidize   To remove oxygen from a material.

Design load   The safe load imposed on a material whereby safe operating conditions prevail.

Diffusion   The mixing of different atoms resulting in the penetration of atoms or molecules into the interstices of other atoms or molecules.

Discontinuity   A metallic separation which may be a crack, hole, or inclusion.

Dislocation   The absence of part of a row of atoms or the presence of extra atoms.

Ductility   The ability of a material to be plastically deformed without rupture as the stress increases beyond the yield strength and below the tensile strength.

Dynamic loading   The rapid application of load such as in impact or shock loading.

Elastic deformation   The temporary dimensional change of a material under load below the yield strength.

Elasticity   The ability of a material to be temporarily deformed as the stress increases up to any value below the elastic limit or the yield strength.

Elastic limit   The strength of a material which allows maximum temporary deformation under load; the value is below the yield strength.

Electrolysis   The passing of an electrical current through a compound of material which causes decomposition of the material.

Electrolyte   A solution which carries an electric current.

Element   A pure substance such as oxygen, gold, or mercury.

Elongation   The ability of a material to stretch temporarily or permanently as the stress increases. The term normally pertains to tensile loading of a material when the stress increases beyond the yield strength, causing an increase in the length dimension.

Elongation %   The ductility factor that is used in forming operations, calculated as follows:

$$\% \, e = \frac{L_f - L_o}{L_o} \times 100$$

where L = length,
$\quad\quad f$ = final length,
$\quad\quad o$ = original length

Endurance limit   The maximum stress in cycles that can be applied to a material without failure.

Endurance strength   The stress at which failure will probably occur after completing a specified number of cycles.

Equiaxed   Pertaining to the stress-free grain in a cast metal that tends to be round in shape but has many facets.

Equilibrium   A condition of balance whereby one force is in balance with an opposing force.

Etch   The application of a chemical onto a prepared specimen's surface for the purpose of microscopic examination.

Eutectic   The alloy in which the lowest structural change takes place from the liquid to the solid on cooling.

Eutectoid   The chemical composition in a series of alloys in which the temperature is the lowest where a structural change takes place on slow cooling.

Extensometer   A device clamped onto a tensile specimen for the purpose of determining the amount of extension of a material under load.

Extruding   The formation of materials by forcing plastic material such as metal through a die of the exact shape of the finished product.

Face-centered cubic   The lattice structure of a metal in which the cubic shaped unit cells have atoms at each face of the cell in addition to the corner atoms; for example, gamma iron and aluminum.

Fatigue The combination of a cyclic load and time on a material which may ultimately lead to material failure, even though the stresses are far below the yield in tension.

Ferrite The solid solution of carbon in alpha iron when the carbon content is below 0.008 percent. (Ferrite and iron are often used interchangeably.)

Ferromagnetic Attracted by magnetic forces as are alpha iron, cobalt, and nickel.

Fine pearlite The very close layers of carbide and ferrite in the microstructure of steel resulting from an intermediate quench from austenite, such as an air or oil quench. The strength of fine pearlite is much greater than that of coarse or regular pearlite.

Flux The magnetic field surrounding a magnet.

Forging A hammering operation which provides granular flow in metals for the purpose of plastic shaping and increased mechanical properties.

Harden To harden a metal by controlled heating and cooling processes.

Hardness The resistance to penetration.

Heat treating The heating and cooling of a metal for the purpose of inducing exact mechanical properties.

High speed steel Excessively high alloy steels which, after heat treatment, resist temperatures up to 1000 °F during machining operations.

Hook's law The equal ratio of stress to strain within the material's elastic limit.

Horsepower 33,000 ft-lb or 396,000 in-lb of work per minute.

Hot working The formation of metals above their recrystallization temperatures.

Hypereutectoid A metallurgical condition in steel in which the carbon content is greater than 0.8 percent.

Hypoeutectoid A metallurgical condition in steel in which the carbon content is less than 0.8 percent.

Inclusion A foreign material trapped in a metal as the metal solidifies.

Ingot A casting which is to be subsequently formed by pressure, such as by rolling.

Inorganic The non–organic materials such as metal.

Intergranular Along the grain boundaries of several grains.

Intermetallic compound An intermediate phase in an alloy system whereby a hard compound is formed from two or more elements.

Intragraular Across a grain or within a grain.

Ionize The division of a material into positive and negative ions with an electric current during electrolysis. The positive ions seek the cathode due to a loss of electrons; negative ions seek the anode due to a gain in electrons.

Iron carbide An extremely hard chemical compound consisting of three parts of iron and one part of carbon, chemically combined; pertains to steels of less than hypereutectoid composition.

Isotope   Radioactive material having an atomic weight different from that of the same nonradioactive material; can be produced by splitting an atom.

Kilogram per square meter (kg/m²)   To find kg/m², multiply psi by 703.1.

Lattice   An organized group of atoms which forms the grains of metals.

Load   The weight, force, or stress imposed on a material.

Malleability   The ability of a metal to be permanently deformed without rupture by plastic flow under compression loading.

Martensite   The solid solution of carbide in iron with a body-centered tetragonal cell arrangement; is extremely hard and brittle and has a very high tensile strength.

Matrix   The base metal in an alloy.

Mechanical properties   The variable properties of a material which relate to engineering properties, such as hardness, strength, ductility, and brittleness.

Metallurgy   The science of metals including the chemical and physical processing.

Metal powder   Specific metals which have been produced in powder form for subsequent pressing and sintering operations.

Mixture   The combination of two or more elements in a metal with each element identifiable under the microscope.

Modulus of ·elasticity   The equal ratio of stress to strain in a material under load when the load is not greater than the elastic limit or does not encroach on the yield property. Formula: $E = S/\epsilon$.

Molecule   A particle consisting of one or more atoms which identifies the material.

Moment   The product of a force and the perpendicular distance from the force to the axis of tending rotation.

Newton per square millimeter (N/mm²)   To find N/mm², multiply psi by 0.006895.

Nitride   Related to steels containing iron nitride (a very hard compound) on their surface.

N/m²   The international symbol for pounds per square inch, or psi. The newton, N, is the kilogram meter per second squared and is the unit of force. N/m^2 is the newton per square meter. The kilogram is equal to 2.204 pounds, and the meter is equal to 39.37 inches. N/m^2 is an extremely small value, being 0.15×10 lb/in².

Normalize   To remove stress in metal by controlled heating and cooling processes.

Organic   Refers to living or fossil materials as opposed to inorganic.

OSHA   Occupational Safety and Health Act, which provides authority in establishing safety and healthful working conditions in industry.

Oxidation   The chemical change in a metal as the result of uniting with oxygen.

Oxide   A compound of oxygen and some material resulting in a brittle material such as iron oxide or scale on the surface of red-hot steel after cooling to room temperature.

Pascal (Pa)   To find Pa, multiply psi by 6894.757.

Pearlite   The mechanical mixture of carbide and alpha iron existing in layer form; soft and ductile as well as low in tensile strength; relative to the family of steels in other microstructural conditions.

Photomicrograph   A photograph of the magnified view of a metal's structure at a magnification greater than $10\times$.

Physical properties   The nonvariable properties of a material which relate to properties such as weight, melting point, color, electrical conductivity, atomic lattice, and magnetic characteristics.

Pig iron   The ferrous alloy produced by the blast furnace for the purpose of making steel, cast iron, or wrought iron; results from processed iron ore.

Plane   A two-dimensional area of a material.

Plastic deformation   The ductile flow of a material under a load beyond its yield strength which results in permanent dimensional change of the material.

Plasticity   The ability of a material to flow without rupture under loads beyond the yield and below the tensile; increasing temperatures usually increase the plasticity factor.

Poisson's ratio   The ratio of lateral strain to axial strain within the elastic limit.

Polymerization   The changing of gaseous materials to liquids and solids resulting in a new material with new mechanical properties and especially applicable to the plastics family of macromolecules.

Polymers   Materials made from large molecules derived from smaller units of matter; common polymers include plastics.

Preheating   The heating of steel to a temperature approximately midway to the austenitic temperaures, or about 700 °F.

Proportional limit   The maximum stress where strain remains proportional.

psi   Pounds per square inch; now being converted to N/m^2 when used with modulus of elasticity, pressure, force, weight, and stress. (See N/m^2.)

Pure metal   An element.

Pure plastic   The necking of a specimen under tensile load whereby the material flows plastically without an increase in load. The tensile follows unless the existing load is removed.

Radiation shielding   The material used to absorb and contain waves of penetrating radiation.

Recoverable energy   The energy stored in a loaded material such as a heavy weight on a coiled compression spring or a material in tension.

Recrystallization   The creation of new and fine grains in a metal at a specific temperature. Grains may be small or large after various heat-treating operations, and they respond to heat with respect to rebirth only at the recrystallization temperature.

Red temperatures   Temperaures ranging from black red (about 1200 °F) to cherry red, to bright red, to salmon (about 1700 °F).

Reduction in area   As a section of material increases in tensile stress, its length increases in dimension and its cross section decreases. Permanent deformation occurs beyond the yield to the tensile. The percent of reduction in area occuring as a result of tensile fracture indicates the material's ductility factor and is mathematically determined as follows:

$$\%RA = \frac{(A_o - A_f \times 100)}{A_o}$$

$$\text{where } A = \text{area},$$
$$o = \text{original},$$
$$f = \text{final}.$$

Refractory material   A material which effectively resists elevated temperatures.

Rigidity   The ability of a material to resist a force. The geometrical shape of a given material influences the rigidity value.

Rolling   The hot or cold forming of metals by rolling actions.

Rupture   Material separation.

SAE   Society of Automotive Engineers.

Safety factor   The failing load or tensile strength divided by a safe operating load which helps assure safe operating conditions in the material.

Sintering   The heating of a compacted briquette for the purpose of bonding the individual surfaces of the powdered metal.

Slag   The foreign matter which floats on the surface of molten metal.

Slip   The movement of metal along a plane of its atoms.

Softness   Opposite of hardness; penetration of a load increases as softness increases, or as hardness decreases.

Solid solution   The combination of two or more elements whereby the elements diffuse into each other.

Specific gravity   Ratio of weight of a volume of material to that of an equal volume of another material; usually water, at 4 °C for liquids and solids, and hydrogen for gases.

Specific heat   Number of calories of heat needed to raise the temperature of one gram of material one degree Celsius.

Spheroidized microstructure   Pertaining to the commercially annealed steels which show spheres of carbides embedded in ferrite when viewed under a microscope. Microstructure is formed as a result of prolonged heating at temperatures just below the steel's critical point or Ac_1.

Spheroidizing   The heating of high carbide steels to temperatures just below the critical temperature for the purpose of causing the carbides to form spheres or balls, and, in turn, maximum malleability is produced.

Static loading   The slow application of a load to a value where load and stress are balanced.

Stock   The raw material used to manufacture parts.

Strain   The change in dimension of a material as a result of a stress. (Stress accompanies a load.) Usually, unless otherwise indicated, the strain is understood to be linear.

Strength   The ability of a material to hold a load to a specific value. Tensile strength is maximum strength during a pulling load. Yield strength is the load which commences failure. Compression strength is maximum strength during a pushing load. Shear strength is maximum strength during a cutting type load. Impact strength is maximum strength during the fast application of a load. All of these strengths are destructive strengths. Safe operating loads always exist below the above values.

Stress   The internal reaction of a material to an external force such as weight, pressure, or load measured in psi.

Stress raiser   The area in a material that allows an over-concentration of stress because of the design of the material or a flaw in the material.

Structure and microstructure   The arrangement of the constituents in a material including the grain structure.

Supersaturated   The condition in a metal whereby all or most of the constituents are retained in solid solution following a quench from the material's solution temperature. As time elapses, some constituents may precipitate from the solution.

Temper   To impart a specific mechanical property in metal by controlled heating and cooling processes conducted below the metal's critical point.

Thermal shock   Development of a steep temperature gradient along with a high stress within a material.

Torque   A force which causes rotation tendency and is a moment.

Torsion   The twisting of a material by force which includes torque.

Toughness   The mechanical property resulting in a material which resists a combination of stresses acting concurrently and includes impact loading.

Transition temperature   The temperature which produces a fracture in a material under load whereby the fracture includes approximately 50 percent cleavage and 50 percent plastic deformation.

Ultimate strength   The maximum strength.

Welding   The fusion of metals.

Wrought   A pressure-forming operation that shapes metal, such as rolling, forging, extruding, stamping, drawing, cupping, and coining.

Young's modulus   The modulus of elasticity of a material.

Bibliography and References for Further Study

Allen, Dell K. *Metallurgy Theory and Practice*. Chicago: American Technical Society, 1969.

American Institute of Timber Construction. *Timber Construction Manual*. New York: John Wiley & Sons, Inc., 1966.

American Iron and Steel Institute. *The Making of Steel*. New York, 1964.

American Society for Metals. *Basic Metallurgy*. Metals Park, Ohio, 1962.

American Society for Metals. *Diffusion in Body-Centered Cubic Metals*. Metals Park, Ohio, 1964.

American Society for Metals. *Forming of Stainless Steels*. Metals Park, Ohio, 1968.

American Society for Metals. *Joining of Stainless Steels*. Metals Park, Ohio, 1967.

American Society for Metals. *Metals Handbook*, Vol. 1–11. Metals Park, Ohio, 1961–76.

American Society for Metals. *Oxidation of Metals and Alloys*. Metals Park, Ohio, 1970.

American Society for Metals. *Solidification*. Metals Park, Ohio, 1970.

American Society for Testing and Materials. "Die-Cast Metals: Light Metals and Alloys," *Annual Book of Standards, Part 6*. Philadelphia, 1971.

American Society for Testing and Materials. *Metal Corrosion in the Atmoshpere*. Philadelphia, 1968.

American Society for Testing and Materials. *Physical and Mechanical Testing of Metals*. Philadelphia, 1973.

American Society for Tool and Manufacturing Engineers. *Handbook of Industrial Metrology*. Englewood Cliffs, N.J.: Prentice-Hall, Inc., 1967.

Anderson, Curtis B., Peter C. Ford, and John H. Kennedy. *Chemistry Principles and Applications*. Lexington, Mass.: D.C. Heath & Co., 1973.

Archer, R. S., J. Z. Briggs, and C. M. Loeb, Jr. *Molybdenum*. New York: Climax Molybdenum Company, 1965.

Avner, Sidney H. *Introduction to Physical Metallurgy*. New York: McGraw-Hill Book Company, 1974.

Betz, Carl E. *Principles of Magnetic Particle Testing*. Chicago: Magnaflux Corporation, 1967.

__________. *Principles of Penetrants*. Chicago: Magnaflux Corporation, 1969.

Bolz, Roger W. *Production Processes*. New York: Industrial Press, 1963.

Bradford, Charles B. *Tool Design*. Chicago: American Technical Society, 1967.

Climax Molybdenum Company. *Atlas, Hardenability of Carburized Steels*. New York, 1960.

Corliss, William R., and Douglas G. Harvey. *Radioisotopic Power Generation*. Englewood Cliffs, N.J.: Prentice-Hall, Inc., 1964.

Davis, Harner E., George E. Troxell, and Clement T. Wiskocil. *The Testing and Inspection of Engineering Materials*. New York: McGraw-Hill Book Company, 1964.

De Garmo, E. Paul. *Materials and Processes in Manufacturing*. New York: Macmillan Publishing Company, 1974.

Freund, John E. *Mathematical Statistics*. Englewood Cliffs, N.J.: Prentice-Hall, Inc., 1962.

Giachino, J. W., William Weeks, and Elmer Brune. *Welding Skills and Practices*. Chicago: American Technical Society, 1971.

Glie, Rowen. *Speaking of Standards*. Boston: Cahners Books, 1972.

Harris, Charles O. *Strength of Materials*. Chicago: American Technical Society, 1963.

__________. *Structural Design*. Chicago: American Technical Society, 1966.

Hansen, Bertrand L. *Quality Control*. Englewood Cliffs, N.J.: Prentice-Hall, Inc., 1963.

Horton, Holbrook L., Paul B. Schubert, and Graham Garratt. *Machinery's Handbook*. New York: Industrial Press, Inc., 1972.

Jensen, Alfred, and Harry H. Chenoweth. *Applied Strength of Materials*. Dallas: McGraw-Hill Book Company, 1975.

Kazanas, H. C., Roy S. Klein, and John R. Lindbeck. *Technology of Industrial Materials*. Peoria, Ill.: Charles A. Bennett Co., Inc., 1974.

Kennedy, Gower A. *Welding Technology*. New York: Howard W. Sams and Company, 1974.

Keyser, Carl A. *Materials Science in Engineering*. Columbus, Ohio: Charles E. Merrill Publishing Company, 1974.

McGannon, Harold E. *The Making, Shaping and Treating of Steel*. Pittsburgh: Herbick and Held, 1971.

McMaster, Robert C. *Nondestructive Testing Handbook.* New York: The Ronald Press Company, 1963.

Munro, Lloyd A. *Chemistry in Engineering.* Englewood Cliffs, N.J.: Prentice-Hall, Inc., 1965.

Nutt, Merle C. *Principles of Modern Metallurgy.* Columbus, Ohio: Charles E. Merrill Publishing Company, 1968.

Pollack, Herman W. *Materials Science and Metallurgy.* Reston, Va.: Reston Publishing Company, Inc., 1973.

Porter, Harold W., Orville D. Lascoe, and Clyde A. Nelson. *Machine Shop.* Chicago: American Technical Society, 1968.

Republic Steel. *Alloy Steels.* Cleveland, 1968.

Richardson, Terry A. *Modern Industrial Plastics.* New York: Howard W. Sams and Company, 1974.

Rusinoff, S. E. *Manufacturing Processes.* Chicago: American Technical Society, 1962.

Schneeman, Justin G. *Industrial X-ray Interpretation.* Evanston, Ill.: Intex Publishing Company, 1968.

Small, Louis. *Hardness, Theory and Practice.* Ferndale, Mich.: Service Diamond Tool Company, 1960.

Smith, Charles O. *The Science of Engineering Materials.* Englewood Cliffs, N.J.: Prentice-Hall, Inc., 1969.

Stokes, Vernon L. *Manufacturing Processes.* Columbus, Ohio: Charles E. Merrill Publishing Company, 1975.

Subbarao, E. C., D. Chakravorty, M. F. Merriam, V. Raghavan and L. K. Singhal. *Experiments in Materials Science.* New York: McGraw-Hill Book Company, 1972.

Swanson, Robert S. *Plastics Technology.* Bloomington, Ill.: McKnight and McKnight, 1965.

Sylvia, J. Gerin. *Cast Metals Technology.* Reading, Mass.: Addison-Wesley Publishing Company, 1972.

Turner, Rufus P. *Metrics for the Millions.* New York: Howard W. Sams, 1974.

United States Steel. *Isothermal Transformation Diagrams.* Pittsburgh, 1963.

Van Vlack, Lawrence H. *Elements of Materials Science.* Reading, Mass.: Addison-Wesley Publishing Company, 1964.

————. *A Textbook of Materials Technology.* Reading, Mass.: Addison-Wesley Publishing Company, 1973.

Weast, Robert C. *Handbook of Chemistry and Physics.* Cleveland: C R C Press, 1973.

Winston, Stanton E. *Machine Design.* Chicago: American Technical Society, 1965.

METRIC CONVERSION TABLE

Multiply	by	to obtain
Kilogram-force (kgf) or kilopound	9.80665	Newton (N)
Kilogram force per square millimeter (kgf/mm^2)	9.806650	Megapascal (MPa) or (MN/m^2)
Newton (N)	0.2248089	Pound-force (lbf)
Newton (N)	0.1019716	Kilogram-force or kilopound (kgf)
Pascal (Pa)	0.1019716	Kilogram per square meter (kg/m^2)
Pascal (Pa)	1.0	Newton per square meter (N/m^2)
Pascal (Pa)	0.02088543	Pound per square foot (lb/ft^2)
Pascal (Pa)	0.0001450377	Pound per square inch (psi)
Pound per square inch (psi)	0.063	Atmosphere
Pound per square inch (psi)	2.036	Inch of Mercury
Pound per square inch (psi)	0.70730697	Kilogram per square centimeter (kg/cm^2)
Pound per square inch (psi)	703.1	Kilogram per square meter (kg/m^2)
Pound per square inch (psi)	6.8948	Kilonewton per square meter (kN/m^2)
Pound per square inch (psi)	51,500	Micron
Pound per square inch (psi)	51.5	Millimeter of Mercury (Torr)
Pound per square inch (psi)	0.6894757	Newton per square centimeter (N/cm^2)
Pound per square inch (psi)	6894.76	Newton per square meter (N/m^2)
Pound per square inch (psi)	0.006895	Newton per square millimeter (N/mm^2)
Pound per square inch (psi)	6894.757	Pascal (Pa)

From "Iron Age," Radnor, Pa.

Index

STATE TECH KNOXVILLE

TA
403
S773

Stokes, Vernon L.

Manufacturing
materials

DATE		

WITHDRAWN

© THE BAKER & TAYLOR CO.